Using R for Numerical Analysis in Science and Engineering

Victor A. Bloomfield
University of Minnesota
Minneapolis, USA

CRC Press is an imprint of the
Taylor & Francis Group, an **informa** business
A CHAPMAN & HALL BOOK

CRC Press
Taylor & Francis Group
6000 Broken Sound Parkway NW, Suite 300
Boca Raton, FL 33487-2742

ISBN-13: 978-1-4398-8448-5 (hbk)

Library of Congress Cataloging-in-Publication Data

Bloomfield, Victor A., author.
Using R for numerical analysis in science and engineering / Victor A. Bloomfield.
pages cm. -- (Chapman & Hall/CRC the R series)
Summary: "This book shows how the free and open-source R environment can be used as a powerful and comprehensive platform for the kinds of numerical analysis that are traditionally employed by MATLAB?. With R code fully integrated, the book offers brief descriptions of basic approaches and emphasizes detailed worked examples. It covers functions in the base installation of R as well as those in contributed packages, which greatly enhance the numerical analysis capabilities of R"-- Provided by publisher.
Includes bibliographical references and index.
ISBN 978-1-4398-8448-5 (hardback)
1. Science--Data processing. 2. Engineering--Data processing. 3. Numerical analysis. 4. R (Computer program language) I. Title.

Q183.9.B56 2014
518.0285'5133--dc23 2014003930

Visit the Taylor & Francis Web site at
http://www.taylorandfrancis.com

and the CRC Press Web site at
http://www.crcpress.com

Chapman & Hall/CRC
The R Series

Series Editors

Aims and Scope

This book series reflects the recent rapid growth in the development and application of R, the programming language and software environment for statistical computing and graphics. R is now widely used in academic research, education, and industry. It is constantly growing, with new versions of the core software released regularly and more than 5,000 packages available. It is difficult for the documentation to keep pace with the expansion of the software, and this vital book series provides a forum for the publication of books covering many aspects of the development and application of R.

The scope of the series is wide, covering three main threads:

- Applications of R to specific disciplines such as biology, epidemiology, genetics, engineering, finance, and the social sciences.
- Using R for the study of topics of statistical methodology, such as linear and mixed modeling, time series, Bayesian methods, and missing data.
- The development of R, including programming, building packages, and graphics.

The books will appeal to programmers and developers of R software, as well as applied statisticians and data analysts in many fields. The books will feature detailed worked examples and R code fully integrated into the text, ensuring their usefulness to researchers, practitioners and students.

Published Titles

Using R for Numerical Analysis in Science and Engineering , *Victor A. Bloomfield*

Event History Analysis with R, *Göran Broström*

Computational Actuarial Science with R, *Arthur Charpentier*

Statistical Computing in C++ and R, *Randall L. Eubank and Ana Kupresanin*

Reproducible Research with R and RStudio, *Christopher Gandrud*

Displaying Time Series, Spatial, and Space-Time Data with R, *Oscar Perpiñán Lamigueiro*

Programming Graphical User Interfaces with R, *Michael F. Lawrence and John Verzani*

Analyzing Baseball Data with R, *Max Marchi and Jim Albert*

Growth Curve Analysis and Visualization Using R, *Daniel Mirman*

R Graphics, Second Edition, *Paul Murrell*

Customer and Business Analytics: Applied Data Mining for Business Decision Making Using R, *Daniel S. Putler and Robert E. Krider*

Implementing Reproducible Research, *Victoria Stodden, Friedrich Leisch, and Roger D. Peng*

Dynamic Documents with R and knitr, *Yihui Xie*

Contents

List of Figures

Preface

The complex mathematical problems faced by scientists and engineers rarely can be solved by analytical approaches, so numerical methods are often necessary. There are many books that deal with numerical methods for scientists and engineers; their content is fairly standardized: Solution of systems of linear algebraic equations and nonlinear equations, finding eigenvalues and eigenfunctions, interpolation and curve fitting, numerical differentiation and integration, optimization, solution of ordinary differential equations and partial differential equations, and Fourier analysis. Sometimes statistical analysis of data is included, as it should be. As powerful personal computers have become virtually universal on the desks of scientists and engineers, computationally intensive Monte Carlo methods are joining the numerical analysis armamentarium.

If there are many books on these well-established topics, why am I writing another one? The answer is to propose and demonstrate the use of a language relatively new to the field: R. My approach in this book is not to present the standard theoretical treatments that underlie the various numerical methods used by scientists and engineers. There are many fine books and online resources that do that, including one that uses R: Owen Jones, Robert Maillardet, and Andrew Robinson. *Introduction to Scientific Programming and Simulation Using R*. Chapman & Hall/CRC, Boca Raton, FL, 2009.

Instead, I have tried to write a guide to the capabilities of R and its add-on packages in the realm of numerical methods, with simple but useful examples of how the most pertinent functions can be employed in practical situations. Perhaps—if it were not for its cumbersomeness—a more accurately descriptive title for this book would be *How To Use R to Perform Numerical Analyses of Interest to Scientists and Engineers*. I believe that the approach I take is the most efficient way to introduce new users to the powerful capabilities of R.

R, with more than two million users worldwide, is well known and widely used among statisticians as a "language and environment for statistical computing and graphics which provides a wide variety of statistical and graphical techniques: linear and nonlinear modeling, statistical tests, time series analysis, classification, clustering, etc." * It runs on essentially all common operating systems: Mac OS, Windows, and Linux.

Less well known than R's statistical prowess is that it has capabilities in the realm of numerical methods very similar to those of excellent but costly commercial

*Comprehensive R Archive Network (CRAN), `http[://cran.r-project.org/`

programs such as MATLAB®, MathCad, and the numerical parts of Mathematica and Maple, with the considerable advantages that it is free and open source. The fact that R is free is important in making its capabilities available to everyone, even if they live in poor countries, do not work in companies or institutions that can afford expensive site licenses, or no longer have student discounts.

R has excellent, publication-quality graphics. It has many useful built-in functions and add-on packages, and can be readily extended with standard programming techniques. For large, computationally demanding projects, R can interface with speedier but more-difficult-to-program languages such as Fortran, C, or C++. It has extensive online help and a large and growing library of books that illustrate its many applications. R is a stable but evolving computational platform, which undergoes continual (but not excessive) development and maintenance, so that it can be relied on over the long term. To quote from the "What Is R?" page `http://www.r-project.org/about.html` linked to the R Project home page at `http://www.r-project.org/`,

> R is an integrated suite of software facilities for data manipulation, calculation and graphical display. It includes
>
> - an effective data handling and storage facility,
> - a suite of operators for calculations on arrays, in particular matrices,
> - a large, coherent, integrated collection of intermediate tools for data analysis,
> - graphical facilities for data analysis and display either on-screen or on hardcopy,
> - a well-developed, simple and effective programming language which includes conditionals, loops, user-defined recursive functions and input and output facilities.
>
> The term "environment" is intended to characterize it as a fully planned and coherent system...

Who should read this book?

I have written this book with the hope of convincing every practicing scientist and engineer that R can be their fundamental computational, graphics, and data analysis tool. As summarized above, and as will be developed throughout the book, R has virtually all the capabilities that are needed for high-level quantitative work in the physical, biological, and engineering sciences. Importantly, that work can be developed and shared in teaching and collaborative efforts, thanks to the free, open-source nature of R.

Readers of this book should have the standard set of introductory undergraduate math courses: differential and integral calculus, linear algebra, and differential equations. Some contact with statistics would be desirable for the last two chapters. Familiarity with basic numerical methods—e.g., trapezoidal or Simpson's rule integration, Euler's method for integrating differential equations, linear least squares fitting of points to a line—would be desirable to provide intuition and motivation. But my aim is to provide a guide to a standard set of high-level numerical analysis tools as implemented in R, without burdening the reader with detailed derivations or rare exceptions: numerical methods that *usually* work (apologies to Forman S. Acton). My goal is to provide a pragmatic guide to these tools, illustrated with suitable examples, to encourage a broad range of scientists and engineers—current practitioners and students—to use them in their work.

Overview of the contents of this book

Chapter 1, Introduction describes how to obtain and install R, how to find help, how to augment R with external packages, and how to learn more about R through books and online resources.

Chapter 2, Calculating lists the basic operators and functions that make R a powerful calculator. It shows how to assign and work with variables, especially the vectors and matrices that are R's core numeric types.

Chapter 3, Graphing introduces the types of plots most useful in science and engineering work. It also shows how to modify axes, add text and math expressions to a plot, combine several plots in a figure, and produce animated graphics.

Chapter 4, Programming and functions introduces the basic programming concepts used in R. It shows how R implements conditional and repetitive execution, explains how users can define their own functions, and describes the wide variety of mathematical functions already available in R.

Chapter 5, Solving systems of algebraic equations discusses how to find the zeros of polynomials and other functions, and how to solve systems of linear equations using matrix methods. It describes special methods for handling sparse matrices, introduces the `Matrix` package that has advantages in dealing with very large systems, and shows how to perform the standard types of matrix decomposition (eigen, SVD, QR, etc.). Chapter 5 concludes with a discussion of some of the R packages and functions for solving systems of nonlinear equations.

Chapter 6, Numerical differentiation and integration begins with a discussion of numerical differentiation both in base R and in some specialized packages. Various algorithms for numerical integration in one dimension are then considered, extending to multidimensional integration where Monte Carlo methods come to the fore. It concludes with a discussion of R's facilities for symbolic differentiation, especially of polynomials, and its interfaces to symbolic packages.

Chapter 7, Optimization begins with a discussion of one-dimensional optimization, and then moves on to the numerous methods for performing multidimensional optimization, both unconstrained and constrained. Finding the global minimum of functions with many local minima is tackled via simulated annealing and genetic algorithm approaches. The chapter concludes with discussions of linear and quadratic programming and mixed-integer linear programming.

Chapter 8, Ordinary differential equations considers problems that lie at the heart of numerical methods in science and engineering. It starts with the simple Euler method for integrating initial value ODEs, but moves rapidly to packages that embody the most powerful methods for solving systems of stiff and non-stiff equations. This chapter also deals with difference equations, delay differential equations, differential algebraic equations, steady-state systems, and boundary value problems. It concludes with a treatment of stochastic differential equations.

Chapter 9, Partial differential equations deals with some of the most common and important types of equations encountered in scientific and engineering work, typified by the diffusion/heat conduction equation, the wave equation, and the Laplace or Poisson equation. The `ReacTran` package deals with all of these, and is particularly useful in solving reaction–diffusion systems.

Chapter 10, Analyzing data introduces topics that are not traditionally part of a "numerical methods" book but that should be part of the armamentarium of every scientist and engineer. The chapter discusses how to get external data into R, how to organize it using data frames, how to analyze data from a single sample and compare two samples, and how to assess correlation between variables. The chapter ends with sections that show how to make sense out of large amounts of data by principal component analysis and cluster analysis.

Chapter 11, Fitting models to data shows how to fit data to linear and nonlinear models, and how to interpolate between measurements. An important section deals with time series, spectrum analysis, and other aspects of signal processing. These last two chapters just skim the surface of the enormous statistical capabilities of the R environment, but are intended to give a useful introduction to these powerful tools.

Obtaining the code used in this book

The code for all examples in this book that are longer than two or three lines is available for downloading at the publisher's website, `http://www.crcpress.com/product/isbn/9781439884485`.

Acknowledgments

I am grateful to Hans Werner Borchert, author of the valuable packages `pracma` and `specfun` and maintainer of the Numerical Mathematics Task View on the CRAN website, for his many contributions to this book. In addition to his overall critiques, he wrote the section on Numerical Integration and several sections in the Optimization chapter. Daniel Beard made insightful comments on an earlier version of this manuscript. My editor Rob Calvert, and his assistants Rachel Holt and Sarah Gelson, kept things running smoothly. Karen Simon efficiently shepherded the production process. My greatest thanks, however, go to the large community of R project contributors—both the core group and the authors of the many packages, manuals, and books—who have given so freely of their time and talent to craft a tool of such immense value.

For product information, please contact:

The MathWorks, Inc.
3 Apple Hill Drive
Natick, MA 01760-2098 USA
Tel: 508 647 7000
Fax: 508-647-7001
E-mail: info@mathworks.com
Web: www.mathworks.com

Chapter 1

Introduction

1.1 Obtaining and installing R

You can download and install R from the CRAN (Comprehensive R Archive Network) website at `http://cran.r-project.org/`. Choose the appropriate link for your operating system (Mac OS X, Windows, or Linux), and follow the (not very complicated) directions. Unless you have some special requirements for customization, you should choose the precompiled binary rather than the source code.

As it comes, R has a plain but serviceable interface. It can be run from the command line or from a set of windows (console, graphics, help, etc.) on MacOS X or Windows. A neater, more streamlined—but perhaps less flexible—integrated development interface can be had by installing the freeware RStudio from `http://www.rstudio.org/`/ide.

1.2 Learning R

The next several chapters of this book are intended to provide a basic introduction to R. The basic manual for learning R is the online *An Introduction to R*, found at `http://cran.r-project.org/` → `Documentation`. The section **Learning more about R** at the end of this chapter lists numerous books and online resources.

1.3 Learning numerical methods

This book tries to lightly sketch the basic ideas of the various numerical methods used, but does not attempt to present their theoretical background or details. Currently the standard reference on numerical methods in science and engineering is *Numerical Recipes* by Press et al. (2007), and there are many other worthwhile books devoted to the field and the various topics within it. Readers are encouraged to consult such references, and/or have recourse to various online sources. A Google search on a given topic will typically lead to a useful *Wikipedia* article, several sets of university-level lecture notes, and often treatments based on MATLAB® or Mathematica. Such online resources may be much more accessible than standard printed references, especially for readers without convenient access to specialized research library collections.

1.4 Finding help

If you know the name of an R object, such as a function, and want to know what it does, what its various arguments (including defaults) are, and what values it returns, with examples, typehelp (function.name) or ?function.name. For example, ?solve tells us that "This generic function solves the equation a%*% x = b for x, where b can be either a vector or a matrix." As one example, it gives inversion of a Hilbert matrix:

```
hilbert <- function(n) {i <- 1:n; 1 / outer(i - 1, i, "+")}
h8 <- hilbert(8); h8
sh8 <- solve(h8)
round(sh8 %*% h8, 3)
```

(Don't worry if you don't understand the code at this time. We will discuss R programming beginning in Chapter 4.)

Often, you may need to be reminded of the name of a function. A very useful "cheat sheet" listing many of the more common R functions is "R Reference Card" by Tom Short, available at http://cran.r-project.org/doc/contrib/Short-refcard.pdf.

If you think that an object or function may be available, and can guess part of its name, try apropos(). For example, if you're interested in spectral analysis, apropos(spec) gives

```
[1] "plot.spec" "plot.spec.coherency" "plot.spec.phase" "spec.ar"
[5] "spec.pgram'' "spec.taper" "spectrum"
```

However, this does not turn up Special, which yields special mathematical functions related to the beta and gamma functions. apropos() allows searches using regular expressions; enter ?apropos to see some examples.

help.search() "allows for searching the help system for documentation matching a given character string in the (file) name, alias, title, concept or keyword entries (or any combination thereof), using either fuzzy matching or regular expression matching." Note that the character string must be in quotes. For example, help.search("spectral") turns up five topics, with descriptions:

- eigen: Spectral Decomposition of a Matrix
- plot.spec: Plotting Spectral Densities
- spec.ar: Estimate Spectral Density of a Time Series from AR Fit
- spec.pgram: Estimate Spectral Density of a Time Series by a Smoothed Periodogram
- spectrum: Spectral Density Estimation

Clicking on any of these topics brings up its help page.

Using regular expressions, help.search(^spec) brings up those help pages containing information about topics whose title, alias, or concept contain words that begin with "spec": Special, specific, spectral, specification, etc.

help.start() opens your web browser to R's online documentation. The manual "An Introduction to R" is the standard online reference to the language. Click on

"Search Engine & Keywords" to search for keywords, function and data names, and concepts, and also for words or phrases within help-page titles. A list of keywords arranged by topics (Basics; Graphics; MASS (the book); Mathematics; Programming, Input/Output, and Miscellaneous; and Statistics) is provided to help target the search.

R has a large and helpful online community, of which you can ask questions if you can't find answers through your own efforts. A very large database (nearly 2700 pages as of the end of 2013) of topics is maintained at `http://r.789695.n4.nabble.com/`. Searching this database can provide leads to existing resources, or show how others have solved puzzling problems.

Two sites for doing Google-type searching of the R language are `http://www.dangoldstein.com/search_r.html` and `http://www.rseek.org/`.

If all else fails, you can ask your own questions by going to `http://www.r-project.org/` > `Mailing Lists`. The third item down is R-help. (The first two are R-announce, "for major announcements about the development of R and the availability of new code" and R-packages, "for announcements ... on the availability of new or enhanced contributed packages.") The posting guide gives important advice about "how to ask good questions that prompt useful answers." Follow that advice to avoid grumpy responses from the experts.

A somewhat haphazard but occasionally enlightening way to learn about various aspects of R is to look at R-bloggers (`http://www.r-bloggers.com/`), which collects "daily news and tutorials about R, contributed by over 450 bloggers."

1.5 Augmenting R with packages

R is to a large extent an environment for packages that perform specialized tasks. The R distribution itself installs a number of packages, some of which are "just there" and need not be loaded explicitly. These include `base`, `graphics`, `stats`, `utils`, `splines`, `datasets`, and several others. Some other packages are "recommended," and are included in all binary distributions of R. Most pertinent for our purposes among these are `Matrix` (which we will discuss in Chapters 2 and 5), `cluster` (functions for cluster analysis), and `nlme` (for nonlinear mixed-effects models). These must be loaded with the `library("package-name")` or `require("package-name")` function. (`library` and `require` can generally be used interchangeably, although `require` is intended for use within other functions and the two will give different messages if the package is not available. See `?library` for details of their usage.)

Much of the real power of R comes from contributed packages (over 5000 as of the end of 2013) that can be downloaded from CRAN and installed in your local copy of R using the command `install.packages("package-name")`. (Mac OS X and Windows users can also install packages via the R menu system.) The packages can then be loaded with `require(package-name)` or `library(package-name)`. Packages are in many ways analogous to the add-ons for other mathematical languages, but are generally free and open source. We will describe and use a number of such packages in this book, including packages for ordinary and partial differential equations, orthogonal polynomials, root-finding, optimization, and more.

A package may contain datasets, functions written in R, and dynamically loaded libraries of C or Fortran code. To find what packages are currently installed in R on your computer, type `library()`. The datasets in some packages can be of use as examples in learning about statistical analysis of data, as we will do in Chapter 10. To get summary help about a package you have installed in R, type `library(help = "package.name")` or `help(package = "package.name")`. Navigating to individual packages in the CRAN archive will give access to their reference manuals and (sometimes) vignettes, as downloadable pdf files.

It is often difficult to find a particular function in R, if it's been implemented in one of the many packages. The Task Views page, accessible from the CRAN home page, groups packages according to the tasks that they help to facilitate. For example, the `ChemPhys` task view refers to packages useful in chemometrics and chemical physics that carry out such tasks as linear and nonlinear regression models, curve resolution, differential equations, optimization, cellular automata, etc. The `NumericalMathematics`, `DifferentialEquations`, `Optimization`, and `TimeSeries` task views are particularly pertinent to the material in this book.

Perhaps the best resource for searching the help pages of contributed packages to find particular functions is the `findFn` function in the `sos` package. Its documentation states "The `sos` package provides a means to quickly and flexibly search the help pages of contributed packages, finding functions and datasets in seconds or minutes that could not be found in hours or days by any other means we know."

The R community site `http://www.inside-r.org/` enables you to search for the packages that contain the keyword(s) of interest, and then to browse the help files of those packages. A similar function is served by the community site crantastic! (`http://crantastic.org/`), which also provides information about new and upgraded packages, and allows reviews by users.

`RSiteSearch("keyword")` at the R prompt opens a web-based interface to search functions, contributed packages, and R-help postings. For example, typing `RSiteSearch("orthogonal polynomials")` yielded 194 documents matching the query within function, package vignette, and task view targets.

If you rely on certain packages, and want to check whether they've been updated, a bit of code written by Karthik Ram will give you a list of changes.

```
installed = installed.packages()
available = available.packages()
ia = merge(installed, available, by="Package") [,c
   ("Package", "Version.x", "Version.y")]
updates = ia[as.character(ia$Version.x) != as.character(ia$Version.y),]
updates
```

To install every available update, enter `update.packages`.

One must be cautious when using contributed packages, however, since they are generally less broadly used—hence less thoroughly vetted—than components of the base R installation. If possible, test them with examples for which you know the correct answers, rather than relying on the examples included with the package, before applying them to problems that matter. Packages with vignettes may show that the author has taken extra care, while packages with old dates may indicate that the code is not being maintained and updated.

1.6 Learning more about R

1.6.1 Books

An extensive list of "Books related to R," with bibliographic information and abstracts, can be accessed from the R-Project home page under Documentation. Some that I have found most helpful include

- Peter Dalgaard. *Introductory Statistics with R.* Second Edition, Springer, New York, 2008.
- William N. Venables and Brian D. Ripley. *Modern Applied Statistics with S.* Fourth Edition. Springer, New York, 2002 (the book to which the MASS package is an adjunct).
- Owen Jones, Robert Maillardet, and Andrew Robinson. *Introduction to Scientific Programming and Simulation Using R.* Chapman & Hall/CRC, Boca Raton, FL, 2009.
- Karline Soetaert and Peter M.J. Herman, *A Practical Guide to Ecological Modelling: Using R as a Simulation Platform*, Springer, New York, 2009.
- Victor Bloomfield. *Computer Simulation and Data Analysis in Molecular Biology and Biophysics: An Introduction Using R.* Springer, New York, 2009.
- Norman Matloff. *The Art of R Programming: A Tour of Statistical Software Design.* No Starch Press, San Francisco, 2011.
- Joseph Adler. *R in a Nutshell.* O'Reilly, Sebastopol, CA, 2010.

The first two of these books are standard references, the next three emphasize scientific programming rather than statistics, the sixth is an excellent survey of programming approaches, and the last is a useful overall reference. The book by Jones et al. is particularly valuable as a complement to the one you are reading, because it presents some of the basic theory behind numerical methods, and implements that theory with explicit R scripts.

1.6.2 Online resources

Extensive documentation about R is online at `http://cran.r-project.org/` → Documentation, where one finds Manuals, FAQs, and Contributed material. Manuals have been created by the R Development Core Team. The basic manual for learning the language is *An Introduction to R*, while *The R Reference Index* "contains all help files of the R standard and recommended packages in printable form. (9MB, approx. 3500 pages)."

FAQs contains general information for users on all platforms (Linux, Mac, Unix, Windows), and also platform-specific information for Mac and Windows.

In keeping with the community spirit underlying R, there is extensive Contributed documentation divided into "Documents with more than 100 pages," "Documents with fewer than 100 pages," and "Short Documents and Reference Cards." The document *Using R for Scientific Computing* by Soetaert in the second category has much material along the lines of this book. The very useful *R Reference Card* at `http://cran.r-project.org/doc/contrib/Short-refcard.pdf` has already been mentioned. There are also "Non-English Documents" in many languages.

The reference sheet *R and Octave* (http://cran.r-project.org/doc/contrib/R-and-octave.txt) translates many commands between R and Octave or MATLAB, and the `pracma` package gives R implementation of many advanced math functions from MATLAB and Octave.

Of the many other online resources for learning R, *Programming in R* by Zoonekynd (http:// zoonek2.free.fr/UNIX/48_R/02.html) and the site of the same name by Girke (http://manuals.bioinformatics.ucr.edu/home/programming-in-r) are particularly useful for our purposes.

Chapter 2

Calculating

2.1 Basic operators and functions

R can be used, without any programming, as a powerful calculator. It has all the standard arithmetic operators and functions, which operate on numeric or complex vectors (including scalars, which are vectors of length 1).

- Arithmetic operators: The binary arithmetic operators are +, -, *, /, ^ (exponentiation), %% (mod), and %/% (integer division).
- Logarithms and exponentials: log (natural log), log10, (base 10 log), log2 (base 2 log), log (x,b) (log of x to base b). log1p(x) computes log(1+x) accurately for $|x| << 1$.
- exp computes the exponential function, and expm1 computes exp(x)-1 accurately for $|x| << 1$.
- Trigonometric functions : cos(x), sin(x), tan(x), acos(x), asin(x), atan(x), atan2(y,x) where angles are in radians and x and y are numeric or complex scalars or vectors. atan2(y,x) = atan(y/x) for positive arguments. The `pracma` package, to which we will refer later, adds more trigonometric functions: cot(x), csc(x), sec(x), acot(x), acsc(x), asec(x).
- Hyperbolic functions : cosh(x), sinh(x), tanh(x), acosh(x), asinh(x), atanh(x). `pracma` adds coth(x), csch(x), sech(x), acoth(x), acsch(x), asech(x).
- Miscellaneous mathematical functions: abs(x), sqrt(x).

R also has Bessel functions and special functions related to the beta and gamma functions, and packages add more special functions, as we shall discuss in Chapter 4. Here is an entirely artificial example that demonstrates some of the functions:

```
> log(sqrt(3.2)*besselJ(0.4,0)*exp(-2)/gamma(7.9))
[1] -9.783098
```

The > at the beginning of the line is the R prompt; it appears automatically. The [1] indicates the first answer on that line. In this case there is only one answer, but if there were dozens or hundreds of values returned, with line breaks every 6–8, the index at the beginning of each line would provide useful orientation.

2.2 Complex numbers

R has the standard operations on complex numbers. To get more information, `?complex`.

```
> (1i)^2 # Complex unit i must be multiplied by a scalar
[1] -1+0i

> (1+2i)-(3+4i)  # Addition and subtraction
[1] -2-2i

> (1+2i)*(3+4i)  # Multiplication
[1] -5+10i

> (1+2i)/(3+4i) # Division
[1] 0.44+0.08i

> (1i + (1i)^2 + (1i)^3 + (1i)^4 + (1i)^5) / (1 + 1i)
[1] 0.5+0.5i

> Re(3+2i); Im(3+2i) # Real and imaginary parts
[1] 3
[1] 2

> Mod(3+2i); Arg(3+2i) # Modulus and argument (radians)
[1] 3.605551
[1] 0.5880026

> Mod((1+2i)*(3+4i))  # Modulus of product = product of moduli
[1] 11.18034

> Mod(1+2i)*Mod(3+4i)
[1] 11.18034

> Conj(3+2i) # Complex conjugate
[1] 3-2i

# Example of Euler's formula exp(i*phi) = cos(phi) + i*sin(phi)
> exp(pi/7*1i)
[1] 0.9009689+0.4338837i

> cos(pi/7) + 1i*sin(pi/7)
[1] 0.9009689+0.4338837i
```

In R, everything after # is a comment, and is ignored by the interpreter, but can be extremely useful to programmers and users.

To get fractional roots of negative numbers, you must include an imaginary part. Otherwise, you get NaN (not a number).

```
> (-8)^(1/3)
[1] NaN

> (-8+0i)^(1/3)
[1] 1+1.732051i
```

2.3 Numerical display, round-off error, and rounding

By default, R displays seven digits in calculations. This can be changed by `options(digits = d)`, where d is the desired number of digits. The option remains in effect until changed or until R is restarted.

```
> Arg(3+2i)
[1] 0.5880026

> options(digits=3)

> Arg(3+2i)
[1] 0.588
```

The `round(number, digits)` function rounds the number to the specified number of decimal places. The default is `digits = 0`. It works with both positive and negative numbers of digits.

```
> options(digits=7)

> round(1234.567)  # Default
[1] 1235

> round(1234.567,-2)
[1] 1200

> round(1234.567,2)
[1] 1234.57
```

The function `signif(number, digits)` rounds the number to the specified number of significant digits (default = 6).

```
> signif(1234.567)  # Default
[1] 1234.57

> signif(1234.567,2)
[1] 1200
```

The functions `ceiling(x)`, `floor(x)`, and `trunc(x)` take a single numeric argument x and return the smallest integer not less than x, the largest integer not greater than x, and the integer formed by truncating x toward zero, respectively. If x is a vector (see below, Section 2.6), these rounding functions work on each element of the vector.

For presentation, it is often desirable to format numbers with a given number of digits, commas or other marks separating intervals before the decimal point, in decimal or scientific format, etc. One can do this using the `formatC` function.

```
> options(digits = 7)

> (x = runif(3)) # The outer parentheses cause output to print
[1] 0.4929854 0.5788660 0.2463871

> (x = x + 123456)
[1] 123456.5 123456.6 123456.2

> formatC(x, digits = 2, big.mark = ",", format = "f")
[1] "123,456.49" "123,456.58" "123,456.25"

> formatC(x, digits = 7, big.mark = ",", format = "E")
[1] "1.2345649E+05" "1.2345658E+05" "1.2345625E+05"
```

Note that these are now character strings, not numbers. See `help(formatC)` and `help(format)` for more information about formatting output.

Because computers work with binary rather than decimal arithmetic, fractions may not be exactly represented. For example

```
> .7-.6-.1
[1] -2.78e-17

> .7/.1-7
[1] -8.88e-16

# but
> .7/.1
[1] 7
```

One can "zap" meaningless values close to zero with the `zapsmall` function:

```
> zapsmall(.7/.1) - 7
[1] 0
```

but, of course, one must be cautious in doing so.

R uses the IEEE standard in representing floating-point numbers in 64-bit double precision. (For details see Jones et al., 2009.) The command `.Machine` tells us a variety of things about this standard. The smallest non-zero floating point number that can be represented is `double.xmin`, $2.225074 \times 10^{-308}$, and the largest floating point number is `double.xmax`, 1.797693×10^{308}. The smallest positive number x such that $1+x$ is not equal to 1 is `double.eps`, 2.220446×10^{-16}. The smallest positive number such that $1-x$ can be distinguished from 1 is `double.neg.eps`, 1.110223×10^{-16}. One must exercise care in testing for exact numerical equality if differences are near `double.eps`. (See the section on relational operators, below.)

If a number cannot be represented meaningfully, `Inf` (infinity) or `NaN` (not a number) will generally be returned according to standard computational arithmetic definitions:

```
> 1/0
[1] Inf

> log(0)
[1] -Inf

> Inf*Inf
[1] Inf

> Inf/Inf
[1] NaN

> 0/0
[1] NaN
```

2.4 Assigning variables

To do much useful work in R or any other computer language, one must define variables and assign values to them. The conventional assignment operator in R is `<-` , but = is also allowed, and is the operator I will use in this book, because it is easier to type, is akin to usage in most other languages, and avoids typographical disasters such as `x< -y`, which will be interpreted as “x is less than minus y” rather than the intended “give x the value of y.” (On the other hand, do not confuse = with ==, which means logical equality.)

```
> theta = pi/4
> st = sin(theta)

> theta
[1] 0.7853982

> asin(st)
[1] 0.7853982
```

Names of variables in R may consist of lowercase or capital letters, numbers, “.”, and “_”. The name must begin with a letter or “.”; and if it begins with “.” the next character cannot be a number. R is case sensitive, so x and X are different variables.

A simple but handy use for named variables is to convert between units. For example, to convert between time units, we can use the definitions (separating multiple assignments on the same line with semicolons):

```
> sec. = 1; min. = 60*sec.; hr. = 60*min.; day. = 24*hr.
> week. = 7*day.; yr. = 365.25*day.; century. = 100*yr.
```

```
> 3*century./sec.
[1] 9.47e+09
```

to calculate the (approximate) number of seconds in three centuries. Note that one divides by the desired unit because the answer is a pure number without units. In this example I have adopted the arbitrary but useful convention that unit names end with a period, to avoid conflicts with other potential uses of these variable names. Another example is to convert between degrees and radians when using trigonometic functions.

```
> degree. = pi/180
> sin(30*degree.)
[1] 0.5
```

R has five sets of built-in constants:

- `pi`
- `LETTERS` (the 26 uppercase letters of the Roman alphabet)
- `letters` (the 26 lowercase letters of the Roman alphabet)
- `month.abb` (the 3-letter abbreviations of the month names in English)
- `month.name` (the month names in English)

Although not strictly prohibited, it is not advisable to name variables as "c," "t," "T," or "F" since these are used in R to combine arguments to form a vector, take the transpose of a vector or matrix, and stand as abbreviations for TRUE and FALSE.

In this book we will deal with numeric, complex, or logical (TRUE/FALSE) variables, but R can also deal with character data. We will mainly consider calculations with vectors and matrices, and occasionally lists (the general form of vectors with different types of elements); but R has other types of objects as well: data frames, factors, and arrays (matrices with more than two dimensions).

2.4.1 Listing and removing variables

To find out what variables are currently defined in the R environment, type `ls()`. To remove variables when they are no longer needed, type `remove(variable.names)` or `rm(variable.names)`. To remove all variables and most other objects, type `rm(list = ls())`.

2.5 Relational operators

R has the familiar operators that allow comparison of values, giving TRUE or FALSE answers: < (less than), > (greater than), <= (less than or equal to), >= (greater than or equal to), == (identical to), and != (not equal to). An example that illustrates the limitations of precise representation of decimal fractions:

```
> .5 == 1/2
1] TRUE

> .3/.1 == 3
```

```
[1] FALSE
```

To avoid such situations with numerical or complex quantities, use `all.equal`, a utility that tests near equality (by default within a tolerance of `.Machine$double.eps`$^{0.5}$) of two R objects:

```
> all.equal(.3/.1,3)
[1] TRUE
```

If the operator == or ! = is applied to vectors (see the next section) with n elements, it will generate a logical vector with n TRUE or FALSE values. If what is wanted is instead a single answer to the question whether the vectors are identical, use the `identical` function, which tests for *exact* equality, instead. (See the subsection on logical vectors below.)

2.6 Vectors

For nearly all numerical calculations in R, one uses vectors and matrices. Vectors are the simplest data structure, consisting of an ordered collection of numbers, characters, or logical values that are separated by commas and bracketed by c(), which stands for "combine" or "concatenate." A typical numerical vector might be

```
> x = c(3.2, 1.7, -11.3, -0.67, 4, 0)
```

A scalar can be thought of as a vector of length 1.

2.6.1 Vector elements and indexes

To select a particular element of a vector, one puts its index in square brackets. For example, to select the third element of x

```
> x[3]
[1] -11.3
```

and to select several elements

```
> x[1:3]; x[c(2,3,5)]
[1] 3.2 1.7 -11.3
[1] 1.7 -11.3 4.0
```

The above code illustrates two features of R that are quite different despite their punctuational similarity. The *colon* separating two numbers `from` and `to` produces a sequence from `from` to `to` in steps of 1 or -1. Thus `x[1:3]` selects elements 1, 2, and 3 of the vector `x`. The *semicolon* separates two assignment statements on the same line.

To change the value of an element

```
> x[3] = 10.0
> x
[1] 3.20 1.70 10.00 -0.67 4.00 0.00
```

To "grow" a vector by adding a new element to its end

```
> x[7] = 4.3
> x
[1] 3.20 1.70 10.00 -0.67 4.00 0.00 4.30
```

If an element is skipped in this process, its value will be given as NA (a logical constant indicating "not available").

```
> x[9] = 9
> x
[1] 3.20 1.70 10.00 -0.67 4.00 0.00 4.30 NA 9.00
```

To remove an element, enter its index with a minus sign.

```
> x[-8]
[1]  3.20  1.70 10.00 -0.67  4.00  0.00  4.30  9.00
```

2.6.2 Operations with vectors

Most of the basic numerical operators act on a vector element by element. For example, using the vector x defined above,

```
> x-1
[1]   2.20   0.70 -12.30  -1.67   3.00  -1.00

> 3*x
[1]   9.60   5.10 -33.90  -2.01  12.00   0.00

> x^2
[1]  10.240   2.890 127.690   0.449  16.000   0.000

> cos(x/2)
[1] -0.0292  0.6600  0.8061  0.9444 -0.4161  1.0000
```

There is also a set of functions that return the length, mean, standard deviation, minimum, maximum, range, etc., of a vector.

```
> length(x)  # Number of elements in the vector
[1] 6

> mean(x)
[1] -0.512

> sd(x)  # Standard deviation
[1] 5.58

> var(x)  # Variance, sd^2
[1] 31.1

> min(x)
[1] -11.3
```

```
> range(x)
[1] -11.3   4.0

> sum(x)
[1] -3.07

> prod(x)
[1] 0

> cumsum(x)  # Cumulative sum
[1]  3.20  4.90 -6.40 -7.07 -3.07 -3.07

> cumprod(x)  # Cumulative product
[1]   3.20   5.44 -61.47  41.19 164.74   0.00
```

The function summary gives the minimum and maximum (hence the range), 1st and 3rd quartiles, median, and mean of vector elements regarded as data; but, unfortunately, not the standard deviation or variance.

```
> summary(x)
   Min. 1st Qu.  Median    Mean 3rd Qu.    Max.
 -11.30   -0.50    0.85   -0.51    2.82    4.00
```

A vector y may act on a vector x. If y is shorter than x, y is recycled until a vector with the length of x is obtained, with a warning message if the recycling is fractional. For example

```
> x1 = c(1,2,3,4,5)
> y1 = c(1,2)
> x1*y1
[1] 1 4 3 8 5

Warning message:
In x1 * y1 :
 longer object length is not a multiple of shorter object length
```

2.6.3 Generating sequences

Vectors are simply sequences of numbers. Sometimes these numbers will be data, but often we will want to generate sequences for simulations. These sequences of numbers might be evenly spaced, e.g., time points in a simulation of chemical reaction kinetics. Or they might be random, as in a Monte Carlo simulation.

2.6.3.1 Regular sequences

The colon operator generates a sequence of numbers separated by 1 or -1.

```
> 1:10
```

```
[1]  1  2  3  4  5  6  7  8  9 10

> 5.7:-3.7
[1]  5.7  4.7  3.7  2.7  1.7  0.7 -0.3 -1.3 -2.3 -3.3
```

A common mistake is to forget that the colon has higher priority than other arithmetic operations.

```
> n = 10
> 1:n-1
[1] 0 1 2 3 4 5 6 7 8 9

> 1:(n-1)
[1] 1 2 3 4 5 6 7 8 9
```

If an increment different from 1 is desired, use `seq(from, to, by)`. If the parameters are given in this order, their names may be omitted, but if a different order is used, the names are required. (This is true of all functions in R.)

```
> seq(3,8,.5)
[1] 3.0 3.5 4.0 4.5 5.0 5.5 6.0 6.5 7.0 7.5 8.0

> seq(by=0.45,from=2.7,to=6.7)
[1] 2.70 3.15 3.60 4.05 4.50 4.95 5.40 5.85 6.30
```

The number of elements may be specified by `length.out`, which is often abbreviated to `length` or simply `len`.

```
> seq(-pi,pi,length.out=12) # 12 values between -pi and pi
[1] -3.142 -2.570 -1.999 -1.428 -0.857 -0.286  0.286  0.857
[9]  1.428  1.999  2.570  3.142
```

2.6.3.2 *Repeating values*

The `rep` function repeats values in a sequence.

```
> y = 2; rep(y,5) # Or rep(y, times=5)
[1] 2 2 2 2 2

> w = c(4,5); rep(w,5)  # Repeats w
[1] 4 5 4 5 4 5 4 5 4 5

> rep(w, each=5)  # Repeats each term in w
[1] 4 4 4 4 4 5 5 5 5 5
```

2.6.3.3 *Sequences of random numbers*

Sequences of random numbers are often used in simulations. R has many different probability distributions from which random numbers can be drawn, but two are most commonly used: uniformly distributed random numbers, and normally distributed random numbers.

A sequence of n random numbers uniformly distributed between min and max is generated by runif(n, min, max). If min and max are not specified, the defaults are 0 and 1.

```
> runif(6,-2,2)
[1]  0.0590 -0.0343 -0.2844 -0.0167  1.3273 -0.8960

> runif(6)
[1] 0.176 0.235 0.316 0.656 0.817 0.636
```

Likewise, a sequence of n random numbers drawn from a normal distribution with mean mean and standard deviation sd is generated by rnorm(n, mean, sd). If mean and sd are not specified, the defaults are 0 and 1.

```
> rnorm(6,9,1.5)
[1]  7.36  9.49 10.79  8.59 11.90  6.44

> rnorm(6)
[1] -1.103 -0.849  1.148  1.460 -0.831  0.919
```

In a common simulation scenario, one wants to generate a sequence of values with normally distributed random error of fixed standard deviation. For example:

```
> x = 1:6
> err = rnorm(6,0,0.1) # mean of error = 0, sd = 0.1

> x + err # "experimental" result
[1] 1.01 1.78 3.07 4.11 5.01 5.80
```

Instead, it may be desired to generate a sequence with given relative error:

```
> x*(1+err)
[1] 1.01 1.56 3.21 4.44 5.06 4.79
```

Usually R sets the seed for a sequence of random numbers based on the system clock. To set a specific integer seed, e.g., to check the reproducibility of a calculation involving random numbers, use the set.seed function. For example

```
> set.seed(123)
> round(rnorm(5),3)
[1] -0.560 -0.230  1.559  0.071  0.129
```

2.6.4 *Logical vectors*

Sometimes we want to pick out those elements of a vector that obey some criterion. Suppose we generate a vector v of random numbers and want to pick out those elements that are greater than 0. (Enclosing the expression below in parentheses causes the result to print.)

```
> (v = runif(8,-3,3))
[1] -1.68 1.97 -1.38 1.87 -2.42 -0.74 -0.16 1.25
```

The statement v > 0 will generate a vector the length of v whose elements indicate whether each element of v obeys the criterion.

```
> v > 0
[1] FALSE TRUE FALSE TRUE FALSE FALSE FALSE TRUE
```

The `which` function gives the index of each element that satisfies the criterion.

```
> which(v > 0)
[1] 2 4 8
```

A vector consisting of only those elements that satisfy the criterion can be constructed as follows.

```
> v[v > 0]
[1] 1.97 1.87 1.25
```

Two vectors will usually be compared using the logical operator ==. But just as we saw above with scalars, round-off error may lead to small differences, hence strict inequality, when fractions are concerned. The `all.equal` test may be more useful.

```
> v1 = seq(.1,.6,.1)/.1; v1
[1] 1 2 3 4 5 6

> w1 = 1:6; w1
[1] 1 2 3 4 5 6

> v1 == w1
[1] TRUE TRUE FALSE TRUE TRUE FALSE

> all.equal(v1,w1)
[1] TRUE
```

2.6.5 *Speed of forming large vectors*

When vectors have relatively few elements, the speed of forming them is generally inconsequential. However, if there are tens of thousands of elements, then there may be significant differences in program execution time depending on how the vector elements are formed. If the vector is initially defined as a scalar, and then expanded one element at a time, R has to reallocate computer memory at each step. On the other hand, if the length of the vector is known ahead of time, memory can be allocated at the beginning, and then the process is much faster.

We compare the timing of two ways of extending a vector using a `for` loop (see Chapter 4 on Functions and Programming) with the function `system.time`.

```
> n=10000  # length of vector

> v = 1 # value of first element, starting as a scalar
# Extending v one element at a time
> system.time(for (i in 2:n) v[i] = i)
   user  system elapsed
0.250   0.203   0.560
```

```
# Allocate n places in memory initially:
> v1 = numeric(n)
# Now fill those places
> system.time(for (i in 2:n) v1[i] = i)
   user  system elapsed
  0.034   0.002   0.068
```

Even faster is using the sequence function:

```
> system.time({v2 = 1:n})
user system elapsed
  0    0    0

> head(v2)
[1] 1 2 3 4 5 6

> tail(v2)
[1] 9995 9996 9997 9998 9999 10000
```

The head and tail functions are useful for checking the beginning and end of very large vectors without printing the entire vector.

2.6.6 *Vector dot product and crossproduct*

Up to this point we have been using the standard R meaning of "vector" as a one-dimensional list of numbers. However, a scientist or engineer might expect "vector" to mean a quantity having direction as well as magnitude, typically specified by a triplet of numbers denoting the projections of the vector along three orthogonal axes, e.g., (x, y, z) in Cartesian coordinates. We will denote these vectors by lower-case bold-face names. Two standard operations on such vectors are dot product and cross product. R does not have built-in functions for these operations, but it is easy to construct them.

Consider two vectors **u** and **v**, defined by the triplets $(u1, u2, u3)$ and $(v1, v2, v3)$ respectively. Their dot product is $\mathbf{u} \cdot \mathbf{v} = u1v1 + u2v2 + u3v3$. This can be written in R as the function

```
> dot = function(u,v) as.numeric(u%*%v)
```

where the as.numeric function is needed to coerce the result from a 1×1 matrix to a scalar. The magnitude or Euclidian norm of **v** can then be written as the function

```
> vecnorm = function(v) sqrt(dot(v,v))
```

For example,

```
> u = c(1,2,3)
> v = c(4,5,6)

> dot(u,v)
[1] 32
```

```
> vecnorm(u)  # sqrt(1^2 + 2^2 +3^2) = sqrt(14)
[1] 3.742
```

Note that the magnitude of **v** might also be called its length, but in R the length of a vector is its number of elements, not its magnitude. R has a function `norm` to calculate any one of several norms of a matrix. (Type `?norm` for details.) We can use it to calculate the Euclidian norm of a vector by converting the vector to a (one-dimensional) matrix and choosing the Frobenius option "F" or "f":

```
> norm(as.matrix(u),"F")
[1] 3.742
```

The crossproduct of two three-dimensional vectors is calculated with the function

```
> cross = function(u,v) {c(u[2]*v[3]-u[3]*v[2],
+ u[3]*v[1]-u[1]*v[3], u[1]*v[2]-u[2]*v[1])}

> cross(u,v)
[1] -3  6 -3
```

Note that `"+"` is added automatically to the beginning of the next line when the preceding line does not form a complete statement, in this case because it ends with a comma.

R has a function `crossprod` that, when operating on vectors, behaves like `dot` (but yields a 1×1 matrix). See below for its use in matrix multiplication, and `?crossprod` for details. To add to the confusion, R already has a function `dot` that, in the `plotmath` package for annotating graphics, yields "x with a dot." It's important not to confuse these usages.

More convenient, since we will be using it in a variety of contexts, may be to install and load the add-on package `pracma`, which contains the expected vector dot and crossproducts.

```
> install.packages("pracma")  # If not already installed
> require(pracma)
Loading required package: pracma
Attaching package: pracma
The following objects are masked _by_ .GlobalEnv:
    cross, dot
```

Note that `pracma` now superimposes its definition of `dot` over the base R definition as well as over the functions we have just defined.

```
> u = c(1,2,3)
> v = c(4,5,6)

> dot(u,v)
[1] 32

> cross(u,v)
[1] -3  6 -3
```

The reader is urged to type `cross` (without the ?) to see how the definition of this function is implemented in `pracma`: essentially the same as above.

2.7 Matrices

Matrices—two-dimensional arrays of numbers—are ubiquitous in numerical analysis, and R has an extensive set of functions for dealing with them.

2.7.1 Forming matrices

A matrix may be constructed in R as follows:

```
> (m = matrix(c(3,-4.2,-7.1,0.95),nrow=2,ncol=2))
      [,1]  [,2]
[1,]  3.0 -7.10
[2,] -4.2  0.95
```

`nrow` is the number of rows, `ncol` the number of columns. Note that R, by default, fills the matrix in column order. If row order is desired, it must be specified by `byrow=TRUE`):

```
> (m = matrix(1:6, nrow=2, byrow=T))
     [,1] [,2] [,3]
[1,]    1    2    3
[2,]    4    5    6
```

Given six elements and two rows, R is smart enough to figure out that there should be three columns.

When printed out, the matrix is flanked by row and column indices. A particular element `i,j` is specified by its row and column indices in square brackets, while whole rows or columns are specified by `[i,]` and `[,j]`, respectively.

```
> m[2,3]
[1] 6

> m[2,]
[1] 4 5 6

> m[,3]
[1] 3 6
```

A matrix may also be formed by binding together row (`rbind`) or column (`cbind`) vectors. For example,

```
> x = 1:3; y = 1:3

> rbind(x,y)
  [,1] [,2] [,3]
x    1    2    3
y    1    2    3
```

```
> cbind(x,y)
     x y
[1,] 1 1
[2,] 2 2
[3,] 3 3
```

A diagonal matrix is constructed using `diag`:

```
> diag(c(4,6,5))
     [,1] [,2] [,3]
[1,]    4    0    0
[2,]    0    6    0
[3,]    0    0    5
```

so an $n \times n$ unit matrix can be constructed by `diag(rep(1,n))`:

```
> diag(rep(1,3))
     [,1] [,2] [,3]
[1,]    1    0    0
[2,]    0    1    0
[3,]    0    0    1
```

Similarly, a 3×3 matrix with all zeros is constructed by

```
> matrix(rep(0,9), nrow=3)
     [,1] [,2] [,3]
[1,]    0    0    0
[2,]    0    0    0
[3,]    0    0    0
```

The `outer()` operator forms an $m \times n$ matrix by combining two vectors of lengths m and n according to a function specified by `FUN`. The default function is "*".

```
> x = 1:3; y = 1:3

> outer(x,y)
     [,1] [,2] [,3]
[1,]    1    2    3
[2,]    2    4    6
[3,]    3    6    9

> outer(x,y,FUN="+")
     [,1] [,2] [,3]
[1,]    2    3    4
[2,]    3    4    5
[3,]    4    5    6
```

Here's an easy way to use `outer()` to make a table (actually a matrix) of powers of integers:

```
> x = 1:9; y = 2:8
> names(x)=x; names(y)=y
> outer(y,x,"^")
  1  2   3    4     5      6       7        8         9
2 2  4   8   16    32     64     128      256       512
3 3  9  27   81   243    729    2187     6561     19683
4 4 16  64  256  1024   4096   16384    65536    262144
5 5 25 125  625  3125  15625   78125   390625   1953125
6 6 36 216 1296  7776  46656  279936  1679616  10077696
7 7 49 343 2401 16807 117649  823543  5764801  40353607
8 8 64 512 4096 32768 262144 2097152 16777216 134217728
```

The kronecker function is useful for constructing block matrices. Given two matrices M1 and M2, kronecker(M1,M2) returns a matrix with dimensions dim(M1)*dim(M2).

```
> (M1 = matrix(1:4,2,2))
     [,1] [,2]
[1,]    1    3
[2,]    2    4

> (M2 = diag(1,2))
     [,1] [,2]
[1,]    1    0
[2,]    0    1

> kronecker(M1,M2)
     [,1] [,2] [,3] [,4]
[1,]    1    0    3    0
[2,]    0    1    0    3
[3,]    2    0    4    0
[4,]    0    2    0    4

> kronecker(M2,M1)
     [,1] [,2] [,3] [,4]
[1,]    1    3    0    0
[2,]    2    4    0    0
[3,]    0    0    1    3
[4,]    0    0    2    4
```

A submatrix can be formed from a larger matrix by putting the desired row and column indices in square brackets.

```
> (m3 = matrix(1:9,3,3,byrow=T))
     [,1] [,2] [,3]
[1,]    1    2    3
[2,]    4    5    6
[3,]    7    8    9
```

```
> m3[1:2,c(1,3)]
     [,1] [,2]
[1,]    1    3
[2,]    4    6
```

If the rows and columns of a matrix arise from a series of measurements, as in a database where each row corresponds to a subject and each column to a particular measurement, it may be convenient to give the rows and columns descriptive names. (See the discussion of `data.frame` in Chapter 10.) `rownames` and `colnames` are used for this purpose. For example, using the matrix `m` defined above:

```
> rownames(m) = c("A","B")
> colnames(m) = c("v1","v2","v3")

> m
  v1 v2 v3
A  1  2  3
B  4  5  6
```

The rows or columns may then be queried individually by name, e.g.,

```
> m[,"v1"]
A B
1 4

> summary(m[,"v1"])
   Min. 1st Qu.  Median    Mean 3rd Qu.    Max.
   1.00    1.75    2.50    2.50    3.25    4.00
```

The names can also be assigned when the matrix is constructed:

```
> m = matrix(1:6, nrow=2, ncol=3, byrow=T,
+ dimnames = list(c("A","B"),c("v1","v2","v3")))
```

2.7.2 *Operations on matrices*

2.7.2.1 *Arithmetic operations on matrices*

As with vectors, simple operations on matrices are applied individually to each element.

```
> m-2
  v1 v2 v3
A -1  0  1
B  2  3  4

> m/5
   v1  v2  v3
A 0.2 0.4 0.6
B 0.8 1.0 1.2
```

```
> options(digits=3)

> sqrt(m)
  v1   v2   v3
A  1 1.41 1.73
B  2 2.24 2.45

> m^(-1)
    v1  v2    v3
A 1.00 0.5 0.333
B 0.25 0.2 0.167
```

If two matrices are combined by simple operations, assuming their row and column dimensions match, the operations are applied to the individual elements.

```
> m+m
  v1 v2 v3
A  2  4  6
B  8 10 12

> m/m
  v1 v2 v3
A  1  1  1
B  1  1  1
```

2.7.2.2 Matrix multiplication

Multiplication of two matrices **M1** and **M2** to get a new matrix **M3** is defined as

$$M3_{1j} = \sum_{k=1}^{n} M1_{ik} M2_{kj} \tag{2.1}$$

where n is the number of columns of **M1**, which must equal the number of rows of **M2**.

The R operator for matrix multiplication is %*%. For example,

```
> M1 = matrix(runif(9),3,3); M1
      [,1]  [,2]  [,3]
[1,] 0.261 0.338 0.176
[2,] 0.402 0.786 0.827
[3,] 0.899 0.765 0.340

> M2 = matrix(runif(9),3,3); M2
       [,1]    [,2]  [,3]
[1,] 0.0844 0.00498 0.478
[2,] 0.8571 0.80753 0.505
[3,] 0.9829 0.18913 0.133
```

```
> M3 = M1 %*% M2; M3
       [,1]  [,2]  [,3]
[1,] 0.485 0.308 0.319
[2,] 1.520 0.793 0.699
[3,] 1.066 0.687 0.861
```

2.7.2.3 *Transpose and determinant*

The transpose of a matrix is denoted by `t`:

```
> t(M1)
       [,1]  [,2]  [,3]
[1,] 0.261 0.402 0.899
[2,] 0.338 0.786 0.765
[3,] 0.176 0.827 0.340
```

and the determinant (of a square matrix) by `det`:

```
> det(M1)
[1] 0.0395
```

2.7.2.4 *Matrix crossproduct*

The twin functions `crossprod` and `tcrossprod` are slightly faster ways of multiplying a matrix by the transpose of another matrix.

```
> A = matrix(1:4,2,2)
> B = matrix(5:8,2,2)

> crossprod(A,B)
     [,1] [,2]
[1,]   17   23
[2,]   39   53

> t(A) %*% B
     [,1] [,2]
[1,]   17   23
[2,]   39   53

> tcrossprod(A,B)
     [,1] [,2]
[1,]   26   30
[2,]   38   44

> A %*% t(B)
     [,1] [,2]
[1,]   26   30
[2,]   38   44
```

Type ?crossprod for more details.

2.7.2.5 Matrix exponential

Both the pracma and Matrix packages provide the function expm that computes the exponential of a matrix, formally defined as the infinite Taylor series $exp(A) = I + A + A^2/2! + A^3/3! + \ldots$ For example,

```
> require(Matrix)

> (A = matrix(1:4,2,2))
     [,1] [,2]
[1,]    1    3
[2,]    2    4

> expm(A)
2 x 2 Matrix of class "dgeMatrix"
      [,1]  [,2]
[1,] 51.97 112.1
[2,] 74.74 164.1
```

dgeMatrix is the "standard" class for dense numeric matrices in the Matrix package. Type help(package = "Matrix") for more details. The matrix exponential function may be used to solve sets of first-order, linear differential equations, whose formal solutions are often sums of exponential functions.

2.7.2.6 Matrix inverse and solve

The functions solve and eigen are central to numerical analysis of linear algebraic systems. These and related matrix functions are handled in R via the standard LAPACK code. According to its website, http://www.netlib.org/lapack/,

> LAPACK is written in Fortran 90 and provides routines for solving systems of simultaneous linear equations, least-squares solutions of linear systems of equations, eigenvalue problems, and singular value problems. The associated matrix factorizations (LU, Cholesky, QR, SVD, Schur, generalized Schur) are also provided, as are related computations such as reordering of the Schur factorizations and estimating condition numbers. Dense and banded matrices are handled, but not general sparse matrices. In all areas, similar functionality is provided for real and complex matrices, in both single and double precision.

The inverse of a square matrix is computed using the solve function.

```
> (M1_inv = solve(M1))
       [,1]   [,2]  [,3]
[1,]  -9.23  0.502  3.57
[2,]  15.34 -1.767 -3.66
[3,] -10.10  2.645  1.75
```

Applying solve to this result should yield the original matrix, which it appears to:

```
> solve(solve(M1))
      [,1]  [,2]  [,3]
[1,] 0.261 0.338 0.176
[2,] 0.402 0.786 0.827
[3,] 0.899 0.765 0.340
```

but equality is not exact due to round-off error.

```
# Test for exact equality
> identical(M1, solve(solve(M1)))
[1] FALSE
# Equality to within machine precision
> all.equal(M1, solve(solve(M1)))
[1] TRUE
```

The most familiar use of matrix inversion is to solve sets of linear algebraic equations, as will be discussed in Chapter 5. Here we give a brief demonstration. Consider the system of three equations in three unknowns:

$$x_1 + \frac{1}{2}x_2 + \frac{1}{3}x_3 = 1 \tag{2.2}$$

$$\frac{1}{2}x_1 + \frac{1}{3}x_2 + \frac{1}{4}x_3 = 0 \tag{2.3}$$

$$\frac{1}{3}x_1 + \frac{1}{4}x_2 + \frac{1}{5}x_3 = 0 \tag{2.4}$$

This can be written in matrix form as

$$\begin{pmatrix} 1 & \frac{1}{2} & \frac{1}{3} \\ \frac{1}{2} & \frac{1}{3} & \frac{1}{4} \\ \frac{1}{3} & \frac{1}{4} & \frac{1}{5} \end{pmatrix} \begin{pmatrix} x_1 \\ x_2 \\ x_3 \end{pmatrix} = \begin{pmatrix} 1 \\ 0 \\ 0 \end{pmatrix} \tag{2.5}$$

or in compact form as

$$\mathbf{Ax} = \mathbf{b} \tag{2.6}$$

Premultiplying both sides by $\mathbf{A}^{-1}$,

$$\mathbf{A}^{-1}\mathbf{Ax} = \mathbf{x} = \mathbf{A}^{-1}\mathbf{b} \tag{2.7}$$

Using R, we solve for x by

```
> A = matrix(c(1,1/2,1/3,
+     1/2,1/3,1/4,
+     1/3,1/4,1/5),
+     nrow=3, byrow=T)
> Ainv = solve(A)

> b = c(1,0,0)

> (x = Ainv %*% b)
```

```
     [,1]
[1,]    9
[2,]  -36
[3,]   30

> A %*% x # Check result
         [,1]
[1,] 1.00e+00
[2,] 8.88e-16
[3,] 0.00e+00
```

R also provides a much simpler and more compact way of doing the same thing, with the added advantage that the result is a vector rather than a 3×1 matrix:

```
> solve(A,b)
[1]   9 -36  30
```

2.7.2.7 Eigenvalues and eigenvectors

The eigenvalues and eigenvectors of a square matrix are obtained by the `eigen` function.

```
> eigen(M1)
$values
[1] 1.63+0.0i -0.12+0.1i -0.12-0.1i
$vectors
         [,1]             [,2]             [,3]
[1,] 0.263+0i -0.292-0.149i -0.292+0.149i
[2,] 0.737+0i  0.695+0.000i  0.695+0.000i
[3,] 0.622+0i -0.620+0.156i -0.620-0.156i
```

Note that `eigen` computes both real and imaginary parts as required. To get the eigenvectors alone, use

```
> eigen(M1)$values
[1]  1.63+0.0i -0.12+0.1i -0.12-0.1i
```

A square, symmetric matrix is an example of a Hermitian matrix. It can be proved, and is important in quantum mechanics, that the eigenvalues of a Hermitian matrix are real and the eigenvectors are orthonormal (mutually perpendicular and of unit length). We can demonstrate these properties with a particular numerical example.

```
> H = matrix(c(1,2,3,
+    2,5,-1,
+    3,-1,7),3,3,byrow=T)

> (Hval = eigen(H)$values)
[1]  8.25  5.81 -1.06

> (Hvec = eigen(H)$vectors)
```

```
        [,1]    [,2]   [,3]
[1,] -0.3676 -0.3422  0.865
[2,]  0.0593 -0.9366 -0.345
[3,] -0.9281  0.0756 -0.365

> Hvec[,1]%*%Hvec[,1]
     [,1]
[1,]    1
> Hvec[,2]%*%Hvec[,2]
     [,1]
[1,]    1
> Hvec[,3]%*%Hvec[,3]
     [,1]
[1,]    1

> Hvec[,1]%*%Hvec[,2]
          [,1]
[1,] -2.91e-16
> Hvec[,1]%*%Hvec[,3]
         [,1]
[1,] 5.55e-17
> Hvec[,2]%*%Hvec[,3]
          [,1]
[1,] -2.15e-16
```

More generally, a square matrix with complex elements is Hermitian if the element in the i-th row and j-th column is equal to the complex conjugate of the element in the j-th row and i-th column, for all indices i and j. In this case, the eigenvalues are real and the eigenvectors are orthonormal with their complex conjugates. For example:

```
> Hi = matrix(c(1,2+7i,3,
+    2-7i,5,-1,
+    3,-1,7),3,3,byrow=T)

> (Hival = eigen(Hi)$values)
[1] 11.35  6.77 -5.12

> (Hivec = eigen(Hi)$vectors)
            [,1]           [,2]           [,3]
[1,] 0.594+0.000i -0.160+0.000i  0.788+0.000i
[2,] 0.127-0.679i -0.395+0.182i -0.176+0.549i
[3,] 0.380+0.156i  0.379+0.801i -0.210+0.045i

> Hivec[,1]%*%Conj(Hivec[,1])
     [,1]
[1,] 1+0i
```

```
> Hivec[,1]%*%Conj(Hivec[,2])
                         [,1]
[1,] -5.55e-17-2.78e-17i
```

Of course, the fact that the imaginary parts that should be zero are not exactly so in these calculations shows the limitations of representing decimal numbers in binary arithmetic.

2.7.2.8 Singular value decomposition

Singular value decomposition, or SVD, is useful for dealing with matrices that are either singular (determinant = 0) or nearly so (ill-conditioned).[1] Applying the R function `svd` to an $m \times n$ matrix $\mathbf{X}$ decomposes it into the product of three matrices, $\mathbf{X} = \mathbf{UDV}^t$ where $\mathbf{X}$ is an $m \times n$ matrix, $\mathbf{D}$ is an $n \times n$ diagonal matrix, and $\mathbf{V}^t$ is the transpose of an $n \times n$ matrix.

If $\mathbf{X}$ is a square matrix, `svd(X)` and `eigen(X)` give essentially identical results. We illustrate with the Hilbert matrix used in the `svd` help example, a square matrix in which the i,j element is $1/(i+j-1)$. The Hilbert matrix is notoriously ill-conditioned for moderate and larger n. It may be defined in R for arbitrary n by the function

```
> hilbert = function(n) {i=1:n; 1/outer(i-1,i,"+")}.
```

(Accept this on faith for now. We shall discuss how to define functions in Chapter 4.)

Let us begin with a small value, $n = 4$.

```
> (h4 = hilbert(4))
      [,1]  [,2]  [,3]  [,4]
[1,] 1.000 0.500 0.333 0.250
[2,] 0.500 0.333 0.250 0.200
[3,] 0.333 0.250 0.200 0.167
[4,] 0.250 0.200 0.167 0.143
> s4 = svd(h4)
> s4
$d
[1] 1.50e+00 1.69e-01 6.74e-03 9.67e-05
$u
       [,1]   [,2]   [,3]    [,4]
[1,] -0.793  0.582 -0.179 -0.0292
[2,] -0.452 -0.371  0.742  0.3287
[3,] -0.322 -0.510 -0.100 -0.7914
[4,] -0.252 -0.514 -0.638  0.5146
$v
       [,1]   [,2]   [,3]    [,4]
[1,] -0.793  0.582 -0.179 -0.0292
[2,] -0.452 -0.371  0.742  0.3287
[3,] -0.322 -0.510 -0.100 -0.7914
```

[1] www.math.umn.edu/~lerman/math5467/svd.pdf

```
[4,] -0.252 -0.514 -0.638  0.5146
# Recover h4:
> s4$u %*% diag(s4$d) %*% t(s4$v)
      [,1]  [,2]  [,3]  [,4]
[1,] 1.000 0.500 0.333 0.250
[2,] 0.500 0.333 0.250 0.200
?[3,] 0.333 0.250 0.200 0.167
[4,] 0.250 0.200 0.167 0.143
```

Compare with the eigenanalysis of h4:

```
> eigen(h4)
$values
[1] 1.50e+00 1.69e-01 6.74e-03 9.67e-05
$vectors
      [,1]   [,2]   [,3]    [,4]
[1,] 0.793  0.582 -0.179 -0.0292
[2,] 0.452 -0.371  0.742  0.3287
[3,] 0.322 -0.510 -0.100 -0.7914
[4,] 0.252 -0.514 -0.638  0.5146
```

Thus svd and eigen give essentially the same numerical results for square, symmetric matrices, though they are not strictly identical because of the different steps taken to arrive at those results:

```
> eigen(h4)$values == svd(h4)$d
[1] FALSE FALSE FALSE FALSE
> all.equal(eigen(h4)$values, svd(h4)$d)
[1] TRUE
```

Whether a matrix is ill-conditioned may be judged from its 2-norm condition number, the ratio of its largest to its smallest singular values (equivalently, the ratio of its largest to smallest eigenvalues). We abbreviate this ratio as cn, and define it as a function for our Hilbert matrices:

```
> cn = function(n) max(svd(hilbert(n))$d)/
  + min(svd(hilbert(n))$d)
> cn(5)
[1] 476607
> cn(6)
[1] 14951059
```

Even matrix dimensions of 5 x 5 and 6 x 6 have high condition numbers, and are dangerously ill-conditioned. This can also be seen from the determinants of the matrices:

```
> det(hilbert(5))
[1] 3.75e-12
> det(hilbert(6))
[1] 5.37e-18
```

where the determinant of hilbert(6) is below machine precision.

If the matrix $\mathbf{X}$ is not square, its eigenvalues and eigenvectors cannot be computed, but singular value decomposition is still applicable. If $m > n$, there are more equations than unknowns and the system is overdetermined. In this case $m - n$ of the singular values will be zero. This is the case in the example for `svd` help, where the matrix `hilbert(9)[,1:6]` has nine rows and six columns. More importantly, it is the case in the least squares fitting of data, where there are typically more measurements than parameters to be fit.

If $m < n$ there are more unknowns than equations and the system is underdetermined. Then there will be no unique solution, but instead an infinite $(n - m)$-dimensional family of solutions. For `svd(hilbert(9)[1:6,])`, R picks the same solution as it did for `svd(hilbert(9)[, 1:6])`, but this will not generally be the case for unsymmetrical matrices.

2.7.3 The `Matrix` package

The `Matrix` package has "recommended" priority. That is, it is included with all recent distributions of R, but needs to be loaded with a `library` or `requires` function. It implements "[c]lasses and methods for dense and sparse matrices and operations on them using Lapack and SuiteSparse." The `Matrix` package preserves the sparseness of matrices as successive operations are performed, which may provide substantial improvement in computational speed and memory utilization for the very large matrices that are frequently encountered in real-world applications. However, for the standard solution of large sets of linear equations involving common tridiagonal, banded, and block matrices, `limSolve` (see Chapter 5) may be a simpler choice.

Consider computing the eigenvalues of the large, sparse matrix `CAex` used as an example in the `Matrix` documentation. `CAex` is a 72×72 symmetric matrix with 216 non-zero entries (4.17% of the total) in five bands. It is stored as a sparse matrix of class `dgCMatrix`.

```
> library(Matrix)
> data(CAex)
> image(CAex)
```

We calculate the eigenvalues of `CAex` with the `eigen` function, using the option `only.values = TRUE` (the default is `FALSE`) since calculating the eigenfunctions takes the majority of the computation time.

```
> CAex.eigval=eigen(CAex, only.values=TRUE)$values
> zapsmall(CAex.eigval)
 [1] 1 1 1 1 1 1 1 1 1 1 1 1 1 1 1 1 1 1 1 1 1 1 1 1 1 1 1 1
[29] 1 1 1 1 1 1 1 1 1 1 1 1 1 0 0 0 0 0 0 0 0 0 0 0 0 0 0
[57] 0 0 0 0 0 0 0 0 0 0 0 0 0 0 0 0
```

For a discussion of relative speeds of matrix calculations in base R and in the `Matrix` package, see `http://cran.r-project.org/web/packages/Matrix/vignettes/Comparisons.pdf`.

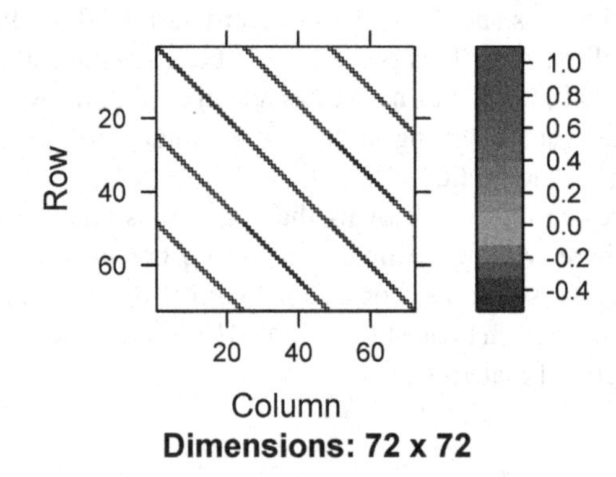

Figure 2.1: *Image plot of sparse banded matrix* `CAex`.

2.7.4 *Additional matrix functions and packages*

The base R installation has the standard QR (`qr`) and Cholesky (`chol`) decomposition functions for matrix manipulation. Since these are mainly used in linear algebra and linear regression applications, we shall delay discussing these functions until Chapter 5.

2.8 Time and date calculations

R has numerous functions to manipulate times and dates. It is frequently desired to determine how long a given process takes. This can be done with the `system.time` function, which is wrapped around the code for the process of interest. For example, to sum the results of evaluations of the sines of ten million uniformly distributed random numbers:

```
> system.time(cumsum(sin(runif(1e7))))
   user  system elapsed
  0.687   0.064   0.746
```

"`user`" is the number of CPU seconds spent running R, "`system`" is the number of CPU seconds spent running the operating system, and "`elapsed`" is the clock time, which may be greater than the sum of "`user`" and "`system`" if the computer is simultaneously running other processes.

The `system.time` function is one of the few instances in which `"="` cannot be substituted for `"<-"` in an assignment statement.

```
> system.time(cs <- cumsum(sin(runif(1e7))))
   user  system elapsed
  0.606   0.015   0.617
> system.time(cs = cumsum(sin(runif(1e7))))
```

```
Error in system.time(cs = cumsum(sin(runif(1e+07)))) :
  unused argument(s) (cs = cumsum(sin(runif(1e+07))))
```

To get the current date, type `Sys.Date()`. To specify some other date, and get the number of days difference,

```
> Earlier.Date = as.Date("1/1/2011", "%m/%d/%Y")
> Sys.Date() - Earlier.Date # As of 10/04/2013
Time difference of 1007 days
```

Descriptions of the classes "`POSIXlt`" and "`POSIXct`" representing calendar dates and times (to the nearest second), and various ways to specify and calculate with these quantities, may be obtained by typing `?DateTimeClasses`. To learn how to compute time differences, type `?difftime`.

Chapter 3

Graphing

It is not traditional for a book on numerical methods to devote much space to graphing of data and functions. Yet graphing is an essential part of scientific and engineering work, and R has very strong graphics tools. Therefore, we include a chapter on the topic here. Two recent books on graphing with R are

- Paul Murrell, *R Graphics*, Second Edition, Chapman & Hall/CRC, 2011
- Hrishi V. Mittal, *R Graphs Cookbook*, Packt Publishing, 2011

3.1 Scatter plots

Perhaps the most common type of plot encountered in the scientific and engineering literature is the scatter plot of a dependent variable y measured or evaluated at a set of discrete points of an independent variable x. As an example, we simulate measurement of the function $y = x^2 e^{-x/2}$ over the x-range 0–10 with 5% normally distributed random error.

```
> set.seed(123)  # Enables reproducible random number generation
> x = 1:10
> y = x^2*exp(-x/2)*(1+rnorm(n=length(x), mean=0, sd=0.05))
> par(mfrow=c(1,2))  # Plots in 1 row, 2 columns
> plot(x,y)
```

This gives a plot (Figure 3.1, left) suitable for initial inspection of the data, but not for formal presentation or even recording in a lab notebook. At a minimum, one will want to label the axes more informatively with `xlab` and `ylab`, and provide a title with `main`. In addition, one may want to change the point symbol with `pch` and make the points smaller with `cex`.

```
> plot(x,y, pch=19, cex=0.7, xlab="Time, sec",
+   ylab = "Signal Intensity", main = "Detector Output")
```

This gives the much better Figure 3.1, right.

By default, R plots (x,y) pairs as points, but lines `(type="l")` and overlays `(type="o")` are also useful (Figure 3.2).

```
> set.seed(123)  # Enables reproducible random number generation
> x = seq(from=0, to=10, by=0.1) # More closely spaced points
> y = x^2*exp(-x/2)*(1+rnorm(n=length(x), mean=0, sd=0.05))
```

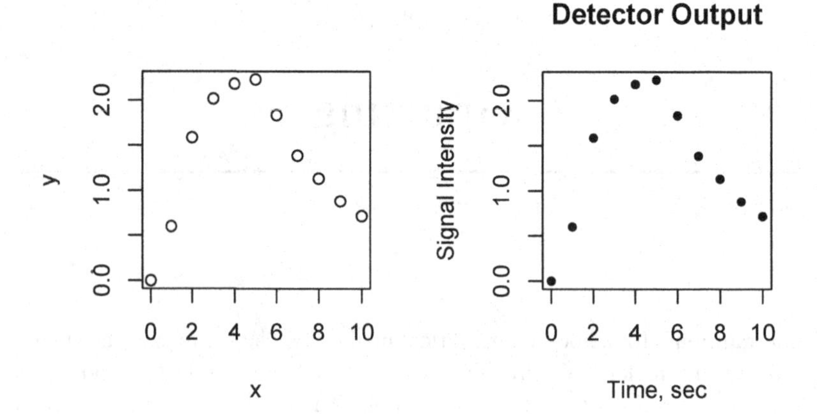

Figure 3.1: *Left: Default data plot; Right: Refined data plot.*

```
> par(mfrow=c(1,2))
> par(mar=c(4,4,1.5,1.5),mex=.8,mgp=c(2,.5,0),tcl=0.3)
> plot(x,y, type = "l")
> plot(x,y, type = "o")
> par(mfrow=c(1,1))
```

As we shall discuss in a few pages, R allows much greater customization of graphs than the few elementary steps we have shown here. However, from now on we will generally use the `par(mar=c(4,4,1.5,1.5),mex=.8,mgp=c(2,.5,0),tcl=0.3)` parameter setting to make the graphs more compact and to place the ticks within the graph, as is standard in most scientific work. Explanation of these parameters will be found in Section 3.4.2.

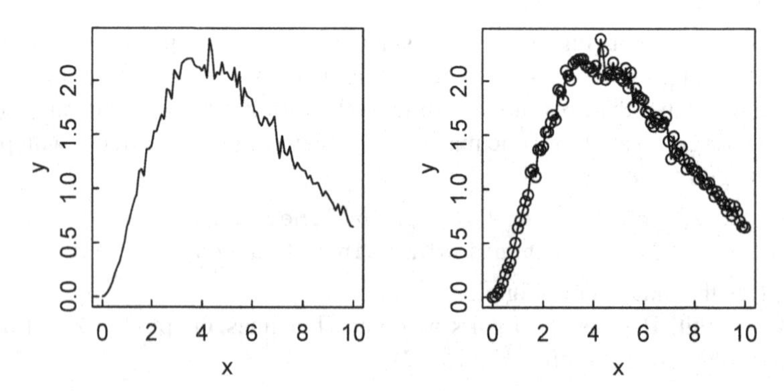

Figure 3.2: *Left:* `plot(x,y,type="l")`; *Right:* `plot(x,y,type="o")`.

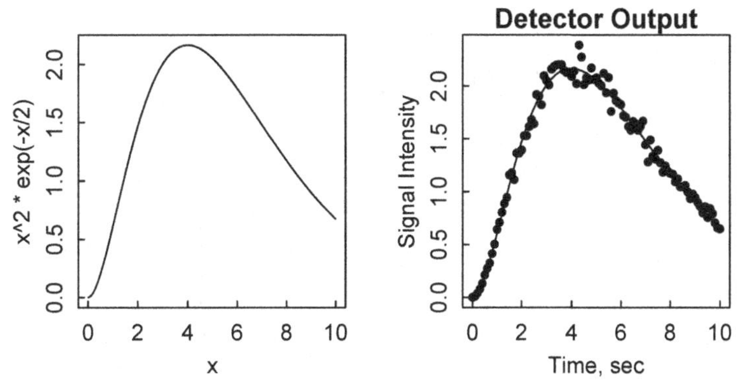

Figure 3.3: *Left: Function plot using* `curve`*; Right: Function plot superimposed on data points.*

3.2 Function plots

R enables plotting of functions of the variable x with the command `curve`. For example, to plot the function used to generate the data above, we write

```
> curve(x^2*exp(-x/2),0,10)
```

with the result shown in Figure 3.3, left. To superimpose the curve on the data points, we plot the data first, then generate the curve with the condition `add = TRUE`:

```
> plot(x,y, pch=19, cex=0.7, xlab="Time, sec",
+   ylab = "Signal Intensity", main = "Detector Output")
> curve(x^2*exp(-x/2),0,10, add=T)
```

as seen in Figure 3.3, right.

The command `curve()` works only with the variable x, and if given a function with no argument will assume that the argument is x, as in

```
> curve(sin,-4*pi,4*pi)
```

where we note that R provides x in the axis labels even though we did not specify it (Figure 3.4).

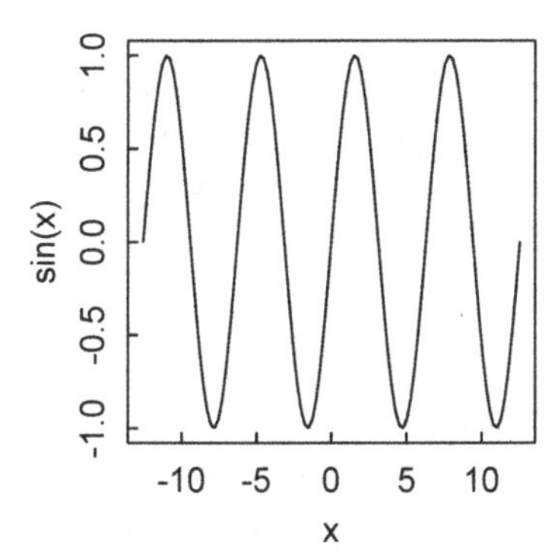

Figure 3.4: *The function sin plotted without specifying the independent variable.*

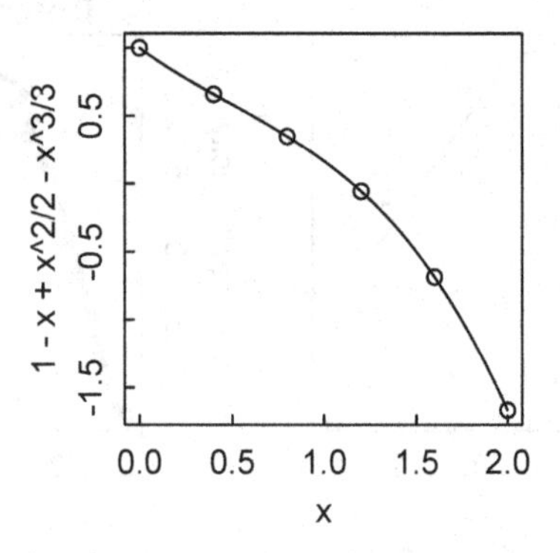

Figure 3.5: *curve plot of a polynomial with points added.*

On the other hand, the code

```
> curve(x)
```

generates the error message

```
Error in eval(expr, envir, enclos) : could not find function "x"
```

because R is unsure whether x is a function or a variable.

Just as we could superimpose a function curve on a scatter plot with the `curve(..., add=T)` command, we can superimpose points on a function plot with the `points()` command. In Figure 3.5, for example, are evenly spaced points imposed on a polynomial curve.

```
> curve(1-x+x^2/2-x^3/3,0,2)
> x = seq(0,2,.4)
> y = 1-x+x^2/2-x^3/3
> points(x,y)
```

In similar fashion, connected line segments could be added to a plot of points with the `lines()` function. See `?lines` for an example using data from the `cars` dataset in the R base package.

3.3 Other common plots

Other common ways of graphically representing scientific and engineering data are bar charts and histograms. In this section we show how to construct such plots in R. We also introduce box plots, which may be less familiar but which provide useful summaries of data.

3.3.1 Bar charts

Bar charts are known in R as `barplots`. For example, suppose a nutritionist is doing a feeding study using feeds A and B. She measures the average weight of two groups of mice, one group fed A and the other B, each week for three weeks, with results as indicated in the following code and shown in Figures 3.6 and 3.7.

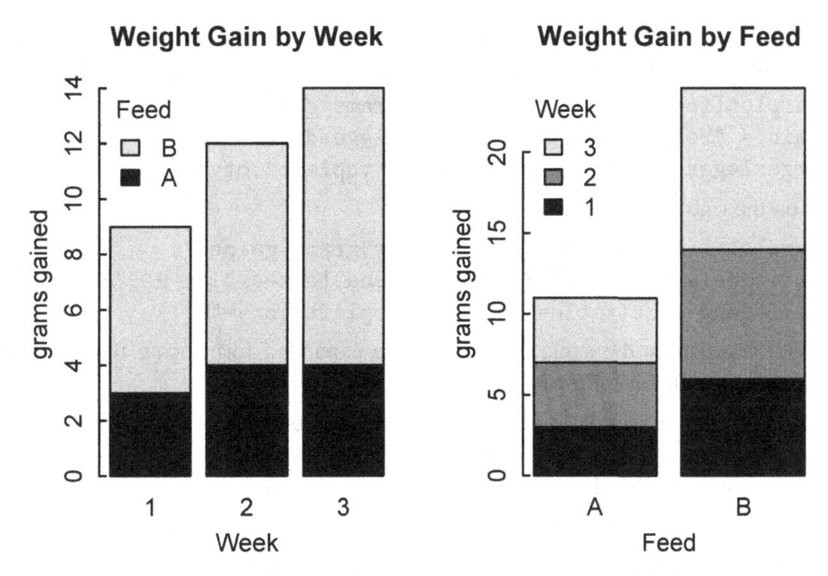

Figure 3.6: *Stacked bar plots using* `beside = FALSE` *default option.*

```
> A = c(3,4,4)
> B = c(6,8,10)
> feed = matrix(c(A,B), nrow=2, byrow=TRUE,
+   dimnames= list(c("A","B"), c("1","2","3")))
> feed  # Check that we've set up the matrix correctly
  1 2  3
A 3 4  4
B 6 8 10
```

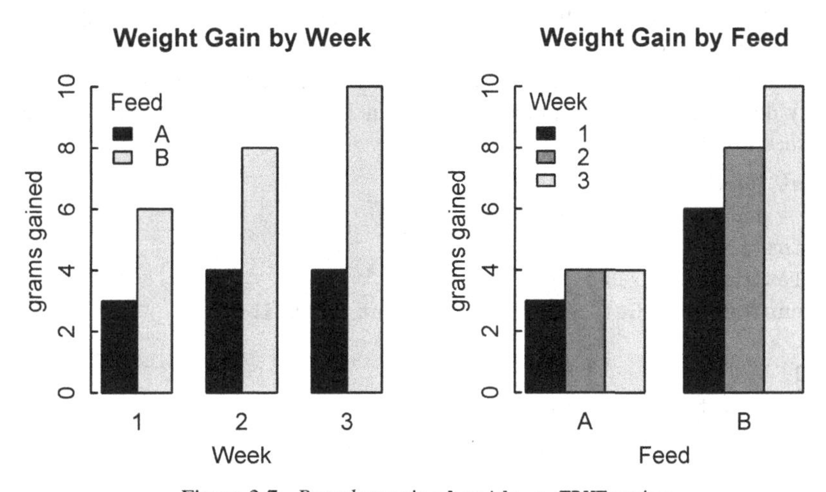

Figure 3.7: *Bar plots using* `beside = TRUE` *option.*

Now plot stacked barplots using the default `beside = FALSE` option, emphasizing, on the left, time as the independent variable:

```
> barplot(feed,xlab="Week",ylab="grams gained",
+ main = "Weight Gain by Week\n", legend.text=c("A","B"),
+ args.legend=list(title="Feed",x="topleft",bty="n"))
```

and, on the right, the feeds:

```
> barplot(t(feed),xlab="Feed",ylab="grams gained",
main = "Weight Gain by Feed\n",legend.text=c("1","2","3"),
args.legend=list(title="Week",x="topleft",bty="n"))
```

We can display the same data in a more expanded form using the `beside = TRUE` option as in the following code:

```
> barplot(feed, beside=T,xlab="Week",ylab="grams gained",
main = "Weight Gain by Week\n", legend.text=c("A","B"),
args.legend=list(title="Feed",x="topleft",bty="n"))
```

and

```
> barplot(t(feed), beside=T,xlab="Feed",ylab="grams gained",
main = "Weight Gain by Feed\n",
legend.text=c("1","2","3"),
args.legend=list(title="Week",x="topleft",bty="n"))
```

In `main` for both of these plots, `\n` is the newline command, introducing an extra line spacing after the main title. There are many options to the `barplot` command, some of which we have used above. See the help page for more details.

3.3.2 Histograms

Histograms are commonly used to display the distribution of repeated measurements. R does this with the `hist` function. If the fraction of measurements falling into each range is desired instead, use `plot(density)`, where the `density` function gives useful numerical data about the distribution. As an example, we generate 1000 normally distributed random numbers with mean 10 and standard deviation 2. Results are shown in Figure 3.8.

```
> set.seed(333)
> x = rnorm(1000,10,2)
> hist(x)
> plot(density(x))
> density(x) # Get information about distribution

Call:
density.default(x = x)
```

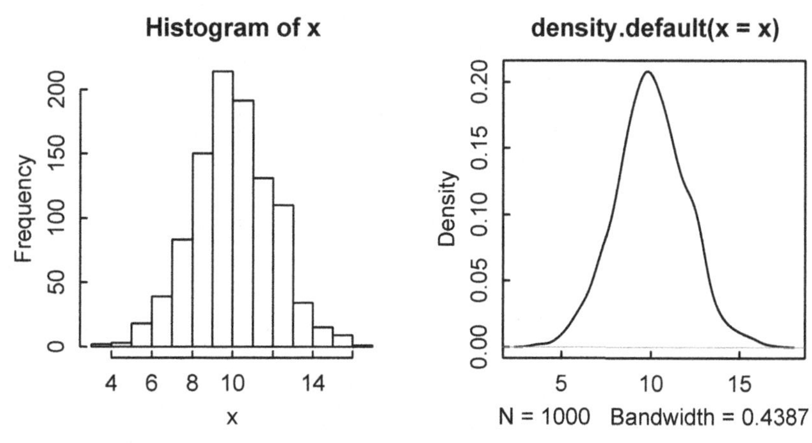

Figure 3.8: *Distribution of 1000 normally distributed random variables with mean = 10 and standard deviation = 2. Left: Histogram; Right: Density plot.*

```
Data: x (1000 obs.); Bandwidth 'bw' = 0.4387

       x                 y
 Min.   : 2.333   Min.   :1.035e-05
 1st Qu.: 6.277   1st Qu.:3.456e-03
 Median :10.220   Median :2.737e-02
 Mean   :10.220   Mean   :6.333e-02
 3rd Qu.:14.164   3rd Qu.:1.183e-01
 Max.   :18.108   Max.   :2.078e-01
```

The `hist` function tends to have a mind of its own when setting breaks between classes. To control this, use the argument `breaks = vecbreaks`, where `vecbreaks` is a vector that explicitly gives the breakpoints between histogram cells.

3.3.3 Box-and-whisker plots

The `boxplot` function gives a very informative graphical representation of a distribution of x in this case (Figure 3.8). The box shows the values of the first and third quartiles, the heavy line in the middle of the box gives the median, and the whiskers give the values of the quartile plus approximately 1.5 times the length of the interquartile range. Values beyond the whiskers are given by points. The plot was generated simply by

```
> boxplot(x, main = "Box Plot")
```

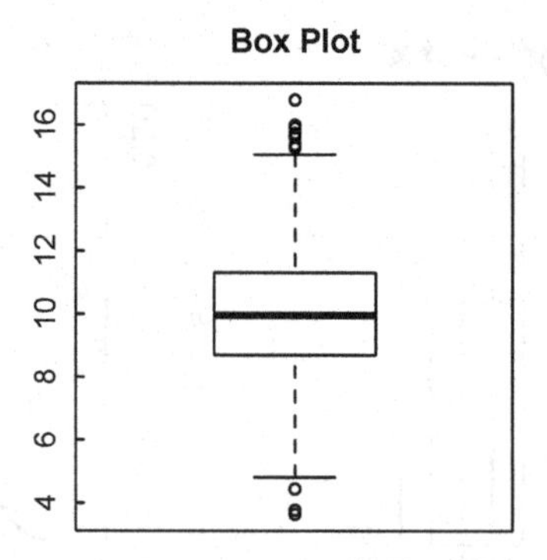

Figure 3.9: *Box plot of distribution of x from Figure 3.8.*

3.4 Customizing plots

3.4.1 Points and lines

R has 26 point styles, numbered from 0 to 25, which can be specified by `pch(n)` (Figure 3.10):

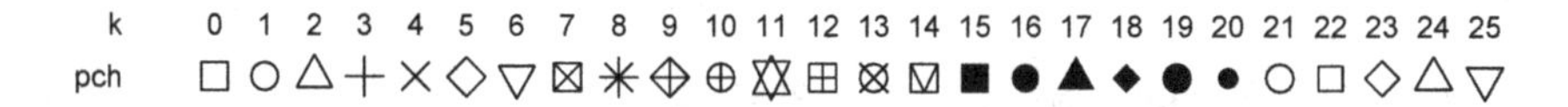

Figure 3.10: *Point characters available in R.*

It also has six line types, numbered from 1 to 6, which can be specified by `lty(n)` (Figure 3.11):

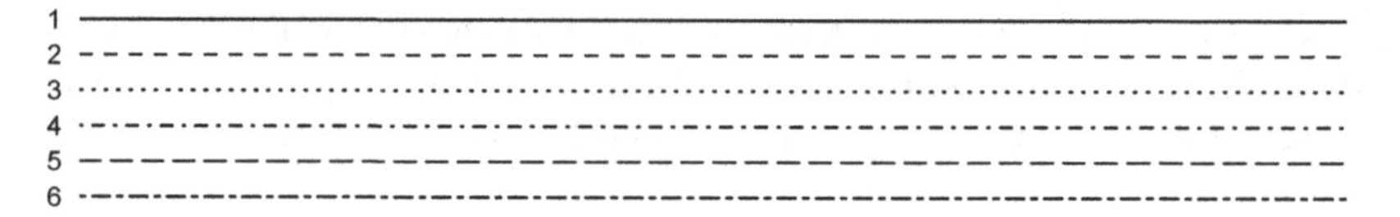

Figure 3.11: *Line types available in R.*

As noted earlier, the size of points, relative to the default size, can be set by `cex`. Similarly, the relative thickness of lines can be set by `lwd`.

3.4.2 Axes, ticks, and `par()`

R automatically chooses the scales of x and y axes, but sometimes you'd like a different choice. This is done with `xlim` and `ylim`, in which you explicitly set the upper and lower limits for the axes. Points and lines can also be colored, using the parameter `col` in `plot` or `curve`. The default, `col=1`, gives black. Integers 2–6 correspond to

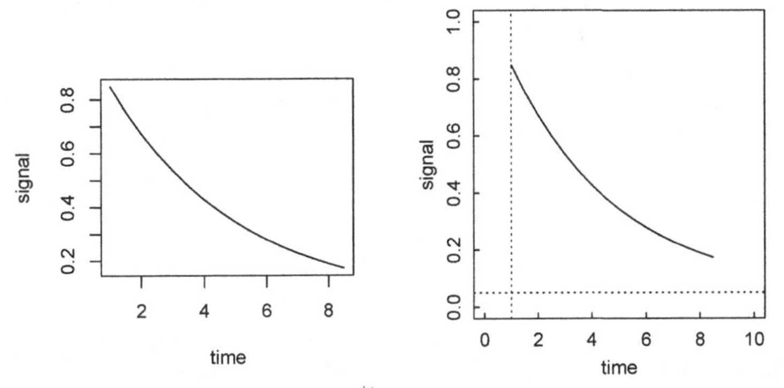

Figure 3.12: *Left: Default plot of* $0.8e^{-t/4}+0.05$*; Right: Plot modified as described in the text.*

red, green, blue, cyan, and magenta, respectively. These colors can also be called by name, e.g., `col ="red"`.Typing `?palette` gives much more information on graphic color capabilities in R.

The axes, text placement, and other aspects of a graph can readily be customized in R. For example, suppose we want a line plot of the function $0.8e^{-t/4}+0.05$ for t values between 1 and 8.5. The result, produced by the code

```
> time = seq(1,8.5,.5)
> signal = 0.8*exp(-(time-1)/4) + 0.05
> plot(time, signal, type="l")
```

is shown in Figure 3.12, left.

We can modify this default result in several ways. For example, suppose we want the ordinate to run from 0 to 10, and the abscissa from 0 to 1. We want the ticks to be inside the axes rather than outside and to be a bit shorter. We want the plot margins to be somewhat smaller, and the axis labels to be closer to the axes, to tighten up the white space. We also want to add some lines to the graph, to emphasize that the signal starts at time = 1 and that it levels off at `signal = 0.05`. These modifications are accomplished by the following code, giving the result shown in Figure 3.12, right.

```
> par(mar=c(4,4,1.5,1.5),mex=.8,mgp=c(2,.5,0),tcl=0.3)
> plot(time, signal, type="l",xlim = c(0,10), ylim = c(0,1))
> abline(h = 0.05, lty=3) # Horizontal line at y = 0.05
> abline(v=1,lty=3) # Vertical line at x = 1
```

The axis limits are specified with `xlim = c(0,10)` and `ylim = c(0,1)`. The tick direction and length are set with `tcl = 0.3` (the default is -0.5, where the minus sign specifies ticks outside the plot, and the number is the fraction of a line height). The internal lines in the plot are drawn with the `abline` command.

Additional aspects of the plot are set with various arguments to the `par` function, which specifies a broad range of graphical parameters. For example, figure margins may be made tighter or looser with the `mar` argument to `par`, (the default is `mar =`

`c(5,4,4,2)+0.1)`, where the numbers are multiples of a line height and are in the order bottom, left, top, right. `mex` determines coordinates in the margins of plots; values less than 1 move the labels toward the margins, thus decreasing white space. The location of the axis labels is set by `mgp` (the default is `mgp = c(3,1,0)`). The new settings of the par parameters are retained until modified by future `par()` statements, or until a new R session is started. Type `?par` for a listing of the many parameters that go into a graph.

3.4.3 Overlaying plots with graphic elements

Once a plot has been formed, various elements may be added to it. We have already learned about `points` and `lines`, and will shortly learn how to add error bars and text or mathematical annotations. In this subsection we demonstrate how to add other basic graphic elements—line segments, rectangles, polygons, arcs, circles, and ellipses—to a plot. The first three of these elements are called from base R, the last four from the `plotrix` package. See Figure 3.13 for the results.

```
> par(mar=c(1.5,1.5,1.5,1.5)) # Reduce margins
> par(mfrow=c(1,2))

> plot.new(); plot.window(c(0,100), c(0,100), asp=1); box()
> x0=c(5,10,15); y0=x0; x1 = x0 + 5; y1 = y0 + 60
> segments(x0=x0,x1=x1,y0=y0,y1=y1, lty=1:3)
> rect(80,25,95,80,border="black",lwd=3)
> polygon(50+25*cos(2*pi*0:8/8), 50+25*sin(2*pi*0:8/8),
+ col=gray(0.8), border=NA)
>
> require(plotrix)  # Assumes the package is already installed
>
> plot.new(); plot.window(c(0,100), c(0,100), asp=1); box()
```

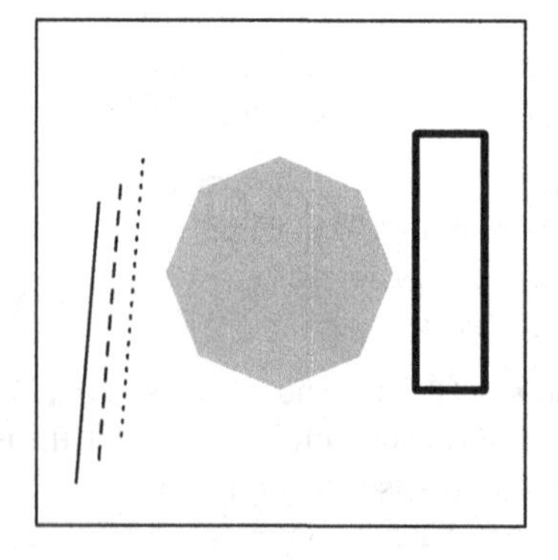

Figure 3.13: *Left: Graphic elements produced with base R; Right: Graphic elements produced with `plotrix` package.*

```
> draw.arc(20, 20, (1:4)*5, deg2 = 1:20*15)
> draw.circle(20, 80, (1:4)*5)
> draw.ellipse(80, 20, a = 20, b = 10, angle = 30, col=gray(.5))
> draw.radial.line(start=2, end = 15, center = c(80,80), angle=0)
> draw.radial.line(start=2, end = 15, center = c(80,80), angle=pi/2)
> draw.radial.line(start=2, end = 15, center = c(80,80), angle=pi)
> draw.radial.line(start=2, end = 15, center = c(80,80), angle=3*pi/2)
>
> par(mfrow=c(1,1))
```

It is evident that these functions could be used to draw simple – or even not so simple – diagrams.

Finally, we show how to color defined areas of a plot with the `polygon()` function. As an example, we distinguish the positive and negative regions of a function with different shades of gray, which might be useful in a pedagogical presentation of how to integrate the function.

Consider integrating the first order Bessel function `besselJ(x,1)` from $x = 0$ to its zero-crossing point near $x = 10$. We first compute the crossing points with the `uniroot` function.

```
> x10 = uniroot(function(x) besselJ(x,1),c(9,11))$root
> x4 = uniroot(function(x) besselJ(x,1),c(3,5))$root
> x7 = uniroot(function(x) besselJ(x,1),c(6,8))$root
```

We compute the value of the function over the desired range, using many steps to give a smooth polygon fill.

```
> x = seq(0,x10,len=100)
> y = besselJ(x,1)
```

Next we construct an empty plot with the desired x and y limits, and add the polygon of the function with a medium gray fill.

```
> plot(c(0,x10),c(-0.5,0.8), type="n", xlab="x", ylab="J(x,1)")
> polygon(x,y,col="gray", border=NA)
```

We then "paint over" the negative region with white, and add a horizontal line at $x = 0$.

```
> rect(0,-0.5,x10,0, col="white", border=NA)
> abline(h=0, col = "gray")
```

We calculate the value of the function in the negative region, again using many steps for smoothness.

```
> xminus = seq(x4,x7,len=50)
> yminus = besselJ(xminus,1)
```

Finally, we cover the negative region with a polygon in a darker gray, and add back the ticks that were painted over in an earlier step.

```
> polygon(xminus,yminus, col=gray(.2), border=NA)
> axis(1, tick=TRUE)
```

The result is seen in Figure 3.14.

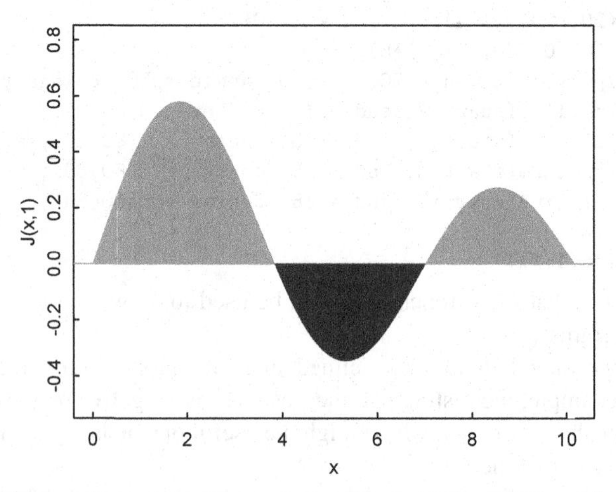

Figure 3.14: *The positive and negative regions of* `besselJ(x,1` *distinguished with different shades of gray using the* `polygon` *function.*

3.5 Error bars

Plots of experimental data or simulations should generally have error bars on the points to indicate uncertainty or statistical variation. Error bars are formed in R using the `arrows` command, as in the following code.

```
> x = 0:20
> y = sin(x)^2 + cos(x/2)
> err.y = 0.1
> err.x = 1
> par(mfrow=c(1,2))
> # Plot just y error bars
> plot(x,y,type="o", pch=19)
> arrows(x,y,x,y+err.y,0.05,90); arrows(x,y,x,y-err.y,0.05,90)
> # Plot both x and y error bars
> plot(x,y,type="o", pch=19)
> arrows(x,y,x,y+err.y,0.05,90); arrows(x,y,x,y-err.y,0.05,90)
> arrows(x,y,x+err.x,y,0.05,90); arrows(x,y,x-err.x,y,0.05,90)
```

with the result shown in Figure 3.15. If each point has a different uncertainty, then `err.y` should be a vector whose length is the same as that of `y`. `arrows` is called with the arguments `(x0,y0,x1,y1,length,angle)` where `(x0,y0)` are the coordinates of the starting point, `(x1,y1)` are the coordinates of the end point, `length` is the length of the arrowhead in inches (default 0.25), and `angle` is the angle of the arrowhead with the shaft. Type `?arrows` for further information.

Error bars or confidence limits can also be plotted with the `plotCI` or `dispersion` functions in the `plotrix` package. See their help pages for details.

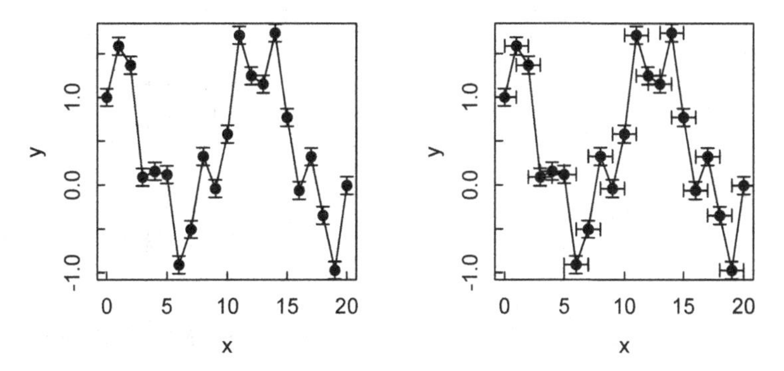

Figure 3.15: *Illustration of error bars using the* `arrows` *command. Left: y error bars only; Right: Both x and y error bars.*

3.6 Superimposing vectors in a plot

An easy way to plot several data vectors in the same plot is to combine the vectors into a matrix with `cbind`, and then use the `matplot` command. Of course, the vectors must be of the same length and refer to the same x-value. As an example we use the `iris` data frame in the base R installation. Type `?iris` for information on this dataset. (Help for `matplot` gives several other examples using `iris`.)

We will use `Sepal.Length` as the independent variable, and plot the other three measured dimensions against it. To simplify the plot, we consider only data on the `setosa` species. Thus we construct the vectors

```
> SL.s = iris$Sepal.Length[iris$Species == "setosa"]
> PL.s = iris$Petal.Length[iris$Species == "setosa"]
> SW.s = iris$Sepal.Width[iris$Species == "setosa"]
> PW.s = iris$Petal.Width[iris$Species == "setosa"]
> setosamat = cbind(SW.s,PL.s,PW.s)
```

The commands above retrieve four row vectors `SL.s`, `PL.s`, `SW.s`, and `PW.s`, from the `iris` dataset distributed with the R package. The command `cbind(SW.s,PL.s,PW.s)` combines the three vectors into the three-column matrix `setosamat`, with each column containing the data in one of the vectors `PL.s`, `SW.s`, and `PW.s`.

The `matplot()` command then plots the data in each of these column vectors against the data in `SL.s`.

```
> matplot(SL.s, setosamat)
```

from which we see that, by default, R labels each point with the number of the vector, and uses the standard "rainbow" colors for further identification. Since this book is in black and white, the colors don't show on the page, but they will on a computer screen (Figure 3.16, left).

The plot can be improved by changing the points to standard symbols and adding better axis labels and a legend (Figure 3.16, right).

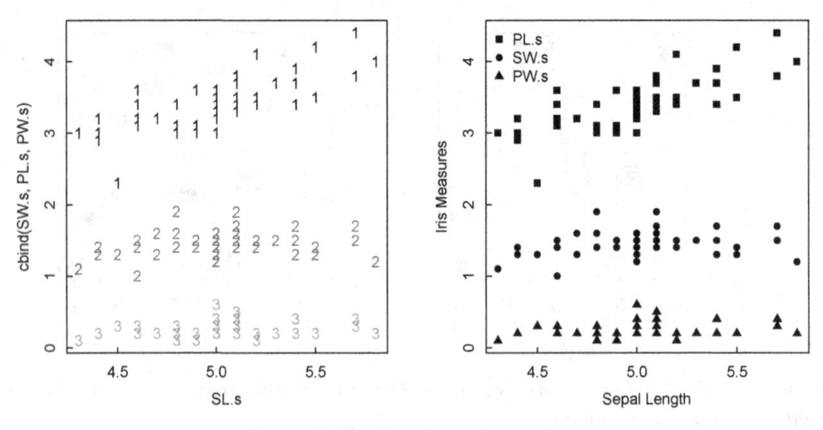

Figure 3.16: *Matplots of* `iris` *data.*

```
> matplot(SL.s, setosamat col=c(1,1,1), pch = 15:17,
+   xlab="Sepal Length", ylab = "Iris Measures")
> legend("topleft", legend=c("PL.s","SW.s","PW.s"),
+  pch=15:17, bty="n")
```

As another example, we define three displaced sine functions, then bind the three y vectors into a matrix **m**, and `matplot` the result, specifying different line types (or we could have used points) and a single line color (1 = black), since by default `matplot` colors the variables successively with `col = c(1,2,3,...)`. To show how the plot parameters work, we modify the y-label, and add main title and subtitle. Once the plot is drawn, we add a legend.

```
> x=seq(-4*pi,4*pi,pi/6)
> y1=sin(x)
> y2=sin(x+pi/6)+0.1
> y3=sin(x+pi/3)+0.2
> m = cbind(y1,y2,y3)
```

```
> matplot(x,m,type="l",ylab="y1,y2,y3",lty=1:3,col="black",
+   main = "Displaced sin functions",
+   sub = "y1 = sin(x), y2 = sin(x+pi/6)+0.1, y3 = sin(x+pi/3)+0.2")
> legend("bottomleft",legend=c("y1","y2","y3"),col=1,lty=1:3,bty="n")
```

The result is shown in Figure 3.17. The x axis was expanded before plotting to allow more room for the legend.

3.7 Modifying axes

Scientific and engineering graphs often need axes other than the default linear axes with ticks on the bottom and left. R has many options for customizing axes, of which we present here a few of the most commonly used.

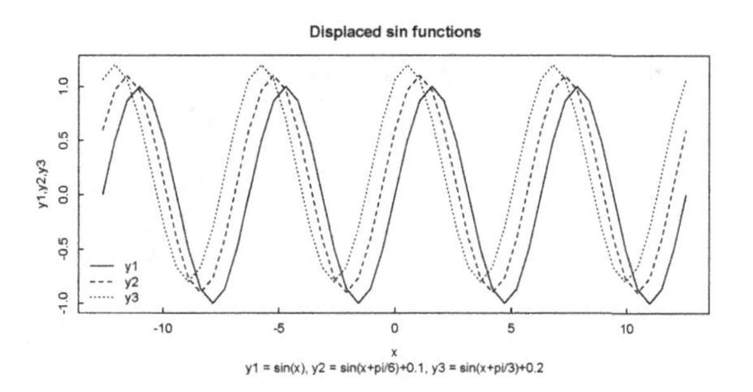

Figure 3.17: *Superimposed vectors using matplot.*

3.7.1 Logarithmic axes

The following code produces a log-log plot of the function $y = 1 + x^{2.3}$ where x runs from 0.2 to 10 (Figure 3.18).

```
> x = seq(.2,10,by=.2)
> y = 1 + x^2.3
> plot(x,y,log="xy", type="l")
```

If the plot call were `plot(x,y,log="x")` then only the x axis would be logarithmic; similarly for y.

3.7.2 Supplementary axes

Sometimes one wants the top and right axes to provide different scales than the bottom and left axes. The axes are numbered 1 (bottom), 2 (left), 3 (top), and 4 (right) . In the following code, we put temperature in celsius on the bottom and in fahrenheit on the top, and energy in kilojoules on the left and kilocalories on the right. The `par` commands online 3 set margins for the plot (Figure 3.19).

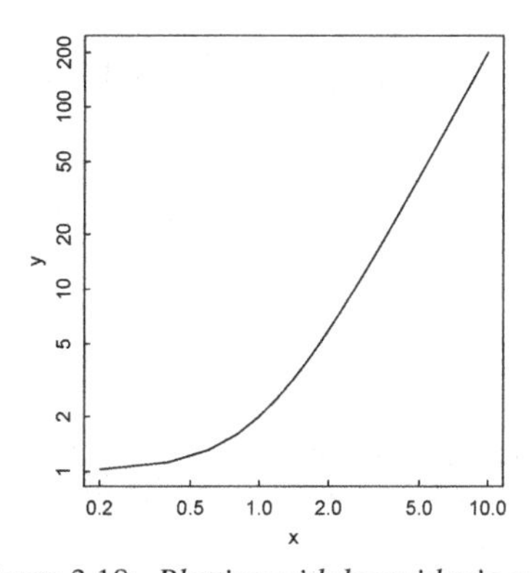

Figure 3.18: *Plotting with logarithmic axes.*

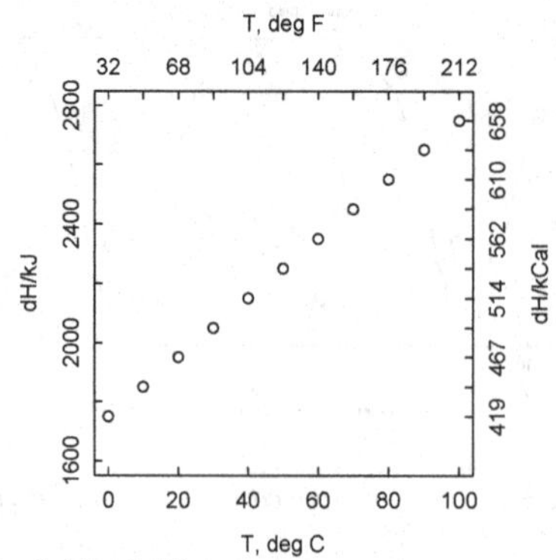

Figure 3.19: *Adding supplementary axes to a graph.*

```
> tC = seq(0,100,10)
> dH = 2000 + 10*(tC - 25)
> par(tcl=0.3, mar=c(3,3,4,4)+0.1, mgp = c(2,0.4,0))
> plot(tC, dH, xlim = c(0,100), ylim = c(1600,2800),
  + xlab="T, deg C",ylab="dH/kJ", tcl=0.3)
> axis(3, at = tC, labels = tC*9/5+32, tcl=0.3)
> mtext(side=3,"T, deg F", line = 2)
> axis(3, at = tC, labels = tC*9/5+32, tcl=0.3)
> axis(4, at = dH, labels = round(dH/4.18,0), tcl=0.3)
> mtext(side=4,"dH/kCal", line = 2)
```

3.7.3 *Incomplete axis boxes*

One may wish to draw some graphs without all four axes. This is controlled by the `bty` (box type) parameter. For example, to draw the graph in Figure 3.20, use the command

```
> curve(log10(x), 0.5,5, bty = "L")
```

since the axes form an L. Likewise, to draw axes 1,2,4 but omit axis 3, use `bty = "U"`; and to draw axes 1,2,3 but omit axis 4, use `bty = C`. To omit all axes, use `bty = "n"` (none). Note that this latter command also prevents the drawing of a box around a legend. (See Figure 3.17 and accompanying code.)

3.7.4 *Broken axes*

Sometimes a dataset will have groups of points with widely different values. If logarithmic axes are not appropriate, one will wish to break the linear axes between the two groups of values, since otherwise the smaller values will be unduly compressed. This may be done with the `axis.break` function in the `plotrix` package, along with reducing the larger values, suppressing the automatic numbering of

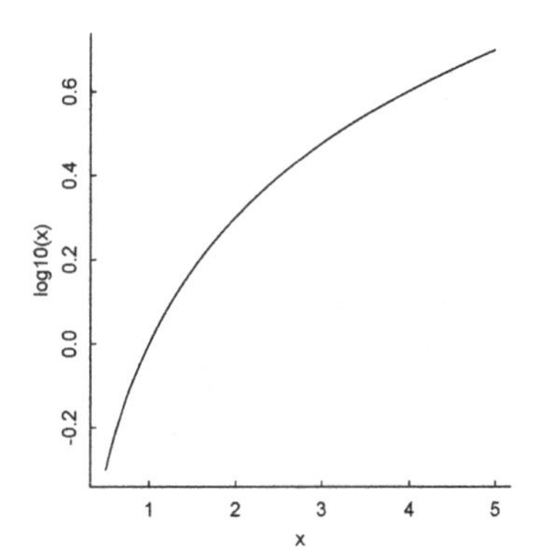

Figure 3.20: *Drawing a graph with only two axes.*

the axes with axes=FALSE, and imposing customized labels with the axis() command. The following code gives an example. Note also that, for illustrative purposes, we have given the x and y axes the two different break styles, zigzag and slash (Figure 3.21).

```
> xstep = 0.1  # Scale factor for x axis
> x1 = seq(0.1,0.5,xstep)
> x2 = seq(5.0,5.3,xstep)
> x2red = x2-min(x2)+max(x1)+2*xstep
> x = c(x1,x2red)
> ystep = 1  # Scale factor for y axis
> y1 = 1:5
> y2 = c(51,53,52,54)
> y2red = y2-min(y2)+max(y1)+2*ystep
> y = c(y1,y2red)
>
> library(plotrix)
>
> plot(x,y,axes=F,xlab="x", ylab="y")
> box() # Draw axes without labels
> axis.break(1,max(x1)+xstep, style="zigzag", brw=0.04)
> axis.break(2,max(y1)+ystep, style="slash", brw=0.04)
>
> lx1 = length(x1); lx2 = length(x2)
> lx = lx1 + lx2
> ly1 = length(y1); ly2 = length(y2)
> ly = ly1 + ly2
>
> axis(1,at=(1:lx1)*xstep,labels=c(as.character(x1)))
> axis(1,at=((lx1+2):(lx+1))*xstep,labels=c(as.character(seq(min(x2),
+   max(x2),by=xstep))))
> axis(2,at=(1:ly1)*ystep,labels=c(as.character(y1)))
> axis(2,at=((ly1+2):(ly+1))*ystep,labels=c(as.character(seq(min(y2),
+  max(y2),by=ystep))))
```

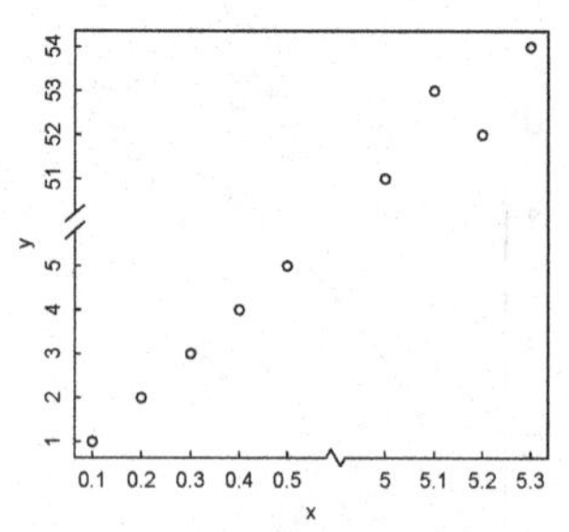

Figure 3.21: *Example of `axis.break()` in `plotrix` to plot data of substantially different magnitudes.*

3.8 Adding text and math expressions

To annotate certain points on a graph, one can use the `arrows` and/or `text` functions.

```
> x = -10:10
> y = sin(x)
> plot(x,y, type="o")
> text(3,1, "Local Max")
> text(-2.5,-1,"Local Min")
> arrows(1,-.1,.1,0,length=0.1,angle=15)
> text(1,-.15,"0,0")
```

In the `arrows` function, length specifies the length of the arrowhead in inches, and angle specifies the angle (degrees) that the arrowhead makes with the shaft (Figure 3.22).

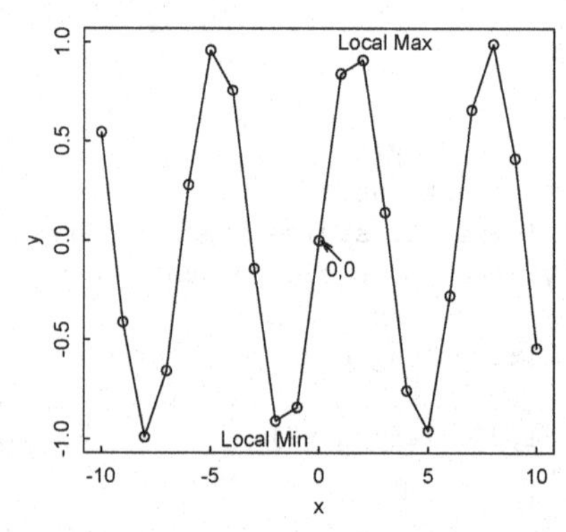

Figure 3.22: *Annotating a graph with text and arrow.*

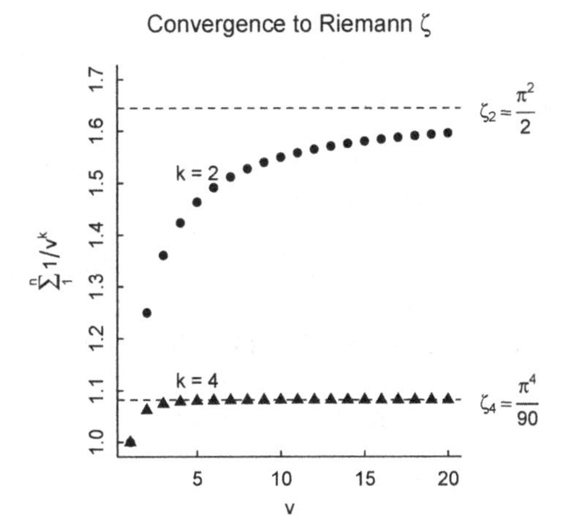

Figure 3.23: *Use of expression() to annotate a graph.*

3.8.1 Making math annotations with expression()

R has the facility to annotate plots with mathematical as well as textual material, using commands similar to those in TEX. Type `demo(plotmath)` or `?plotmath` for a complete list. Here we give an example, using the convergence of sums to the Riemann zeta functions of orders 2 and 4, of how to use `expression()` to produce useful annotations of a plot (Figure 3.23). The function `mtext` enables placing text outside the boundaries of the plot. `side=4` specifies the right-hand vertical axis, `at` gives the vertical placement of the text, `line` specifies the line (counting from 0) parallel to the axis on which the test is to be placed, and `las = 2` indicates that the text is to be written perpendicular to the axis (the default `las = 1` gives parallel orientation).

```
> n = 20
> v = 1:n
> v2 = v^(-2)
> s2 = cumsum(v2)
> v4 = v^(-4)
> s4 = cumsum(v4)
> s = cbind(s2,s4)
> # par(mar=c(5,5,4,4)+0.1)
> par(mar=c(4,5,4,5)+0.1,mex=.8,mgp=c(2,.5,0),tcl=0.3)
> matplot(v,s, type="p", pch=c(16,17), col=1,ylim=c(1,1.7), bty="L",
+   xlab="v", ylab = expression(sum(1/v^k,v=1,n)),
+   main = expression(paste("Convergence to Riemann ",zeta)))
> abline(h=pi^2/6,lty=2)
> abline(h=pi^4/90,lty=2)
> mtext(side=4,at=pi^2/6,text=expression(zeta[2] ==
+   frac(pi^2,2)),line=1,las=2)
> mtext(side=4,at=pi^4/90,text=expression(zeta[4]==frac(pi^4,90)),
+   line=1,las=2)
```

```
> text(5,1.52,"k = 2")
> text(5,1.12,"k = 4")
```

3.9 Placing several plots in a figure

Sometimes one needs to produce a figure that combines several plots. One may do so using the command `par(mfrow=c(n1,n2))` to place the plots in n1 rows and n2 columns, with `mfrow` specifying that they are placed in row order. If column order is desired, use `mfcol`. We have done this already many times in this chapter, to put two plots side-by-side with `par(mfrow=c(1,2))`. Expanding this idea, here is sample code, showing four ways of plotting 1000 normally distributed random numbers, leading to Figure 3.24. The statement `par(mfrow = c(1,1))` at the end reestablishes the default of a single plot in a figure.

```
> n = 1000
> x = 1:n
> set.seed(333)
> y1 = rnorm(n,0,1)
> y2 = rnorm(n,0,2)
> par(mfrow = c(2,2))
> plot(y1)
> boxplot(y1,y2)
> hist(y1)
> plot(density(y2))
> par(mfrow = c(1,1))
```

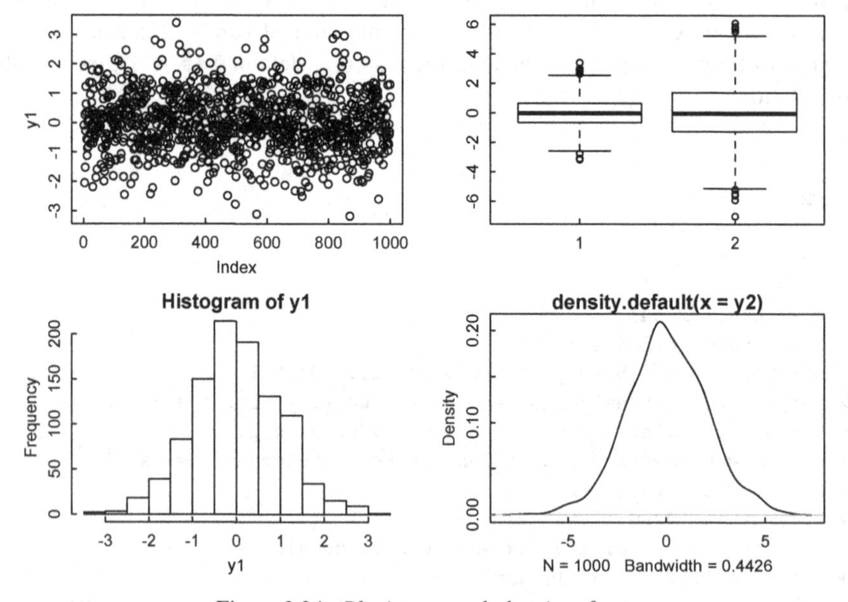

Figure 3.24: *Placing several plots in a figure.*

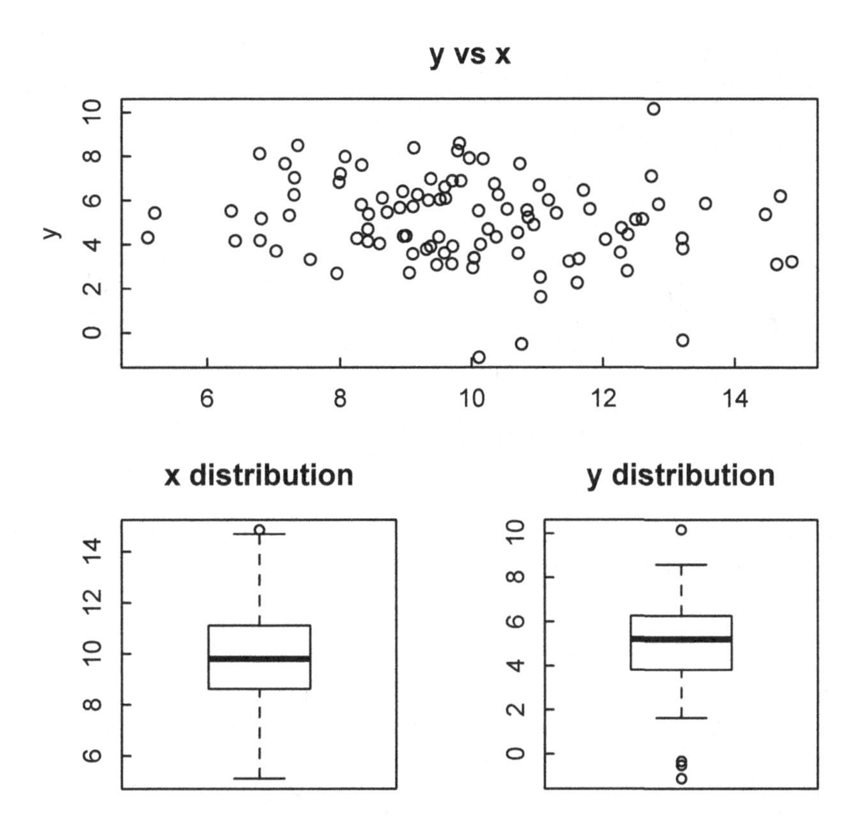

Figure 3.25: *Using `layout` to create a scatter plot with accompanying box plots.*

For more complex layouts, in which the plots are of different sizes, use `layout()`. As a simple example, we construct a figure in which the first plot uses all of the top row, while the second and third plots share the bottom row.

```
> # Divide the device into two rows and two columns
> # Allocate figure 1 all of row 1
> m = matrix(c(1,1,2,3),ncol=2,byrow=TRUE)
> layout(m)
> x = rnorm(100,10,2)
> y = rnorm(100,5,2)
> par(mar=c(2,3,3,2))
> plot(x,y,main="y vs x")
> boxplot(x,main="x distribution")
> boxplot(y,main="y distribution")
```

which gives Figure 3.25.

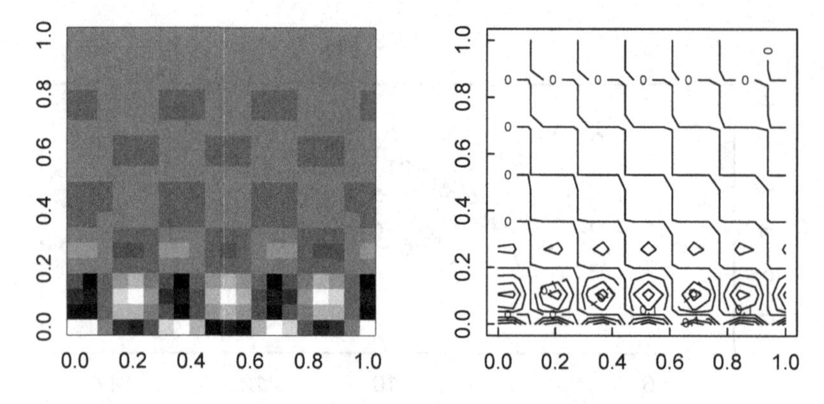

Figure 3.26: *Left: Image plot; Right: Contour plot.*

To see a more complex illustration, run the code to create a scatter plot with marginal histograms in the example file for `help(layout)`.

3.10 Two- and three-dimensional plots

R can produce two-and three-dimensional plots of multidimensional data. For two-dimensional projections, use `image` or `contour`. These functions generally use colors (see help for sample palettes), but in this book we use shades of gray (Figure 3.26).

```
> n = 1:20
> x = sin(n)
> y = cos(n)*exp(-n/3)
> z = outer(x,y)
> par(mar=c(3,3,1.5,1.5),mex=.8,mgp=c(2,.5,0),tcl=0.3)
> par(mfrow=c(1,2)) # Two plots side-by-side
> image(z,col=gray(1:10/10)) # First plot
> contour(z) # Second plot
> par(mfrow=c(1,1)) # Reset to default single plot
```

A variant of `contour` is `filled.contour`, which produces a contour plot with the areas between the contours filled in solid color. A key showing how the colors map to z values is shown to the right of the plot. See `?filled.contour` for details and examples.

A three-dimensional perspective projection is achieved with the `persp()` function (Figure 3.27). Viewing angles are adjusted with `theta` and `phi`. The figure facets can be shaded with the `shade` parameter. With the same x,y,z values as in the previous example, we write

```
> par(mar=c(1,1,1,1))
> par(mfrow=c(1,2))
```

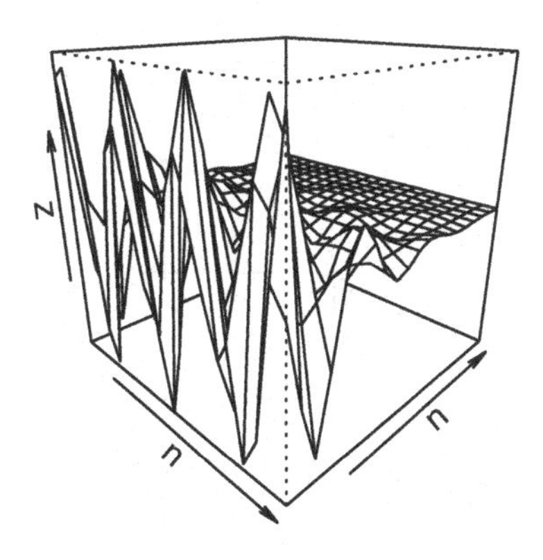

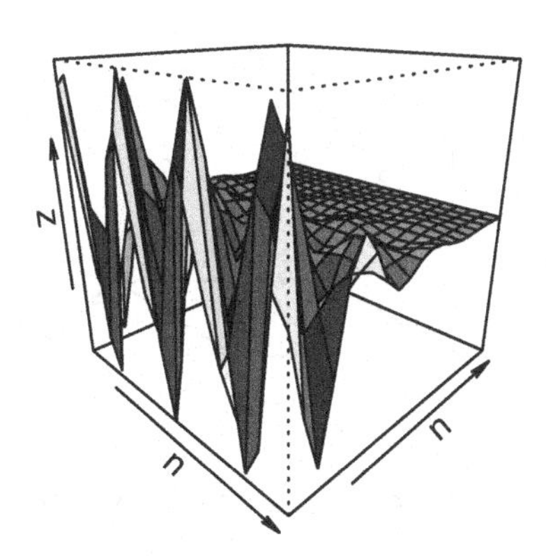

Figure 3.27: *Left: Perspective plot of the outer product of sin(n) and cos(n)$e^{-n/3}$; Right: The same plot with shade applied.*

```
> persp(n,n,z, theta=45,phi=20)
> persp(n,n,z, theta=45,phi=20, shade=0.5)
> par(mfrow=c(1,1))
```

to get Figure 3.27.

The rglpackage has numerous functions for plotting 3D graphics, including persp3d() which is similar to persp(), but does a better job of handling hidden surface removal and produces rotatable plots.

The package scatterplot3d enables plotting of multivariate data in three dimensions. This package is not included in the base R installation, so it must be installed and then loaded with library() or require().

```
> install.packages("scatterplot3d")
> library(scatterplot3d) # Or require(scatterplot3d)
```

Simple examples of its use, with two plot types (Figure 3.28), are obtained from the code below.

```
> x=1:20
> y=1:20
> set.seed(17)
> z = runif(20)
> par(mfrow=c(1,2))
> par(mar=c(1,1,1,1))
> scatterplot3d(x,y,z)
> scatterplot3d(x,y,z, type="h")
```

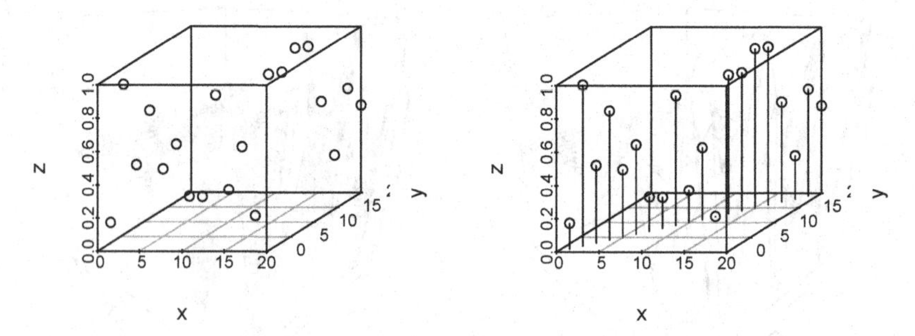

Figure 3.28: *`scatterplot3d` plots (left) default (`type = ''p''`); Right: (`type =''h''`).*

The help page for `scatterplot3d()` lays out the many optional arguments to the function and shows some interesting examples of its use to draw 3D geometrical figures as well as data points.

3.11 The `plotrix` package

The add-on `plotrix` package contains a number of plotting functions that may be of interest to scientists and engineers. We survey some of them in this section. Of course, the package must be installed and loaded with

```
install.packages("plotrix")
require(plotrix)
```

before its functions can be used.

3.11.1 `radial.plot` and `polar.plot`

Data are sometimes more conveniently displayed in a polar coordinate system than a Cartesian one. `plotrix` has `radial.plot` and `polar.plot` functions to enable such displays (Figure 3.29). The two functions are nearly identical, except for the values of the angular variables (radians and degrees) and the names of those variables (`radial.pos` and `polar.pos`, respectively). The following code shows three examples of the use of these functions, using the three types of plots available: polygon (convenient for function plotting), symbol, and radial line.

```
> require(plotrix)
> par(mfrow=c(1,3)) # 3 plots in a row
> par(cex.lab=0.7)  # reduce the size of interior labels
>
> angle = seq(0, 2*pi, len = 50)
> radius = 10*cos(angle)^2 + cos(angle + pi/6)
> radial.plot(lengths = radius, radial.pos = angle,
```

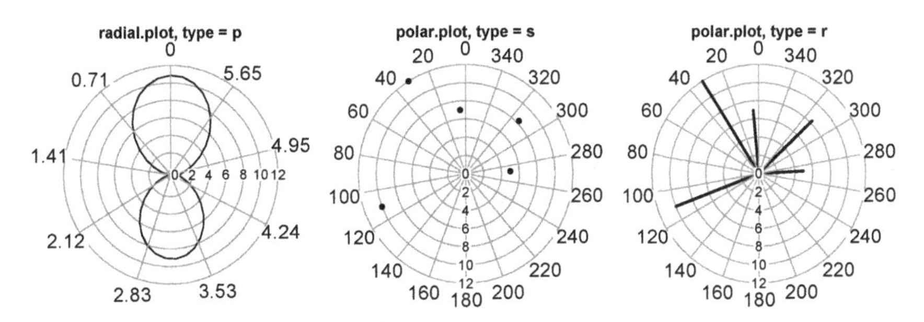

Figure 3.29: *Radial (left) and polar (center, right) plots using (p)polygon, (s)ymbol, and (r)adial line representations.*

```
+   start=pi/2, rp.type = "p",
+   main = "radial.plot, type = p")
>
> angle = c(0,5,32,111,273,314)
> distance = c(0,7,12,10, 5, 8.3)
> polar.plot(lengths = distance, polar.pos = angle,
+   start=90, rp.type = "s", point.symbols=16,
+   show.grid.labels=1,
+   main = "polar.plot, type = s")
>
> angle = c(0,5,32,111,273,314)
> distance = c(0,7,12,10, 5, 8.3)
> polar.plot(lengths = distance, polar.pos = angle,
+   start=90, rp.type = "r", lwd=2,
+   show.grid.labels=1,
+   main = "polar.plot, type = r")
```

Parameters such as grids, labels, and angular start position can be adjusted in the function calls, as detailed in the help pages. The help page for `radial.plot` is more comprehensive than that for `polar.plot`.

3.11.2 Triangle plot

In disciplines that study mixtures, such as soil science or thermodynamics of alloys, triangle plots that characterize ternary mixtures are useful. The `plotrix` package has the `triax.plot` function for constructing such plots. The help page shows how plot details like grids, labels, and symbols can be adjusted. Here is an example for a set of four three-component alloys (Figure 3.30).

```
> require(plotrix)
>
> alloy1 = c(20, 75, 5)
```

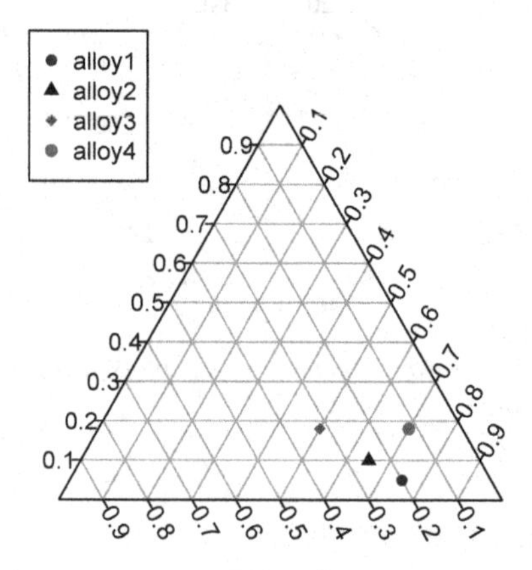

Figure 3.30: *Triangle plot of alloy composition.*

```
> alloy2 = c(25, 65, 10)
> alloy3 = c(32,50, 18)
> alloy4 = c(12,70, 18)
> alloys = rbind(alloy1, alloy2, alloy3, alloy4)
>
> triax.return = triax.plot(alloys, main = "Alloy Composition",
+   show.grid = TRUE, show.legend = TRUE,
+   col.symbols = gray(2:5/10), pch = 16:19, bty="n")
>
> par(triax.return$old.par) # Change parameters back
```

The functions `triax.points` and `triax.abline` enable placing points and drawing straight lines in a triangle plot. `triax.frame` draws the triangle outline, axis labels, and title without data. `triad.fill` fills a triangle plot with smaller triangles. See the help pages of these functions for details and examples.

3.11.3 Error bars in `plotrix`

As mentioned earlier (Section 3.5), `plotrix` contains the function `plotCI` which, given a set of x and y values and upper and lower bounds, plots the points with error bars. Depending on the situation, this function may enable greater control of the plot details with less coding than using the `arrows` function approach.

Likewise, the function `dispersion` in `plotrix` will display either vertical lines with horizontal caps (i.e., error bars) or lines (filled if desired) that form a "confidence band" around the line connecting the data points. See the help pages for details.

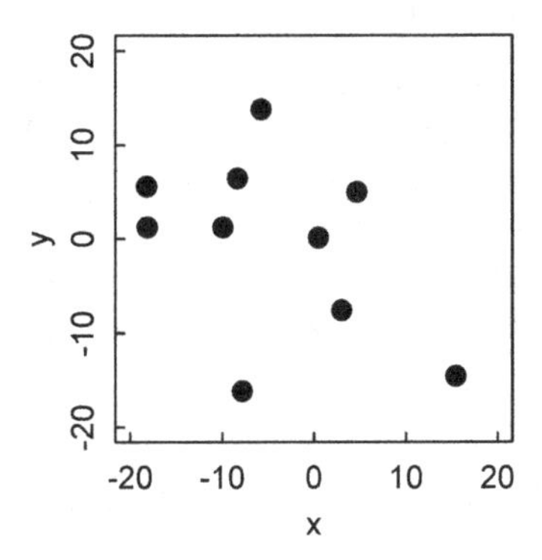

Figure 3.31: *Result of Brownian motion animation after 100 steps with 10 particles.*

3.12 Animation

Animation is often useful in scientific or engineering calculations, to visualize the time course of complex events. The `animation` package in R presents "A Gallery of Animations in Statistics and Utilities to Create Animations." The results can be shown either in an R graphics window or in an HTML browser window. Here we present an adaptation of the `brownian.motion` demo, with a figure (Figure 3.31) showing the final state of 10 particles making 100 random steps in two dimensions. Of course, a static book page cannot represent an animated graphics window, so the reader should run the code.

```
> install.packages("animation")
> require(animation)
> brownian.motion = function(n = 10, xlim = c(-20, 20),
+  ylim = c(-20, 20), ...) {
+     x = rnorm(n)
+     y = rnorm(n)
+     interval = ani.options("interval")
+     for (i in seq_len(ani.options("nmax"))) {
+         plot(x, y, xlim = xlim, ylim = ylim, ...)
+        # text(x, y)
+         x = x + rnorm(n)
+         y = y + rnorm(n)
+         Sys.sleep(interval)
+ }
+     invisible(NULL)
+ }
> # Change options from default (interval = 1, nmax = 50)
> oopt = ani.options(interval = 0.05, nmax = 100)
> brownian.motion(pch = 16, cex = 1.5)
> ani.options(oopt) # Restore default options
```

3.13 Additional plotting packages

A great deal of effort has been put by developers into devising useful functions for graphing almost any kind of data or functions. We have already mentioned `plotrix`, `scatterplot3d`, `rgl`, and `animation`. The graphics task view (http://cran.r-project.org/web/views/ Graphics.html) summarizes the many other packages available. In particular, the `lattice` package (included in the R installation) and the `ggplot2` package provide important facilities for producing more complex graphics than we have considered here.

Chapter 4

Programming and functions

R is a full-fledged programming language, with standard programming constructs like conditional execution, loops, and subroutines (called functions in R). In this chapter we show how these constructs are implemented.

4.1 Conditional execution: `if` and `ifelse`

Computer programs can choose different computations depending on whether a logical condition evaluates to true or false. The relational operators in R that perform such evaluation are listed in Section 2.5. The logical operators are & (and), | (or), and ! (not).

The `if` statement operates on logical vectors of length 1. A formal construction is

```
if(condition 1) {
    result 1
  } else if (condition 2) {
    result 2
  } else {
    result 3
}
```

For example

```
> x = -3
> if(x < -1) {
+   y = -1
+   } else if (x < 0) {
+     y = 0
+   } else {
+     y = 1
+ }
> y
[1] -1
```

Be careful not to insert a line feed between } and `else`, because R will interpret everything up to the } as a complete command and will return a result prematurely. This is a general rule: If you are forced to break a line because it is too long, be sure to break it so as to yield an incomplete command. R will then insert a + at the beginning of the next line.

In simple cases, the `if-else` construction can be written on a single line:

```
> x = -1
> if (x > 0) 1 else 0
[1] 0
```

Likewise for more than two choices, with `else if`:

```
> x = 0
> if (x < 0) 0 else if (x == 0) 0.5 else 1
[1] 0.5
```

A conditional construction may be used inside another construction, e.g.,

```
> x = if(y > 0) pi else pi/2
```

To apply conditional execution to each element of a vector, use the function `ifelse`:

```
> set.seed(333)
> x = round(rnorm(10),2)
> y = ifelse(x>0, 1, -1)
> x
 [1] -0.08  1.93 -2.05  0.28 -1.53 -0.27  1.23  0.63  0.35 -0.56
> y
 [1] -1  1 -1  1 -1 -1  1  1  1 -1
```

R also has a switch function, `switch(EXPR, cases)`, which evaluates `EXPR` and accordingly chooses one of the cases.

```
> set.seed(123)
> x = rnorm(10, 2, 0.5)
> y = 3
> switch(y, mean(x), median(x), sd(x))
[1] 0.476892
```

4.2 Loops

Computations often must repeat certain steps either a given number of times, or until some condition is met. R, like other programming languages, has the looping functions to deal with such situations, though it is often possible, and usually desirable, to avoid loops by taking advantage of vectorization.

4.2.1 *for loop*

To repeat an operation through a given range of elements of a vector, use the `for` construction.

```
> sum = 0
> for (i in 1:10) sum = sum + i
```

```
> sum
 [1] 55
```

An alternative that uses vectorization rather than a loop:

```
> sum(1:10)
[1] 55
```

For a slightly more complicated example, sum over only even numbers:

```
> sum = 0
> for (i in 1:10) if (i%%2 == 0) sum = sum + i
> sum
[1] 30
```

The alternative that takes advantage of vectorization, though it is hardly necessary for such a short vector:

```
> sum(c(2,4,6,8,10))
[1] 30
```

If the number of repetitions of the loop operation gets very large, `for` loops become notably slower than vectorized operations. Consider calculation of a million sine functions:

```
> n = 1000000
> x=0
> system.time(for (i in 1:n) x = sin(i/n))
   user  system elapsed
  0.739   0.005   0.742
>
> i = 1:n # vectorized
> system.time(sin(i/n))
   user  system elapsed
  0.057   0.004   0.061
```

However, if the value of an element of the vector is changed according to the values of other elements, then vectorization is not possible. A common example is numerical integration of a differential equation using Euler's method. Consider radioactive decay, in which the rate of decrease in the number of radioactive atoms N is proportional to the number remaining (Figure 4.1):

$$\frac{dN}{dt} = -kN \tag{4.1}$$

with initial number N_0 at $t = 0$.

```
# Set up initial conditions and define variables
> tmin = 0; tmax = 100; dt = 1
> n = (tmax - tmin)/dt + 1 # 101 time values from 0 to 100
> time = seq(tmin, tmax, by = dt)
> k = .03 # Decay rate constant
> N0 = 100 # Initial number of atoms
> N = N0 # Initialize N
```

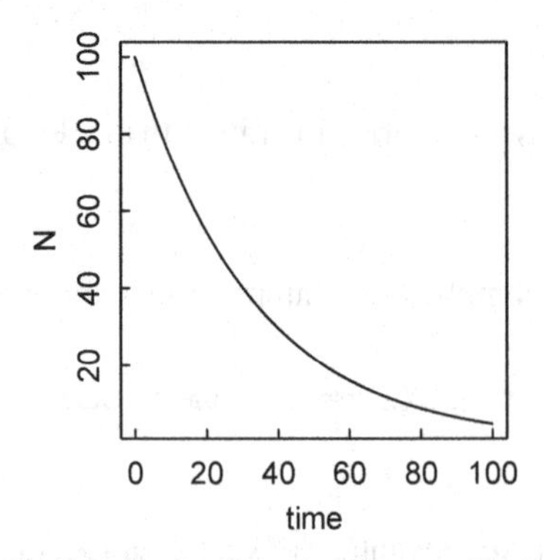

Figure 4.1: *Simulation of radioactive decay using Euler's method.*

```
# Run for loop
> for (i in 2:n) {
+    dN = -k*N[i-1]*dt
+ N[i] = N[i-1] + dN}
# Plot results
> plot(time, N, type = "l")
```

Since N_i depends on N_{i-1}, vectorization is not possible and a `for` loop is required.

4.2.2 *Looping with* `while` *and* `repeat`

Sometimes the number of repetitions in a loop is not known beforehand, but the loop is to be terminated when some condition is met. Then a `while` or `repeat` loop is called for.

Consider, for example, simulation of a particle undergoing a biased random walk in one dimension, starting at the origin and taking steps of mean length 0.5 and standard deviation 1. We use a `while` loop to report how many steps it takes to get to position greater than 10, and what that position is.

```
> x=0
> n=0
> set.seed(333)
> while (x <= 10) {
+    n=n+1
+    x=x+rnorm(1,mean=.5,sd=1)
+    }
>
> print(paste ("n = ", n, ", x = ",round(x,2) ))
[1] "n =  26 , x =  11.05"
```

A similar result can be achieved with a `repeat` loop, with `break` terminating the loop if the condition is met. Without `break`, one would have an infinite loop. As an added feature, next stops the current iteration if the result is unacceptable (too large a negative step in this case) and proceeds to the next iteration.

```
> x=0
> n=0
> set.seed(333)
> repeat {
+   n=n+1
+   dx=rnorm(1,mean=0.5,sd=1)
+   if (dx < -1) next # Reject large negative steps
+   x=x+dx
+   if (x > 10) break
+   }
> print(paste ("n = ", n, ", x = ",round(x,2) ))
[1] "n =  19 , x =  10.55"
```

4.3 User-defined functions

R allows user-defined functions, which might be called subroutines in some other programming languages. The general form of a function definition is
`f = function(x,y,...) expression involving x, y, ...`
The result of the function will be the last evaluated expression, unless return() is called (see sqrt_N below). In this section we give a few simple examples of user-defined functions; more will be used throughout the book.

Here's a simple function that calculates the first three powers of a vector and arranges the result as a matrix.

```
> powers = function(x) {matrix(c(x,x^2,x^3),nrow=length(x),ncol=3)}
> v = 1:5
> powers(v)
     [,1] [,2] [,3]
[1,]    1    1    1
[2,]    2    4    8
[3,]    3    9   27
[4,]    4   16   64
[5,]    5   25  125
```

Here's a function of two variables for doing modular division.

```
> modcalc = function(x,y) c(x%/%y, x%%y)
> modcalc(17,3)
[1] 5 2
```

Functions may have options, which may in turn have default values. The Newton method for finding square roots includes a loop and a conditional in the function, and we have defined default values for the tolerance (relative change per iteration) and maximum number of iterations.

```
> sqrt_N = function(a, tol=2.22e-16, max.iter=50) {
```

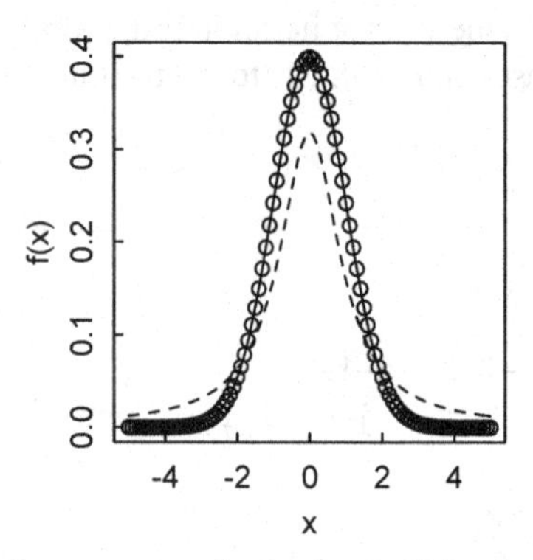

Figure 4.2: *Overlay of* gaussian(x,0,1) *(solid line),* dnorm *(points), and* lorentzian(x,0,1) *(dotted line) functions.*

```
+    x=a
+    iter = 0
+    xdiff = Inf
+    while (xdiff > tol) {
+      iter = iter + 1
+      xold = x
+      x = (x + a/x)/2
+      xdiff = abs(x-xold)/abs(x)
+      if (iter > max.iter) {print(paste("Not converged after",
+        iter,"cycles."))
+        break
+    }}
+ return(print(paste("sqrt(",a,")=",x,", xdiff=", xdiff,",
+ iterations=",iter)))  }
> sqrt_N(52.3)
[1] "sqrt( 52.3 )= 7.23187389270582 , xdiff= 0 , iterations= 9"
```

The Gaussian and Lorentzian functions are frequently encountered in spectroscopy and other scientific areas. gauss(x,0,1) is equivalent to dnorm, the probability density function for the normal distribution built into R (Figure 4.2).

```
> gauss = function(x,x0,sig) {1/sqrt(2*pi)*sig*exp(-(x-x0)^2/(2*sig^2))}
> lorentz = function(x,x0,w) {w/pi/((x-x0)^2 + w^2)}
>
> curve(gauss(x,0,1), -5,5, ylab = "f(x)", main = "Distributions")
> curve(dnorm,-5,5,type="p", add=T)
> curve(lorentz(x,0,1), xlim = c(-5,5), lt = 2, add=T)
```

We shall construct and use many functions throughout this book. Consider, for example, this variation on the one-dimensional random walk theme. Define the function randwalk(N), in which we start at 0 and generate N steps of unit length taken randomly to either the left or the right. We pick the direction of each step

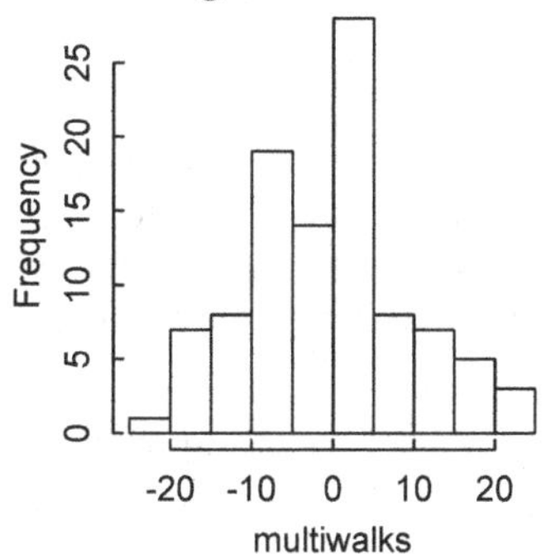

Figure 4.3: *Histogram of displacements of 100 one-dimensional random walks.*

by generating a uniformly distributed random number x between 0 and 1, moving in the negative direction if $x \leq 0.5$ and in the positive direction if $x > 0.5$. The function returns the displacement of the walk from its starting point.

```
> randwalk = function(N) {
+   walk = rep(0,N+1)  # Initialize the vector of steps
+   for (i in 2:(N+1)) {
+   x = runif(1)
+   if (x <= 0.5) walk[i]=walk[i-1]-1 else walk[i]=walk[i-1]+1
+   }
+   return(walk[N]) # End point
+ }
```

We then repeat the random walk many times (100 in the example below), generating one component of the displacement vector `multiwalks` for each repetition, and compute the mean and standard deviation of the displacement for this ensemble of random walks. Since steps to left and right are equally probable, the mean should be zero, while the standard deviation should be $\sqrt{N}$, or 10 in this case. We also plot the histogram of the displacements (Figure 4.3).

```
> multiwalks = c()
> for (k in 1:100) {  # 100 walks
+  multiwalks[k] = randwalk(100) # endpoint of k-th walk
+ }
> mean(multiwalks) # Should be near 0
[1] -0.08
> sd(multiwalks)    # Should be near 10
[1] 9.92261
> hist(multiwalks)
```

The mean and standard deviation accord well with expectations.

A faster and more elegant way to generate the steps of a random walk is with the `sample()` function:

```
sample(x, size, replace = FALSE, prob = NULL)
```

which takes a sample of the specified size from the elements of the vector `x` using the probabilities given in `prob`. The default is sampling without replacement, but in this case we want sampling with replacement. In our example, `x` has the elements 1 and -1, each with probability 0.5, and we want a sample of size `N`.

```
> randwalk2 = function(N) {
+ sum(sample(c(1,-1),size=N, replace=TRUE, prob=c(0.5,0.5)))
+ }
```

Then proceed as before:

```
> multiwalks2 = c()
> for (k in 1:100) {
+  multiwalks2[k] = randwalk2(100)
+ }
> mean(multiwalks2)
[1] -0.78
> sd(multiwalks2)
[1] 9.727167
```

4.4 Debugging

Things can often go wrong when writing functions or other code. Indeed, it may frequently be the case that more time will be spent on debugging than on writing the original code. Therefore, methods for tracking and correcting errors are important. Perhaps the simplest method is to insert `print()` or `cat` statements at intermediate points in the program. For example, here is code for a function that takes a vector of numbers *x*, squares each element, subtracts 4, and takes the natural logarithm of the result. Of course, if the x value is too small, the final operation will be taking the log of zero or a negative number, which gives `NaN`. We put a `print` statement in the function to see whether such numbers appear before logs are taken.

```
> f1 = function(x) {
+ xsq = x^2
+ xsqminus4 = xsq - 4; print(xsqminus4)
+ log(xsqminus4-4)
+ }
> f1(6:1)
[1] 32 21 12  5  0 -3
[1] 3.332205 2.833213 2.079442 0.000000      NaN      NaN
Warning message:
In log(xsqminus4 - 4) : NaNs produced
```

Alternatively, we can omit the `print` statement in the function, and use the debug function to step through the program one instruction at a time, invoking `print` from the Browser prompt to see intermediate results where we suspect a problem may arise.

```
> debug(f1)
> f1(1:6)
debugging in: f1(1:6)
debug at #1: {
    xsq = x^2
    xsqminus4 = xsq - 4
    log(xsqminus4 - 4)
}
Browse[2]>
debug at #2: xsq = x^2
Browse[2]>
debug at #3: xsqminus4 = xsq - 4
Browse[2]>
debug at #4: log(xsqminus4 - 4)
Browse[2]> print(xsqminus4-4)
[1] -7 -4  1  8 17 28
Browse[2]>
exiting from: f1(1:6)
[1]       NaN       NaN 0.000000 2.079442 2.833213 3.332205
Warning message:
In log(xsqminus4 - 4) : NaNs produced
```

To end debugging, type undebug(f1). Type ?debug for more information.

4.5 Built-in mathematical functions

At the beginning of Chapter 2 we noted that many mathematical functions—in addition to the standard logarithmic, trigonometric, and hyperbolic functions—are built in to the base installation of R. In these functions, as in others we shall discuss later, some options are given default values. Other values might be chosen, e.g., expon.scaled = TRUE. Details of implementation and examples of usage of these operators and functions can be obtained by entering help(function.name) or ?function.name at the R prompt. Here we give just a few examples of usage.

4.5.1 Bessel functions

- besselI(x, nu, expon.scaled = FALSE)
- besselK(x, nu, expon.scaled = FALSE)
- besselJ(x, nu), besselY(x, nu)

We can readily plot these functions to get a quick grasp of their behavior. For example, In Figure 4.4 it is apparent that the zeros in $J(x,1)$ coincide with the maxima and minima of $J(x,0)$, consistent with the relation $J(x,0)\prime = J(x,1)$.

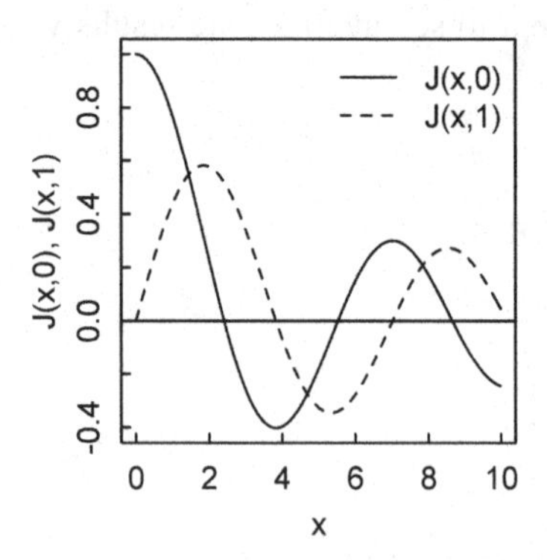

Figure 4.4: *Bessel functions J(x,0) and J(x,1).*

```
> curve(besselJ(x,0),0,10, ylab = "J(x,0), J(x,1)",
  + main = "Bessel Functions")
> curve(besselJ(x,1),lty=2,add=T)
> abline(0,0)
> legend("topright", legend = c("J(x,0)", "J(x,1)"), lty = 1:2,
  + bty="n")
```

The `Bessel` package contains additional Bessel functions `BesselI()` and `BesselH()` (Hankel function), computations for real and complex numbers, and asymptotic approximations for large arguments.

4.5.2 Beta and gamma functions

- `beta(a, b)`, `lbeta(a, b)` The beta function and its natural logarithm.
- `gamma(x)`, `lgamma(x)` The gamma function and the natural logarithm of its absolute value.
- `factorial(x)`, `lfactorial(x)` For non-negative integer x, gamma(x+1) and lgamma(x+1).
- `digamma(x)`, `trigamma(x)` The first and second derivatives of lgamma(x)
- `psigamma(x, d)` The d-th ($d >= 0$) derivative of digamma(x); sometimes called the polygamma function. The default is $d = 0$.

```
> gamma(7)
[1] 720
> trigamma(10.3)
[1] 0.102
> psigamma(10.3,1)
[1] 0.102
```

4.5.3 Binomial coefficients

- `choose(n, k)`, `lchoose(n, k)` Binomial coefficient and log of its absolute value.

```
> choose(7,5)
[1] 21
> factorial(7)/(factorial(5)*factorial(2))
[1] 21
```

4.6 Special functions of mathematical physics

4.6.1 The `gsl` package

The foremost package for special functions in mathematical physics is `gsl`. The `gsl` package provides an R wrapper for the special functions and quasi random number generators of the Gnu Scientific Library (http://http://www.gnu.org/software/gsl/).

Many special functions of mathematical physics are covered in the Gnu Scientific Library. Functions included (each with all the standard variants) are: `Airy, Bessel, Clausen, Coulomb, Coupling, Dawson, Debye, Dilog, Ellint, Elljac, Error, Expint, Fermi-Dirac, Gamma, Gegenbauer, Hyperg, Laguerre, Lambert, Legendre, Log,Poly, Powint, Psi, zeta`, and trigonometric functions.

As stated in the `gsl` package documentation, documentation is generally limited to a pointer to the GSL reference book and reproductions of some tables and figures in Abramowitz and Stegun (1964).

As an example, we use the `gsl` package to plot the Laguerre polynomials of order 2–5 (Figure 2.5).

```
> install.packages("gsl")
> library(gsl)
> x = seq(from=0,to=6,len=100)
> par(mar=c(4,4,1.5,1.5),mex=.8,mgp=c(2,.5,0),tcl=0.3)
> plot(x,laguerre_n(2,0,x),xlim=c(0,6),ylim=c(-2,3),
+   type="l",bty="L",xlab="x",ylab="Laguerre(n,x)")
> lines(x,laguerre_n(3,0,x), lty=2)
> lines(x,laguerre_n(4,0,x), lty=3)
> lines(x,laguerre_n(5,0,x), lty=4)
> legend("topleft", legend=c(2:5), lty=c(1:4), bty="n")
```

Readers are urged to explore this rich resource to find the special functions of interest to them. Special functions of mathematical physics, especially the various orthogonal polynomials, are also available in the `orthopolynom` package discussed later in this chapter.

4.6.2 Special functions in other packages

One interesting package is `hypergeo`, which provides very accurate calculations of the hypergeometric and generalized hypergeometric functions, applying different

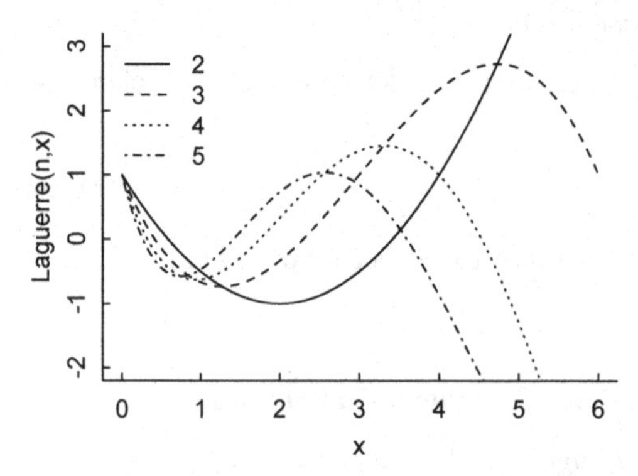

Figure 4.5: *Laguerre polynomials.*

representations such as power series, Euler's integral, or continued fractions. These different representations guarantee an accuracy that one computational method alone will not reach. `hypergeo` depends on two other packages: `contfrac` and `elliptic`; the latter did not have a binary available for OS X at the time of this writing.

Examples of two of the rare cases where the hypergeometric function returns a rational value:

```
> hypergeo(1/3, 2/3, 5/6, 27/32)
[1] 1.6+0i                         # [1] 1.6 = 8/5
> hypergeo(1/4, 1/2, 3/4, 80/81)
[1] 1.8+0i                         # [1] 1.8 = 9/5
```

The `pracma` package provides many special functions of use in science and engineering. These include (function names in parentheses):

- Chebyshev polynomials (chebPoly)
- Dirichlet eta function (eta)
- Elliptic integrals (ellipke,ellipj)
- Error functions (erf, erfc, erfcinv, erfcx, erfi, erfinv)
- Exponential integral (expint)
- Fresnel integrals (fresnelS, fresnelC)
- Legendre functions, associated, first kind (legendre)
- Gamma function, incomplete (gammainc)
- Gamma function, complex (gammaz)
- Lambert W function (lambertWp)
- Polylogarithm function (polylog)
- Riemann zeta function (zeta)

Here, for example, we use `pracma` to visualize the Fresnel sine and cosine integrals, $S(x)$ and $C(x)$, over the range (0,5) (Figure 4.6).

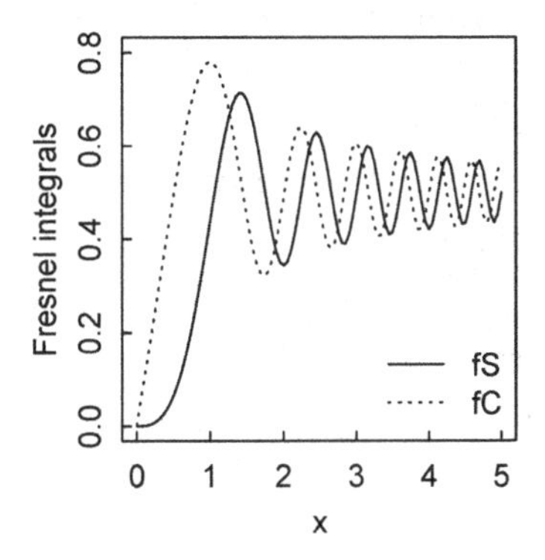

Figure 4.6: *Fresnel sine and cosine integrals.*

```
> require(pracma)
> fS = function(x) fresnelS(x)
> fC = function(x) fresnelC(x)
> curve(fS,0,5,ylim=c(0,0.8),ylab="Fresnel integrals")
> curve(fC,0,5,add=T,lty=3)
> legend("bottomright",legend=c("fS","fC"),lty=c(1,3), bty="n")
```

There are several equivalent definitions of the Fresnel integrals, differing in normalization and scaling: pracma uses

$$S(x) = \int_0^x \sin\left(\frac{\pi}{2}t^2\right) dt,\ C(x) = \int_0^x \cos\left(\frac{\pi}{2}t^2\right) dt \tag{4.2}$$

Other definitions (Abramowitz and Stegun, 1965, p. 300) are

$$S_1(x) = \sqrt{\frac{2}{\pi}} \int_0^x \sin(t^2) dt,\ C_1(x) = \sqrt{\frac{2}{\pi}} \int_0^x \cos(t^2) dt \tag{4.3}$$

and

$$S_2(x) = \frac{1}{\sqrt{2\pi}} \int_0^x \frac{\sin(t)}{\sqrt{t}} dt,\ C_2(x) = \frac{1}{\sqrt{2\pi}} \int_0^x \frac{\cos(t)}{\sqrt{t}} dt \tag{4.4}$$

Relations between these definitions are

$$S(x) = S_1\left(\sqrt{\frac{\pi}{2}}x\right) = S_2\left(\frac{\pi}{2}x^2\right) \tag{4.5}$$

with similar equations for the cosine integral. We shall use S_2 and C_2 in a calculation at the end of this chapter.

The specfun package is another package devoted to special functions in mathematics and physics. Functions included are:

- Gamma, Beta, Airy, and Psi functions
- Legendre functions of first and second kind

- Bessel and modified Bessel functions, spherical Bessel functions, and integrals of Bessel functions
- Kelvin, Struve, and Mathieu functions
- Hypergeometric (and confluent hypergeometric) functions
- Parabolic cylinder functions
- Spheroidal wave functions
- Error functions and Fresnel integrals
- Elliptic integrals and Jacobian elliptic functions
- Cosine and sine integrals
- Exponential integrals

One aspect of this package is that all functions, where appropriate, accept complex arguments, whereas most special functions in R and other packages only work for real arguments.

Because names like "beta" or "gamma" are so often used in R, all function names in `specfun` are prepended with a "sp." to avoid name clashes. For example the Beta and Gamma functions are called `sp.beta()` and `sp.gamma()`, respectively.

4.7 Polynomial functions in packages

Univariate polynomials (polynomials in a single variable) play an important role in science and engineering calculations, and R has three add-on packages for dealing with them: `polynom`, `PolynomF` and `orthopolynom`. The relations between these three is a bit complicated. `PolynomF` is the successor to `polynom`, while `orthopolynom` requires `polynom`, which contains numerous functions that have the same names as those in `PolynomF`. If there are name conflicts, the most recently loaded package takes precedence (a behavior known as "shadowing" or "masking"), so to avoid confusion only one should be loaded in a session.

We first install the three packages:

```
> install.packages(c("PolynomF","polynom","orthopolynom"))
```

If now we load `PolynomF` first, and then `orthopolynom`, we get

```
> require(PolynomF)
Loading required package: PolynomF
> require(orthopolynom)
Loading required package: orthopolynom
Loading required package: polynom

Attaching package: polynom

The following object(s) are masked from package:PolynomF:

    as.polylist, change.origin, GCD, integral, is.polylist, LCM,
    poly.calc, poly.from.roots, poly.from.values, poly.from.zeros,
    poly.orth, polylist
```

If we quit and restart the session, reversing the loading order, we get

```
> require(orthopolynom)
Loading required package: orthopolynom
Loading required package: polynom
> require(PolynomF)
Loading required package: PolynomF

Attaching package: PolynomF

The following object(s) are masked from package:polynom:

    as.polylist, change.origin, GCD, integral, is.polylist, LCM,
    poly.calc, poly.from.roots, poly.from.values, poly.from.zeros,
    poly.orth, polylist
```

4.7.1 PolynomF package

We consider PolynomF first. Having installed it, we load it and explore some of its capabilities. Solving for the roots of polynomials, and differentiating and integrating them, will also be dealt with in later chapters; but it is useful to recognize that the PolynomF (and orthopolynom) packages have these functions in convenient form for polynomials.

Note that below we use the function solve in quite a different context than we used it previously to find the inverse of a matrix or to solve a set of linear algebraic equations. This is an example of *overloading*, in which several functions or methods may be defined with the same name, but which differ in the type of the input and the output of the function.

```
> require(PolynomF)
Loading required package: PolynomF
> # Define a polynomial
> x = polynom() # Make x an object of class polynom
> p = x^3 - 3*x^2 - 2*x + 7
> # Ask the class of p. It should inherit the class of x
> class(p)
[1] "polynom"
> # Ask the modes of x and p. Both are functions.
> mode(x); mode(p)
[1] "function"
[1] "function"
1] "function"
```

A vector of the coefficients of the polynomial, with indices in ascending order, can be generated with the coef function in PolynomR.

```
> coef(p)
[1]  7 -2 -3  1
```

PolynomF can calculate with only one polynomial variable at a time, hence the restriction to *univariate* polynomials. For example, we can define other polynomials y and q(y), but combining them with x and p(x) leads to a polynomial in x only.

```
> y = polynom()
> class(y)
[1] "polynom"
> q = y^2 +2*y
> class(q)
[1] "polynom"
> p + q
7 - 2*x^2 + x^3
```

Using PolynomF we can perform arithmetic on polynomials:

```
> q = x^2 + 2 # Redefine q from example above
> p+q
9 - 2*x - 2*x^2 + x^3
> p-q
5 - 2*x - 4*x^2 + x^3
> p*q
14 - 4*x + x^2 - 3*x^4 + x^5
> round(p*q/5) # Each coefficient is rounded.
3 - x - x^4
> q^2
4 + 4*x^2 + x^4
```

The PolynomF package enables differential and integral calculus on polynomial functions.

```
> dpdx = deriv(p,"x") # Differentiate p with respect to x
> dpdx
-2 - 6*x + 3*x^2
```

We plot p and its derivative, observing that dpdx passes through 0 at the maxima and minima of p (Figure 4.7):

```
> p = x^3 - 3*x^2 - 2*x + 7
> dpdx = deriv(p,"x") # Differentiate p with respect to x
> dpdx
-2 - 6*x + 3*x^2
> curve(p,-2,3, ylab = "p(x), dp/dx")
> curve(dpdx, lty=2, add=T)
> abline(0,0, col = gray(.7))
> dpdx_zeros = solve(dpdx)
> dpdx_zeros
[1] -0.2909944  2.2909944
> abline(v=dpdx_zeros[1], col=gray(.7))
```

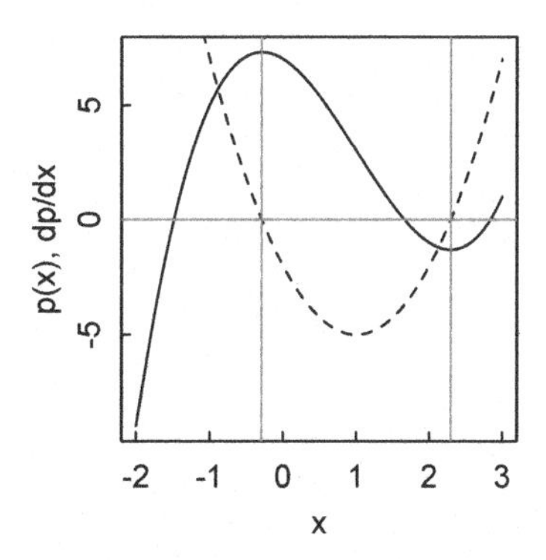

Figure 4.7: *Plot of a polynomial and its first derivative.*

```
> abline(v=dpdx_zeros[2], col=gray(.7))
```

We can integrate the derivative, recovering the polynomial except for the constant term, and we can evaluate the definite integral of p:

```
> integral(dpdx)
-2*x - 3*x^2 + x^3
> integral(p,limits = c(-2,3))
[1] 11.25
```

Given a set of zeros, we can calculate the polynomial that has those zeros:

```
> poly.from.zeros(c(-1,-1))
1 + 2*x + x^2
> poly.from.roots(c(-1,1)) # Alias for poly.from.zeros
-1 + x^2
```

Polynomials are often used to fit to (x,y) data. We shall discuss this further in the chapter on fitting functions to data, but we note that `PolynomF` has the function `poly.calc` that performs this fit (Figure 4.8).

```
> x = -3:3
> y = sin(x)
> polyfit = poly.calc(x,y) # Fit the sine function to a
                             polynomial in x
> polyfit
0.9941212*x - 0.1585776*x^3 + 0.005927377*x^5
> plot(x,y)
> curve(polyfit,add=T)
```

We can shift the origin of the polynomial, perhaps enabling a simplification. If q and p are polynomials and `q = change.origin(p,a)`, then $q(x) = p(x+a)$.

```
> x = polynom()
> p = 1 + 2*x + x^2
> change.origin(p, -1)
```

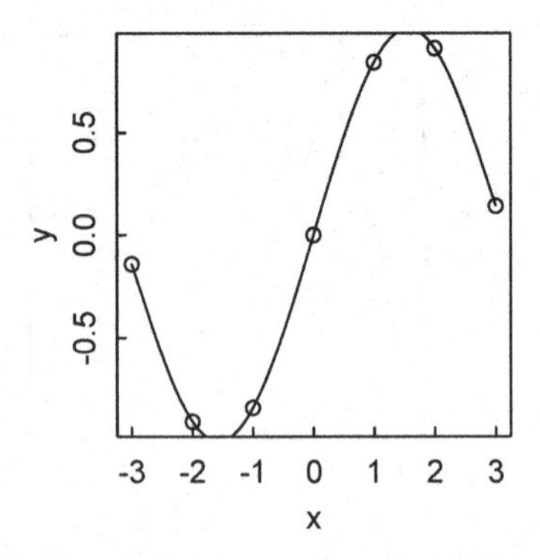

Figure 4.8: *Fitting data to a polynomial with* `poly.calc`.

```
x^2
```

We can use recursion relations to generate higher order orthogonal polynomials from the zeroth and first terms. (Note that the nth item in the list below is the (n-1)st order polynomial.) We use the Hermite polynomial example from the `PolynomF.pdf` reference manual, denoting them as $He(n)$ to differentiate them from the differently scaled Hermite polynomials $H(n)$ used by physicists (e.g., Abramowitz and Stegun, 1965):

```
> x = polynom()
> He = polylist(1, x)
> for(j in 2:10) He[[j+1]] = x*He[[j]] - (j-1)*He[[j-1]]
> He
List of polynomials:
[[1]]
1
[[2]]
x
[[3]]
-1 + x^2
[[4]]
-3*x + x^3
[[5]]
3 - 6*x^2 + x^4
[[6]]
15*x - 10*x^3 + x^5
[[7]]
-15 + 45*x^2 - 15*x^4 + x^6
[[8]]
-105*x + 105*x^3 - 21*x^5 + x^7
[[9]]
```

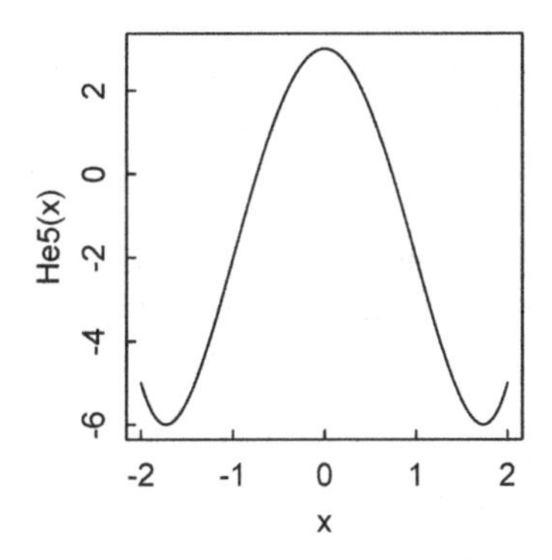

Figure 4.9: *Plot of 5th Hermite polynomial.*

```
105 - 420*x^2 + 210*x^4 - 28*x^6 + x^8
[[10]]
945*x - 1260*x^3 + 378*x^5 - 36*x^7 + x^9
[[11]]
-945 + 4725*x^2 - 3150*x^4 + 630*x^6 - 45*x^8 + x^10
```

Then individual polynomials may be picked out of the list by their indices, and manipulated as functions (Figure 4.9).

```
> He5 = He[[5]]
> curve(He5,-2,2)
```

4.7.2 `orthopolynom` *package*

Generating orthogonal polynomials from their recursion relations, and then manipulating them, is the special province of the `orthopolynom` package, to which we will turn next. `orthopolynom` is a collection of functions to construct sets of orthogonal polynomials and their recurrence relations. Additional functions are provided to calculate the derivative, integral, value, and roots of lists of polynomial objects. (F. Novometsky, Package orthopolynom, CRAN pdf, 2011). The polynomials included are

- Chebyshev: first (T, C) and second (U, S) kinds, and shifted
- Gegenbauer
- Hermite (H, He) and generalized Hermite
- Laguerre and generalized (associated) Laguerre
- Jacobi (P, G)
- Legendre and shifted
- monic (descending powers, with leading coefficient unity)
- spherical

- ultraspherical

For each type of polynomial, functions are given to calculate the inner product (norm squared), a list of polynomials for orders from 0 to n, a coefficient vector of recurrence relations, and the weight function for each order. Additional functions are provided to calculate values for a given x, derivatives and indefinite integrals (in symbolic form), powers, and roots, and to coerce the polynomials to functions.

As an example, consider the generalized Laguerre polynomials, perhaps better known in quantum mechanics as the associated Laguerre polynomials that govern the radial behavior of the electron in the hydrogen atom. By the code

```
> require(orthopolynom)
Loading required package: orthopolynom
Loading required package: polynom
> p.list = glaguerre.polynomials(3,1, normalized=FALSE)
> p.list
[[1]]
1
[[2]] 2-x
[[3]]
3 - 3*x + 0.5*x^2
[[4]]
4 - 6*x + 2*x^2 - 0.1666667*x^3
```

we get a list that contains the generalized Laguerre polynomials of degrees 1–4 and order 1. In terms of the orbital description of the hydrogen atom, these correspond to the s orbitals: 1s, 2s, 3s, and 4s. The p orbitals are of order 2, and would be generated by `glaguerre.polynomials(3,2)`. (The default is `normalized=FALSE`, so this option need not be stated explicitly.), The number 3 would be given if we wanted the first four degrees. (The convention in R is that numbering of vectors and lists starts with 1, but the convention in polynomials is that the list begins with order zero.)

In `p.list`, x is just a symbol. If we want to turn `p.list` into a function of x that can be used for calculation or graphing, we have to apply `polynomial.functions`.

```
> # Pick the third item in the list
> L31 = polynomial.functions(p.list)[[3]]
> L31(1.5)
[1] -0.375
```

We use this approach to plot the radial components of the 2s and 2p orbitals of the hydrogen atom (Figure 4.10), where the associated Laguerre polynomials $L_2^1(x)$ and $L_3^3(x)$ are squared and weighted by the factor $x^2 e^{-x}$ to account for normalization and the volume element in spherical polar coordinates.

```
> orb2s = polynomial.functions(glaguerre.polynomials(2,1))[[2]]
> orb2p = polynomial.functions(glaguerre.polynomials(3,2))[[3]]
> curve(orb2p(x)^2*x^2*exp(-x),0,25,lty=2, xlab="r (reduced)",
ylab ="Electron Density", main = "2s and 2p Orbitals of H Atom")
> curve(orb2s(x)^2*x^2*exp(-x),0,25,add=T)
> legend("topright",legend=c("2s","2p"),lty=1:2,bty="n")
```

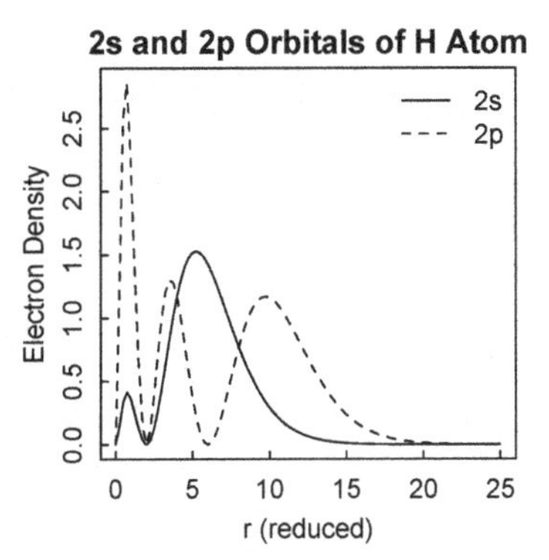

Figure 4.10: *Normalized associated Laguerre polynomials used to calculate the electron densities of the 2s and 2p orbitals of the hydrogen atom.*

The functions `polynomial.derivatives` and `polynomial.integrals` act on a list of polynomials.

```
> polynomial.derivatives(p.list)
[[1]]
0
[[2]] -1
[[3]] -3 + x
[[4]]
-6 + 4*x - 0.5*x^2
> polynomial.integrals(p.list)
[[1]]
x
[[2]]
2*x - 0.5*x^2
[[3]]
3*x - 1.5*x^2 + 0.1666667*x^3
[[4]]
4*x - 3*x^2 + 0.6666667*x^3 - 0.04166667*x^4
```

and again a particular item can be picked out of the list:

```
> polynomial.derivatives(p.list)[[4]]
-6 + 4*x - 0.5*x^2
```

Although `polynomial.integrals` does not give a constant of integration, the polynomial packages give R significant elements of a useful symbolic mathematics capability.

4.8 Case studies

4.8.1 Two-dimensional random walk

Earlier in this chapter we considered various models for one-dimensional random walks. Things get considerably more interesting when considering random walks in two (or more) dimensions. The walk can be taken in free space, as below, or on various geometries of lattice, and can be either self-intersecting (as in diffusion of a particle) or non-intersecting (as in the path of a linear polymer chain). This last situation, sometimes called the "excluded volume effect," leads to some deep theoretical and computational challenges. Unfortunately, it would take us too far out of our way to consider them here.

The model we consider in the code below is of a particle diffusing in random directions, but with steps of constant length = 1, on a surface bounded by a circular wall of radius `Rmax`. If a step leads to the particle colliding with the wall, it is rejected and another random direction is chosen. This process is carried out in the `while` loop of the code below, in which a test radius `Rtest > Rmax` is initially set, and the generation of the next random step is repeated while `Rtest` remains greater than R.

```
> Rmax = 10  # radius of boundary
> N = 200  # number of steps
> coords = matrix(nrow=N+1,ncol=2) # x and y coordinates
> coords[1,1] = coords[1,2] = 0  # start at origin
> twopi = 2*pi
> for (i in 2:(N+1)) {
+  Rtest = 1.1*Rmax
+  while (Rtest > Rmax) {
+    xold = coords[i-1,1]
+    yold = coords[i-1,2]
+    theta = runif(1,0,twopi) # Random angle for next step
+    xstep = cos(theta)  # x and y coords of next step
+    ystep = sin(theta)
+    xnew = xold + xstep  # New trial x and y coords
+    ynew = yold + ystep
+    R = sqrt(xnew^2 + ynew^2)  # New distance from origin
+    # If inside Rmax, reset Rtest to exit while loop
+    if (R < Rmax) Rtest = R
+  }
+ coords[i,1] = coords[i-1,1] + xstep  # New x, y coords
+ coords[i,2] = coords[i-1,2] + ystep
+ }
```

After `N` successful steps have been taken, we plot the path of the particle inside the circular boundary (Figure 4.11).

```
> plot.new() # New plot frame with no axes
> par(mar=c(0,0,0,0)) # Minimize margins
> # Make a square plot window with no box outlining it
```

Figure 4.11: *Path of a two-dimensional random walk confined to a circular domain.*

```
> plot.window(c(-Rmax,Rmax), c(-Rmax,Rmax), asp=1,bty="n")
> # Draw a polygon with 50 sides, approximating a circle
> polygon(Rmax*cos(2*pi*(0:50/50)), Rmax*sin(2*pi*(0:50/50)))
> # Plot (x,y) of the particle path
> lines(coords[,1],coords[,2])
```

Note how the particle "bounces off" the wall if it tries to go too far.

4.8.2 *Eigenvalues of a polymer chain*

We conclude this chapter with a calculation that brings together a number of the topics we have covered in this and previous chapters: functions in packages and user-defined functions, matrix construction, loops, conditionals, complex numbers, eigenvalue calculations, and rounding and sorting of numerical results. The code below recapitulates work done by Zimm et al. (1956) to calculate the eigenvalues of a matrix arising in the theory of polymer dynamics. The eigenvalues are proportional to the bending frequencies of the chain. In hydrodynamics, it is the lowest frequencies (slowest motions), hence the smallest eigenvalues, that are of greatest interest.

It might well be argued that the hardest work came in developing the analytical expressions for the matrix elements, but in 1956—before the general availability of digital computers and the numerical analysis codes whose development they facilitated—the numerical work was difficult, tedious, and fraught with potential for error. Now it takes just a few lines of code and a few milliseconds of personal computer time.

The expressions for the matrix elements involve the Fresnel sine and cosine integrals. Zimm et al. used the definition of the Fresnel integrals in Equation 4.4. We use Equation 4.5 to convert to $S_2(x)$ and $C_2(x)$ from the definitions for `fresnelS` and `fresnelC` used in the `pracma` package, and plot the comparison curves (Figure 4.12).

```
> require(pracma)
> S2 = function(x) fresnelS(sqrt(2*x/pi))
```

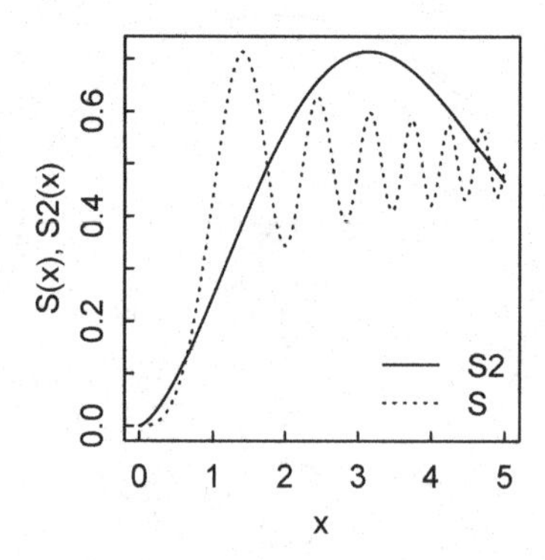

Figure 4.12: *Comparison of S2 and S function definitions for Fresnel sine integral.*

```
> C2 = function(x) fresnelC(sqrt(2*x/pi))
> curve(S2,0,5)
> curve(fresnelS,0,5,lty=2,add=T)
```

The matrix **G** whose eigenvalues we wish to find has elements

$$G_{0m} = 2\pi m^{1/2}(-1)^{(m/2+1)}S_2(\pi m), \tag{4.6}$$

$$G_{lm} = 2\pi(-1)^{(m-l+2)/2}\frac{m^2}{l^2-m^2}[l^{1/2}S_2(\pi l) - m^{1/2}S_2(\pi m)],\ l \neq 0, m, \tag{4.7}$$

$$G_{kk} = \frac{\pi k^{1/2}}{2}[2\pi k C_2(\pi k) - S(\pi k)]. \tag{4.8}$$

These equations are implemented in the R code below. Note that the factors of the form $(-1)^{m/2+1}$ will give the result `NaN` if m is odd and an imaginary part not exactly 0 if m is even. Therefore we have used the formulation `Re(round((-1+0i)^((m1/2)+1), 0))` to get the appropriate values of 0 if m is odd and ± 1 if m is even. We follow Zimm et al. in calculating only the first eight rows and columns of the matrix, which is sufficient to calculate the lowest frequencies. However, since their matrix indices started at 0, while R's start at 1, we must subtract 1 from the indices before evaluating the elements.

```
> G = matrix(nrow=8,ncol=8)
> for (m in 1:8) { # First row
+  m1 = m-1
+  v = Re(round((-1+0i)^((m1/2)+1), 0))
+  G[1,m] = 2*pi*m1^(1/2)*v*S2(pi*m1)
+ }
>
> for (m in 1:8) {
+  m1 = m-1
+  for (l in 1:8) {
+  l1 = l-1
```

```
+   if (l1 == m1) next
+       v = Re(round((-1+0i)^((m1-l1+2)/2), 0))
+       vw = 2*pi*v*m1^2/(l1^2-m1^2)
+       G[l,m] = vw*(sqrt(l1)*S2(pi*l1)-sqrt(m1)*S2(pi*m1))
+  }
+ }
>
> for (k in 1:8) { # Diagonal
+   k1 = k-1
+   G[k,k] = pi*sqrt(k1)/2*(2*pi*k1*C2(pi*k1)-S2(pi*k1))
+ }
```

We display the matrix elements rounded to three decimals.

```
> round(G,3)
     [,1]   [,2]   [,3]   [,4]   [,5]   [,6]   [,7]   [,8]
[1,]    0  0.000  3.052  0.000 -4.875  0.000  6.284  0.000
[2,]    0  4.098  0.000  2.653  0.000 -4.113  0.000  5.347
[3,]    0  0.000 12.867  0.000  2.432  0.000 -3.637  0.000
[4,]    0  0.295  0.000 24.271  0.000  2.485  0.000 -3.527
[5,]    0  0.000  0.608  0.000 37.914  0.000  2.536  0.000
[6,]    0 -0.165  0.000  0.895  0.000 53.413  0.000  2.632
[7,]    0  0.000 -0.404  0.000  1.127  0.000 70.606  0.000
[8,]    0  0.109  0.000 -0.648  0.000  1.343  0.000 89.314
```

Finally, we calculate the eigenvalues, round the results to three decimals, and sort the values in ascending order.

```
> sort(round(eigen(G)$values,3))
[1]  0.000  4.035 12.779 24.200 37.892 53.412 70.715 89.449
```

The results agree with those of Zimm et al. to within ± 1 in the last decimal.

Chapter 5

Solving systems of algebraic equations

Solving equations is central to numerical analysis, and R has numerous tools for doing so. We begin by extending our consideration of polynomials from the previous chapter.

5.1 Finding the zeros of a polynomial

The base installation of R has the function `polyroot` to find the zeros of a real or complex polynomial, specified by the vector of its coefficients in ascending order. This function uses the algorithm of Jenkins and Traub (Jenkins and Traub (1972) TOMS Algorithm 419. Comm. ACM, 15, 9799). For example,

```
> polyroot(c(4,5,6))
[1] -0.4166667+0.7021791i -0.4166667-0.7021791i
```

Using the package `PolynomF` , the code is a bit more cumbersome

```
> x = polynom()
> (q = solve(4 + 5*x + 6*x^2)) # Solve and display in one line
[1] -0.4166667-0.7021791i -0.4166667+0.7021791i
```

but `PolynomF` has the advantage that the polynomial can be calculated from the roots, to within a normalization factor:

```
> poly.calc(q)
0.6666667 + 0.8333333*x + x^2
```

On the other hand, `polyroot` can solve for the zeros of polynomials with complex coefficients, while `PolynomF` cannot. For example, the roots of the complex polynomial $1+2ix+(3-7i)x^2$ are found as follows:

```
> polyroot(c(1, 2i, 3-7i))
[1] -0.1019883+0.2473059i  0.3433676-0.3507542i
```

To see how close to zero the polynomial is at the roots:

```
> sol = polyroot(c(1, 2i, 3-7i))
> p = function(x) 1 + 2i*x + (3-7i)*x^2
> p(sol[1])
[1] 0+2.775558e-17i
> p(sol[2])
[1] -6.661338e-16-1.110223e-16i
```

There are certain "pathological" polynomials for which finding roots should be easy, but isn't. The prime example is Wilkinson's polynomial (http://en.wikipedia.org/wiki/Wilkinson's polynomial; Acton, p. 201):

$$W(x) = \prod_{i=1}^{20}(x-i) = (x-1)(x-2)\ldots(x-20)$$

Obviously the roots are the integers 1 to 20 and they are well separated, but there are limitations in the precision available to the root-solving functions that undermine the process, both with `polyroot` and with `solve`, which uses an eigenvalue computation. Jenkins–Traub is usually the most reliable, but in this case neither works.

```
> require(PolynomF)
> x = polynom()
> W=(x-1)
> for (j in 2:20) W = W*(x-j)
> solve(W)
 [1]  1.000000  2.000000  3.000000  4.000000  5.000000  6.000000
 [7]  6.999973  8.000284  8.998394 10.006060 10.984041 12.033449
[13] 12.949056 14.065273 14.935356 16.048275 16.971132 18.011222
[19] 18.997160 20.000325
```

Now try with `polyroot`, after getting the polynomial coefficients with `coif(W)`.

```
> polyroot(coef(W))
 [1]  1.000000+0.000000i  2.000000+0.000000i  3.000000-0.000000i
 [4]  4.000000+0.000000i  5.000000-0.000000i  7.000005-0.000014i
 [7]  5.999999+0.000002i  9.000227-0.000165i  7.999960+0.000059i
[10] 11.002737-0.000451i 11.993971+0.000451i  9.999080+0.000322i
[13] 13.986821+0.000200i 15.013075-0.000118i 13.010246-0.000335i
[16] 17.005442-0.000049i 17.997884+0.000022i 15.990103+0.000082i
[19] 19.000507-0.000005i 19.999943+0.000000i
```

5.2 Finding the zeros of a function

A common task in numerical analysis is to determine the roots of a function, the places where its value equals zero. Base R has the function `uniroot` for doing this, but we shall begin by coding two simpler methods that are often used.

5.2.1 Bisection method

The bisection method brackets an interval in which a root of the function $f(x)$ must lie, then repeatedly bisects the interval until a root is found within the desired precision. The initial guesses for the lower and upper limits, `xmin` and `xmax`, of the interval must give function values of opposite signs, since then a zero crossing is guaranteed to lie between them. In the code below, the default value of the desired precision is set to `tol = 1e-5`; this can be changed as desired.

```
> bisectionroot = function(f, xmin, xmax, tol=1e-5) {
+   a = xmin; b = xmax
+   # Check inputs
+   if (a >= b) {
+     cat("error: xmin > xmax \n")
+     return(NULL)
+   }
+   if (f(a) == 0) {
+     return(a)
+   } else if (f(b) == 0) {
+     return(b)
+   } else if (f(a)*f(b) > 0) {
+     cat("error: f(xmin) and f(xmax) of same sign \n")
+     return(NULL)
+   }
+   # If inputs OK, converge to root
+   iter = 0
+   while ((b-a) > tol) {
+     c = (a+b)/2
+     if (f(c) == 0) {
+       return(c)
+     } else if (f(a)*f(c) < 0) {
+       b = c
+     } else {
+       a = c
+     }
+     iter = iter + 1
+     }
+     return(c((a+b)/2, iter, (b-a))) # root, iterations, precision
+ }
```

We use `bisectionroot` to find a root of the function $f(x) = x^3 - \sin(x)^2$, obtaining

```
> f = function(x) x^3 - sin(x)^2
> bisectionroot(f,0.5,1)
[1] 8.028069e-01 1.600000e+01 7.629395e-06
```

5.2.2 *Newton's method*

A second commonly used algorithm for root-finding is Newton's method, also known as the Newton–Raphson method. This method obtains an improved estimate x_1 for the root from an initial guess x_0 according to the equation

$$x_1 = x_0 - \frac{f(x_0)}{f'(x_0)}, \tag{5.1}$$

iterating until the desired precision is reached or the maximum number of iterations is exceeded. The method is implemented by the following code:

```
> newtonroot = function(f, df, x0, tol=1e-5, maxit = 20) {
+   root = x0
+   for (jit in 1:maxit) {
+     dx = f(root)/df(root)
+     root = root - dx
+     if (abs(dx) < tol) return(c(root, jit, dx))
+   }
+   print(" Maximum number of iterations exceeded.")
+ }
```

We test the code with the same function as before, but supply the required first derivative as well.

```
> f = function(x) x^3 - sin(x)^2
> df = function(x) 3*x^2 - 2*cos(x)*sin(x)
> newtonroot(f,df,1)
[1] 8.028037e-01 5.000000e+00 4.275506e-08
```

Note that `newtonroot` required 5 iterations to converge, while `bisectionroot` required 16. The latter is generally slower than other methods, but is guaranteed to converge, while alternatives may sometimes not do so.

5.2.3 *uniroot and uniroot.all*

The base installation of R has the function `uniroot()` to search for a root of a function `f` in a specified interval. If successful, it yields the root, the value of `f` at the root, the number of iterations to achieve the desired tolerance, and the estimated precision of the root. The help page for `uniroot` says that it uses the Brent method. According to *Numerical Recipes in Fortran 77*, 2nd ed., pp. 353-4, "Brent's method combines root bracketing, bisection, and inverse quadratic interpolation to converge from the neighborhood of a zero crossing. ... [It thereby] combines the sureness of bisection with the speed of a higher-order method when appropriate."

Consider the function $f(x,a) = x^{1/3}\sin(5x) - a\sqrt{x}$. (Figure 5.1) We treat a as a parameter, rather than a constant, to demonstrate how to treat a parameter in plotting and root-finding contexts.

```
> f = function(x,a) x^(1/3)*sin(5*x) - a*x^(1/2)
> curve(f(x,a=0.5),0,5)
> abline(h=0, lty=3)

> uniroot(f,c(.1,1),a=0.5)
$root
[1] 0.5348651
$f.root
[1] -2.762678e-05
$iter [1] 7
```

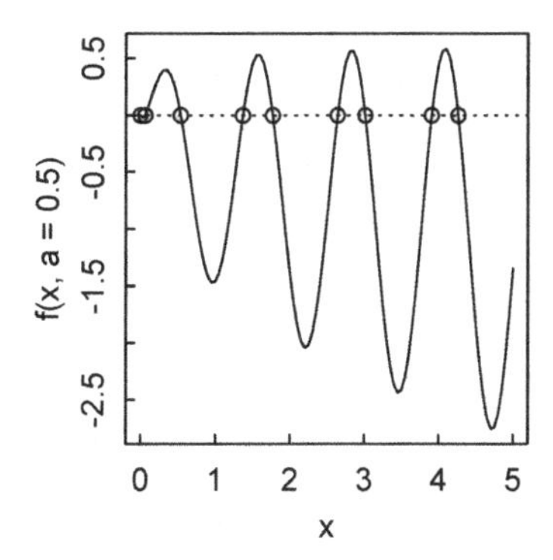

Figure 5.1: *The function f(x,a) with a = 0.5. Roots are located by the* `points` *command once they have been calculated by* `uniroot.all`*.*

```
$estim.prec
[1] 6.103516e-05
```

In this example, we started the root search in the region `c(0.1,1)` rather than `c(0,1` because the function must be of opposite signs at the beginning and end of the interval. If not, an error message is generated.

```
> uniroot(f,c(.1,.5),a=0.5)
Error in uniroot(f, c(0.1, 0.5)) :
f() values at end points not of opposite sign
```

If the function has several zeros in the region of interest, the function `uniroot.all` from the package `rootSolve` (which uses `uniroot`) should find all of them, though success is not guaranteed in pathological cases.

```
> require(rootSolve)
Loading required package: rootSolve
> zpts=uniroot.all(f,c(0,5),a=0.5)
> zpts
[1] 0.00000000 0.06442212 0.53483060 1.36761623 1.76852629
[6] 2.63893168 3.01267402 3.90557382 4.26021380
> yz=rep(0,length(zpts))
> points(zpts,yz)  # Locate roots on graph of function
```

Note that `uniroot.all` does not provide the information about convergence and precision that `uniroot` does. Note also the differences in how to deal with the parameter a in the calls to `curve` and in `uniroot` or `uniroot.all`.

`uniroot` will not work if the function only touches, but does not cross, the x axis, unless one end of the search range is exactly at the root. For example,

```
> ff = function(x) sin(x)+1
> uniroot(ff,c(-pi,0))
Error in uniroot(ff, c(-pi, 0)) :
  f() values at end points not of opposite sign
```

```
# but
> uniroot(ff,c(-pi,-pi/2))
$root
[1] -1.570796
$f.root
[1] 0
$iter
[1] 0
$estim.prec
[1] 0
```

Of course, if the position of the root is already known, there is no need to do the calculation. In general, however, it may be best to seek the minimum of such a function by procedures discussed later in this book in the chapter on optimization.

5.3 Systems of linear equations: matrix `solve`

The need to solve systems of linear equations arises in nearly all fields of science and engineering. Such equations can be formulated as $\mathbf{A}x = B$ where $\mathbf{A}$ is a square $n \times n$ matrix, B is a vector of length n, and x is the vector (length n) to be solved for. Formally, $x = \mathbf{A}^{-1}B$ where $\mathbf{A}^{-1}$ is the inverse of $\mathbf{A}$. However, computing the inverse and then multiplying is inefficient and prone to inaccuracy. R uses the finely honed routines in LAPACK (Linear Algebra PACKage), the standard software library for numerical linear algebra. It invokes these routines with the `solve` function.

The mechanics can be illustrated simply with a 4x4 random matrix **m** and 4-vector b.

```
> options(digits=3)
> set.seed(3)
> m = matrix(runif(16), nrow = 4)
> m
[,1] [,2] [,3] [,4]
[1,] 0.168 0.602 0.578 0.534
[2,] 0.808 0.604 0.631 0.557
[3,] 0.385 0.125 0.512 0.868
[4,] 0.328 0.295 0.505 0.830
> b = runif(4)
>b
[1] 0.111 0.704 0.897 0.280
 > solve(m,b)
[1]  0.528 -3.693  5.850 -2.121
> m%*%solve(m,b) # Should recover b
      [,1]
[1,] 0.111
[2,] 0.704
[3,] 0.897
```

```
[4,] 0.280
> solve(m)%*%b  # Same: multiply b by inverse of m
       [,1]
[1,]  0.528
[2,] -3.693
[3,]  5.850
[4,] -2.121
```

5.4 Matrix inverse

It is rarely necessary to calculate the inverse of a matrix, but if it is so desired it is readily obtained with `solve()`.

```
> set.seed(333)
> M = matrix(runif(9), nrow=3)
> M
           [,1]       [,2]       [,3]
[1,] 0.46700066 0.57130558 0.60939363
[2,] 0.08459815 0.02011937 0.30671935
[3,] 0.97348527 0.72355739 0.06350984
> Minv = solve(M)
> Minv
           [,1]      [,2]       [,3]
[1,] -2.4561314  4.504248  1.8140634
[2,]  3.2638470 -6.273329 -1.0205667
[3,]  0.4633475  2.429467 -0.4334035
> Minv%*%M
             [,1]          [,2]         [,3]
[1,] 1.000000e+00 -2.220446e-16 0.000000e+00
[2,] 0.000000e+00  1.000000e+00 6.938894e-17
[3,] 5.551115e-17  5.551115e-17 1.000000e+00
> zapsmall(Minv%*%M)
     [,1] [,2] [,3]
[1,]    1    0    0
[2,]    0    1    0
[3,]    0    0    1
```

5.5 Singular matrix

In the code below, matrix `A.sing` is singular because columns 2 and 4 are proportional to each other. In this case the system of equations cannot be solved.

```
> A.sing = matrix(c
(1,2,-1,-2,2,1,1,-1,1,-1,2,1,1,3,-2,-3),nrow=4,byrow=T)
> A.sing
[,1] [,2] [,3] [,4]
[1,] 1 2-1-2
```

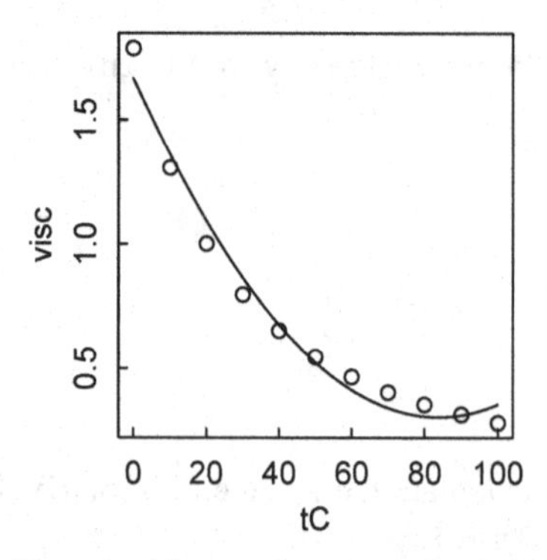

Figure 5.2: *Viscosity of water fit to a quadratic in temperature.*

```
[2,] 2 1 1 -1
[3,] 1 -1 2 1
[4,] 1 3-2-3
> B = c(-1,4,5,-3)
> solve(A.sing,B)
Error in solve.default(A.sing, B) :
  LAPACK  routine dgesv: system is exactly singular
```

5.6 Overdetermined systems and generalized inverse

A matrix has an inverse as usually defined only when it is square. If it has more rows than columns, this is equivalent to the overdetermined system $\mathbf{A}y = b$ where there are more equations than unknowns. The unknowns y may be solved in the least-squares sense using one of several methods in base R or its packages. Consider, for example, three ways of fitting the viscosity of liquid water to a quadratic in Celsius temperature (Figure 5.2).

```
> options(digits=3)
> tC = seq(0,100,10)  # Temperatures between freezing and boiling
> visc = c(1.787,1.307,1.002,0.798,0.653,0.547,0.467,
+    0.404,0.355,0.315,0.282)
> plot(tC,visc)
> const = rep(1,length(tC)) # For proper representation of quadratic
> tC_sq = tC^2
> A = cbind(const,tC,tC_sq)
```

(1) `qr.solve` in base R:

```
> qr.solve(A,visc)
    const          tC      tC_sq
 1.665685 -0.032501  0.000194
> # or equivalently
> solve(qr(A,LAPACK=TRUE),visc)
[1]  1.665685 -0.032501  0.000194
```

(2) the generalized inverse function ginv defined in the MASS package.

```
> require(MASS)
> gv = ginv(A)%*%visc
> gv
          [,1]
[1,]  1.665685
[2,] -0.032501
[3,]  0.000194
# Define a function with the calculated coefficients
> g = function(x) gv[1,1] + gv[2,1]*x + gv[3,1]*x^2
# Superimpose the function plot on the data points
> curve(g(x),0,100,add=T)
```

(3) the Solve function in the limSolve package, which must first be installed, and which automatically loads three other packages on which it depends. Solve also uses the generalized inverse function ginv from MASS.

```
> install.packages("limSolve")
> require(limSolve)
Loading required package: limSolve
Loading required package: quadprog
Loading required package: lpSolve
Loading required package: MASS
> Solve(A,visc)
const         tC      tC_sq
 1.665685 -0.032501  0.000194
```

As we shall see in a later chapter, we would normally do such data fitting using a linear model, which would give estimates of the uncertainties in the parameters.

If, on the other hand, there are fewer equations than unknowns, there are no unique solutions. If there are N unknowns and M equations, there will generally be an $N-M$-dimensional family of solutions. Singular value decomposition can find the subspace of solutions. See svd later in this chapter in the section on matrix decompositions.

5.7 Sparse matrices

Sparse matrices are ones in which only a small fraction of the entries are non-zero. Modern computers are so fast that special treatment is usually needed only for very large sparse matrices, but the R packages limSolve, Matrix, and SparseM provide such capability when needed.

5.7.1 Tridiagonal matrix

Perhaps the most commonly encountered type of sparse matrix is the tridiagonal matrix, in which only the main diagonal and the diagonals just above and below it have non-zero entries. Such matrices may arise when considering interactions between neighbors to the right and left. We first set up and solve a small problem (Hanna and Sandall, pp. 40-43) using the solve function in the base R installation.

```
> n = 11 # Size of matrix
> m = matrix(n,n,data=0) # Set up square matrix

> # Put ones below the diagonal
> aa = rep(1,n)
> aa[1] = aa[n] = 0 # Except first and last element
> aa = aa[-1] # Trim aa to fit below the diagonal

>  # Set up diagonal
> bb = rep(-1.99,n)
> bb[1] = bb[n] = 1

> # Put ones above the diagonal
> cc = rep(1,n)
> cc[1] = cc[n] = 0 # Except first and last element
> cc = cc[-n] # Trim cc to fit above the diagonal

> # Define rhs of linear system
> d = rep(0,n)
> d[1] = 0.5
> d[n] = 0.69

> # Assemble matrix
> m[1,1:2] = c(bb[1],cc[1])
> m[n,(n-1):n] = c(aa[n-1],bb[n])
> for (i in 2:(n-1)) m[i,(i-1):(i+1)] = c(aa[i-1],bb[i],cc[i])

> options(digits=3)

> # Solve
> soln = solve(m,d)
> soln
 [1] 0.500 0.547 0.589 0.625 0.655 0.678 0.694 0.704
 [9] 0.706 0.702 0.690
```

Now suppose that the set of equations to be solved gets 100 or 1000 times bigger. On my 2012 laptop, for $n = 1001$,

```
   user  system elapsed
  0.331   0.004   0.332
```

and for $n = 10001$

```
   user  system elapsed
 341.84    8.24  371.07
```

This is close to expected since the `solve` algorithm goes as n^3. Now try with `Solve.tridiag` from the `limSolve` package. This algorithm goes as n. We need to provide just the vectors, not the matrix **m**.

```
> require(limSolve)
> n = 1001

> # Above-diagonal vector
> aa = rep(1,n)
> aa[1] = aa[n] = 0
> aa=aa[-1]

> # Diagonal vector
> bb = rep(-1.99,n)
> bb[1] = bb[n] = 1

> # Below-diagonal vector
> cc = rep(1,n)
> cc[1] = cc[n] = 0
> cc=cc[-n]

> # rhs of system
> d = rep(0,n)
> d[1] = 0.5
> d[n] = 0.69

> system.time(tri.soln <- Solve.tridiag(aa,bb,cc,d))
   user  system elapsed
      0       0       0
```

For $n = 10001$

```
   user  system elapsed
  0.006   0.001   0.010
```

The time saved is huge, a reduction from minutes to milliseconds.

5.7.2 *Banded matrix*

Less frequently encountered, but still worth considering, are banded matrices. These have non-zero entries only on the main diagonal and on `nup` diagonals above and `nlow` diagonals below the main diagonal. `limSolve` provides the `Solve.banded` function for such matrices.

```
> require(limSolve)
> options(digits=3)

> set.seed(333)
```

```
> n = 500 # 500 x 500 matrix

> # Lower diagonals
> dn1 = runif(n-1)
> dn2 = runif(n-2)

> # Diagonal
> bb = runif(n)

> # Upper diagonals
> up1 = runif(n-1)
> up2 = runif(n-2)

> # Assemble matrix
> abd = rbind(c(0,0,up2),c(0,up1),bb,c(dn1,0),c(dn2,0,0))

> B = runif(n) # rhs of system

> system.time(Band <- Solve.banded(abd, nup=2, nlow=2,B))
   user  system elapsed
      0       0       0
> Band[1:5]   # Show the first five values in solution vector
[1]  11.847 -21.239   0.246  -2.005  -9.015
```

We compare with `solve`, which gives the same result and is not much slower for n = 500.

```
> bndmat = matrix(nrow=n,ncol=n,data=rep(0,n*n))
> diag(bndmat) = bb
> for(i in 1:(n-2)) bndmat[i+2,i] = dn2[i]
> for(i in 1:(n-1)) bndmat[i+1,i] = dn1[i]
> for(i in 1:(n-1)) bndmat[i,i+1] = up1[i]
> for(i in 1:(n-2)) bndmat[i,i+2] = up2[i]
> system.time(bnd <- solve(bndmat,B))
   user  system elapsed
  0.036   0.000   0.036
> bnd[1:5]
[1]  11.847 -21.239   0.246  -2.005  -9.015
```

5.7.3 *Block matrix*

A block matrix may be viewed as a matrix of distinct smaller matrices, typically arrayed on or near the diagonal of the full matrix. They may be encountered, for example, in input-output tables where the inputs fall into discrete clusters. The `limSolve` package has the function `Solve.block` that "solves the linear system A*X=B where A is an almost block diagonal matrix of the form:

`TopBlock`

```
... Array(1) ... ... ...
... ... Array(2) ... ...
...
... ... ... Array(Nblocks)...
... ... ... BotBlock''
```

As one of many examples of R routines calling a faster compiled language, `Solve.block` uses the FORTRAN subroutine `colrow`, whose "method is based on Gauss elimination with alternate row and column elimination with partial pivoting, producing a stable decomposition of the matrix A without introducing fill-in." We illustrate with the example from the help page for `Solve.block`.

```
> # Define matrix dimensions, set elements to 0
> AA = matrix (nr= 12, nc=12, 0)

> # Enter matrix elements
> AA[1,1:4]  = c( 0.0,  -0.98, -0.79, -0.15)
> AA[2,1:4]  = c(-1.00,  0.25, -0.87,  0.35)
> AA[3,1:8]  = c( 0.78,  0.31, -0.85,  0.89, -0.69, -0.98, -0.76, -0.82)
> AA[4,1:8]  = c( 0.12, -0.01,  0.75,  0.32, -1.00, -0.53, -0.83, -0.98)
> AA[5,1:8]  = c(-0.58,  0.04,  0.87,  0.38, -1.00, -0.21, -0.93, -0.84)
> AA[6,1:8]  = c(-0.21, -0.91, -0.09, -0.62, -1.99, -1.12, -1.21,  0.07)
> AA[7,5:12] = c( 0.78, -0.93, -0.76,  0.48, -0.87, -0.14, -1.00, -0.59)
> AA[8,5:12] = c(-0.99,  0.21, -0.73, -0.48, -0.93, -0.91,  0.10, -0.89)
> AA[9,5:12] = c(-0.68, -0.09, -0.58, -0.21,  0.85, -0.39,  0.79, -0.71)
> AA[10,5:12] = c( 0.39, -0.99, -0.12, -0.75, -0.68, -0.99,  0.50, -0.88)
> AA[11,9:12] = c( 0.71, -0.64,   0.0,  0.48)
> AA[12,9:12] = c( 0.08, 100.0, 50.00, 15.00)
>
> AA  # Show matrix
      [,1]  [,2]  [,3]  [,4]  [,5]  [,6]  [,7]  [,8]  [,9]  [,10] [,11]
 [1,]  0.00 -0.98 -0.79 -0.15  0.00  0.00  0.00  0.00  0.00   0.00  0.00
 [2,] -1.00  0.25 -0.87  0.35  0.00  0.00  0.00  0.00  0.00   0.00  0.00
 [3,]  0.78  0.31 -0.85  0.89 -0.69 -0.98 -0.76 -0.82  0.00   0.00  0.00
 [4,]  0.12 -0.01  0.75  0.32 -1.00 -0.53 -0.83 -0.98  0.00   0.00  0.00
 [5,] -0.58  0.04  0.87  0.38 -1.00 -0.21 -0.93 -0.84  0.00   0.00  0.00
 [6,] -0.21 -0.91 -0.09 -0.62 -1.99 -1.12 -1.21  0.07  0.00   0.00  0.00
 [7,]  0.00  0.00  0.00  0.00  0.78 -0.93 -0.76  0.48 -0.87  -0.14 -1.00
 [8,]  0.00  0.00  0.00  0.00 -0.99  0.21 -0.73 -0.48 -0.93  -0.91  0.10
 [9,]  0.00  0.00  0.00  0.00 -0.68 -0.09 -0.58 -0.21  0.85  -0.39  0.79
[10,]  0.00  0.00  0.00  0.00  0.39 -0.99 -0.12 -0.75 -0.68  -0.99  0.50
[11,]  0.00  0.00  0.00  0.00  0.00  0.00  0.00  0.00  0.71  -0.64  0.00
[12,]  0.00  0.00  0.00  0.00  0.00  0.00  0.00  0.00  0.08 100.00 50.00
     [,12]
 [1,]  0.00
 [2,]  0.00
 [3,]  0.00
 [4,]  0.00
 [5,]  0.00
 [6,]  0.00
 [7,] -0.59
 [8,] -0.89
 [9,] -0.71
[10,] -0.88
[11,]  0.48
[12,] 15.00
```

The vector B (right-hand side of the system of equations) is

```
B = c(-1.92,-1.27,-2.12,-2.16,-2.27,-6.08,-3.03,-4.62,-1.02,
      -3.52,0.55,165.08)
```

The matrix AA is divided into blocks as follows:

```
> Top = matrix(nr=2, nc=4, data=AA[1:2,1:4])
> Top
     [,1]  [,2]  [,3]  [,4]
[1,]    0 -0.98 -0.79 -0.15
[2,]   -1  0.25 -0.87  0.35

> Bot = matrix(nr=2, nc=4, data=AA[11:12,9:12])
> Bot
     [,1]   [,2] [,3]  [,4]
[1,] 0.71  -0.64    0  0.48
[2,] 0.08 100.00   50 15.00

> Blk1 = matrix(nr=4, nc=8, data=AA[3:6,1:8])
> Blk1
      [,1]  [,2]  [,3]  [,4]  [,5]  [,6]  [,7]  [,8]
[1,]  0.78  0.31 -0.85  0.89 -0.69 -0.98 -0.76 -0.82
[2,]  0.12 -0.01  0.75  0.32 -1.00 -0.53 -0.83 -0.98
[3,] -0.58  0.04  0.87  0.38 -1.00 -0.21 -0.93 -0.84
[4,] -0.21 -0.91 -0.09 -0.62 -1.99 -1.12 -1.21  0.07

Blk2 = matrix(nr=4, nc=8, data=AA[7:10,5:12])
> Blk2
      [,1]  [,2]  [,3]  [,4]  [,5]  [,6]  [,7]  [,8]
[1,]  0.78 -0.93 -0.76  0.48 -0.87 -0.14 -1.00 -0.59
[2,] -0.99  0.21 -0.73 -0.48 -0.93 -0.91  0.10 -0.89
[3,] -0.68 -0.09 -0.58 -0.21  0.85 -0.39  0.79 -0.71
[4,]  0.39 -0.99 -0.12 -0.75 -0.68 -0.99  0.50 -0.88
```

We combine the inner blocks into a $4 \times 8 \times 2$ array `AR`, since each is of dimension 4×8, and there are two of them.

```
> AR = array(dim=c(4,8,2),data=c(Blk1,Blk2))
```

The quantity `overlap` is the sum of the number of rows of `Top` and `Bot`. Combining these results, we find that

```
> Solve.block(Top,AR,Bot,B,overlap=4)
```

yields a vector of 12 ones.

5.8 Matrix decomposition

The code underlying the matrix algorithms embodied in `solve`, `limSolve`, and `Matrix` uses various decompositions of a matrix: factorization of the matrix into some canonical form. We shall not discuss these decompositions in detail, since our interest is in using R functions to solve sets of equations, rather than delving into how the functions work. However, we note the most common decompositions here.

5.8.1 QR decomposition

The **QR decomposition** of an $m \times n$ (not necessarily square) matrix factors the matrix into an orthogonal $m \times m$ matrix $\mathbf{Q}$ and an upper triangular matrix $\mathbf{R}$. It is invoked in the base R installation with `qr()` and used to solve overdetermined systems in a least-square sense with `qr.solve()`, being therefore useful in computing regression coefficients and applying the Newton–Raphson algorithm. In the `Matrix` package, `x = "dgCMatrix"` gives the QR decomposition of a general sparse double-precision matrix.

We give two examples, starting with an overdetermined system with 4 equations and 3 unknowns.

```
> set.seed(321)
> A = matrix((1:12)+rnorm(12),nrow=4)
> b = 2:5
> qr.solve(A,b)  # Solution in a least-squares sense
[1]  0.625  1.088 -0.504
```

The QR decomposition of A, itself, is simply obtained by

```
> qr(A)
$qr
       [,1]     [,2]    [,3]
[1,] -5.607 -13.2403 -21.515
[2,]  0.230  -3.9049  -4.761
[3,]  0.485   0.4595   1.228
[4,]  0.692  -0.0574   0.515

$rank
[1] 3

$qraux
[1] 1.48 1.89 1.86

$pivot
[1] 1 2 3

attr(,"class")
[1] "qr"
```

If, on the other hand, there are 3 equations and 4 unknowns, we have an underdetermined system.

```
> set.seed(321)
> A = matrix((1:12)+rnorm(12),nrow=3)
```

```
> b = 3:5

> qr.solve(A,b) # Default LAPACK = FALSE uses LINPACK
[1] -0.1181  0.8297  0.0129  0.0000

> solve(qr(A, LAPACK = TRUE),b)
[1]  0.0387  0.0000 -0.4514  0.6756
```

5.8.2 Singular value decomposition

The **singular value decomposition** `svd()` in the base installation decomposes a rectangular matrix into the product $\mathbf{UDV_H}$, where $\mathbf{D}$ is a nonnegative diagonal matrix, $\mathbf{U}$ and $\mathbf{V}$ are unitary matrices, and $\mathbf{V_H}$ denotes the conjugate transpose of $\mathbf{V}$ (or simply the transpose if $\mathbf{V}$ contains real numbers only). The singular values are the diagonal elements of $\mathbf{D}$. For square matrices, `svd()` and `eigen()` give equivalent eigenvalues. In fact, the routines that R uses to calculate eigenvalues and eigenfunctions, based on LAPACK and its predecessor EISPACK, are based on SVD calculations.

An example of singular value decomposition of a matrix with 6 rows and 5 columns, yielding a diagonal matrix of 5 singular values, which would be eigenvalues if the matrix were square:

```
> set.seed(13)
> A = matrix(rnorm(30), nrow=6)
> svd(A)
$d
[1] 3.603 3.218 2.030 1.488 0.813

$u
       [,1]    [,2]    [,3]   [,4]   [,5]
[1,] -0.217 -0.4632  0.4614  0.164  0.675
[2,] -0.154 -0.5416  0.0168 -0.528 -0.444
[3,]  0.538 -0.1533  0.5983 -0.290 -0.124
[4,]  0.574 -0.5585 -0.5013  0.319  0.070
[5,]  0.547  0.3937  0.0449 -0.261  0.285
[6,]  0.104  0.0404  0.4190  0.664 -0.496

$v
       [,1]    [,2]   [,3]   [,4]  [,5]
[1,]  0.459 -0.0047  0.712 -0.159 0.507
[2,] -0.115 -0.5192 -0.028  0.758 0.377
[3,]  0.279  0.7350 -0.355  0.352 0.363
[4,]  0.333 -0.4023 -0.604 -0.448 0.402
[5,] -0.766  0.1684  0.039 -0.275 0.554
```

An interesting and insightful article about the geometric interpretation of the SVD in terms of linear transformations, its theory, and some applications, is "A Singularly Valuable Decomposition: The SVD of a Matrix" by Dan Kalman.[1]

5.8.3 *Eigendecomposition*

The familiar process of finding the eigenvalues and eigenvectors of a square matrix can be viewed as **eigendecomposition**. It factors the matrix into $\mathbf{VDV^{-1}}$, where $\mathbf{D}$ is a diagonal matrix formed from the eigenvalues, and the columns of $\mathbf{V}$ are the corresponding eigenvectors.

A familiar example from physics textbooks is a system of 3 masses of mass m attached to parallel walls by 4 springs of force constant k. Analysis of this system (e.g., Garcia, 2000, pp. 164–5) leads to the matrix equation

$$\begin{pmatrix} 2 & -1 & 0 \\ -1 & 2 & -1 \\ 0 & -1 & 2 \end{pmatrix} \mathbf{a} = \lambda \mathbf{a} \tag{5.2}$$

We wish to solve this equation for the eigenvalues $\lambda = m\omega^2/k$ leading to the characteristic frequencies ω, and for the eigenvectors $\mathbf{a}$. The analytical solutions, readily obtained in this simple case, are $\lambda = 2, 2+\sqrt{2}, 2-sqrt2$ with eigenvectors

$$\mathbf{a}_0 = \begin{pmatrix} 1/\sqrt{2} \\ 0 \\ -1/\sqrt{2} \end{pmatrix}, \quad \mathbf{a}_\pm = \begin{pmatrix} 1/2 \\ \mp 1/\sqrt{2} \\ 1/2 \end{pmatrix} \tag{5.3}$$

These results agree with the numerical values obtained by the R code

```
> options(digits=3)
> M = matrix(c(2,-1,0,-1,2,-1,0,-1,2), nrow=3, byrow=TRUE)
> eigen(M)
$values
[1] 3.414 2.000 0.586

$vectors
       [,1]      [,2]  [,3]
[1,] -0.500 -7.07e-01 0.500
[2,]  0.707  1.10e-15 0.707
[3,] -0.500  7.07e-01 0.500
```

5.8.4 *LU decomposition*

The **LU decomposition** factors a square matrix into a lower triangular matrix $\mathbf{L}$ and an upper triangular matrix $\mathbf{U}$. It can be called from the `Matrix` package with the function `lu()`. LU decomposition is commonly used to solve square systems

[1] www.math.umn.edu/~lerman/math5467/svd.pdf

of linear equations, since it is about twice as fast as QR decomposition. Here is an example from the LU (dense) Matrix Decomposition help page.

```
> options(digits=3)
> set.seed(1)
> require(Matrix)
> mm = Matrix(round(rnorm(9),2), nrow = 3)
> mm
3 x 3 Matrix of class "dgeMatrix"
      [,1]  [,2] [,3]
[1,] -0.63  1.60 0.49
[2,]  0.18  0.33 0.74
[3,] -0.84 -0.82 0.58
> lum = lu(mm)
> str(lum)
Formal class 'denseLU' [package "Matrix"] with 3 slots
  ..@ x   : num [1:9] -0.84 0.75 -0.214 -0.82 2.215 ...
  ..@ perm: int [1:3] 3 3 3
  ..@ Dim : int [1:2] 3 3
> elu = expand(lum)
> elu # three components: "L", "U", and "P", the permutation
$L
3 x 3 Matrix of class "dtrMatrix" (unitriangular)
     [,1]    [,2]    [,3]
[1,]  1.0000       .       .
[2,]  0.7500  1.0000       .
[3,] -0.2143  0.0697  1.0000

$U
3 x 3 Matrix of class "dtrMatrix"
     [,1]   [,2]   [,3]
[1,] -0.840 -0.820  0.580
[2,]      .  2.215  0.055
[3,]      .      .  0.860

$P
3 x 3 sparse Matrix of class "pMatrix"

[1,] . | .
[2,] . . |
[3,] | . .
```

5.8.5 *Cholesky decomposition*

The **Cholesky decomposition** is a special case of the LU decomposition for real, symmetric, positive-definite square matrices. It is invoked from base or `Matrix` with `chol()`. `chol2inv` in base R computes the inverse of a suitable matrix from its Cholesky decomposition.

For example, the matrix **M** in the eigendecomposition section is real, symmetric, and positive-definite. Its Cholesky decomposition is

```
> chol(M)
     [,1]   [,2]   [,3]
[1,] 1.41 -0.707  0.000
[2,] 0.00  1.225 -0.816
[3,] 0.00  0.000  1.155
```

5.8.6 Schur decomposition

The **Schur decomposition** is available in the `Matrix` package. To quote from its help page:

> "If A is a square matrix, then A = Q T t(Q), where Q is orthogonal, and T is upper block-triangular (nearly triangular with either 1 by 1 or 2 by 2 blocks on the diagonal) where the 2 by 2 blocks correspond to (non-real) complex eigenvalues. The eigenvalues of A are the same as those of T, which are easy to compute. The Schur form is used most often for computing non-symmetric eigenvalue decompositions, and for computing functions of matrices such as matrix exponentials."

See `help(Schur)` for some examples.

5.8.7 `backsolve` and `forwardsolve`

If a decomposition into triangular form has been achieved, the base functions `backsolve()` and `forwardsolve()` solve systems of linear equations where the coefficient matrix is upper or lower triangular. For example, if the right-hand side of the equation of motion for the mass and spring system is the vector $(0,1,0)$, the system of equation may be solved as

```
> backsolve(chol(M), x=c(0,1,0))
[1] 0.408 0.816 0.000
```

5.9 Systems of nonlinear equations

5.9.1 `multiroot` in the `rootSolve` package

To solve for the roots of systems of nonlinear equations, one may use the `multiroot()` function in the `rootSolve` package. It employs the Newton–Raphson method, as described in any standard text on numerical analysis.

As a first example, consider the cubic equation

$$s^3 - 3s^2 + 4\rho = 0 \tag{5.4}$$

which arises when Archimedes' principle is used to calculate the ratio s of the height submerged to the radius of a sphere in a fluid, where the ratio of sphere density to

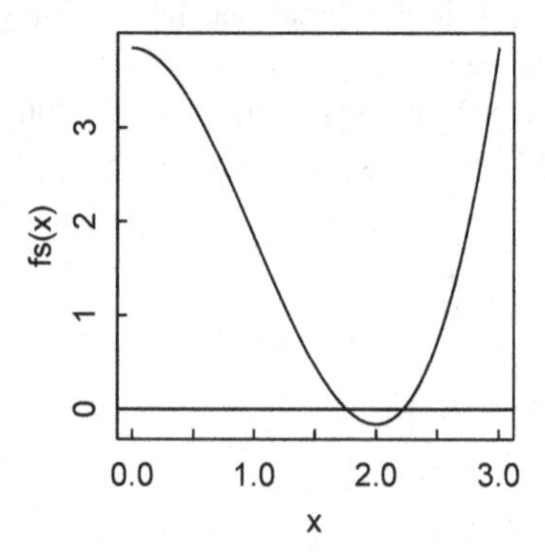

Figure 5.3: *Plot of the lhs of Equation 5.4.*

fluid density is ρ. Suppose we want to use this equation to calculate the fraction of the height of an iceberg (modeled as a sphere) that is submerged in water just above freezing. The density ratio of ice to water near 0 °C is about 0.96. We plot the equation and find that the physically sensible root is a little below 2. (The maximum ratio of depth to diameter is 1, so the maximum ratio of depth to radius is 2.)

```
> fs = function(s) s^3 - 3*s^2 + 4*rho
> rho = 0.96
> curve(fs(x),0,3); abline(h=0)
```

Thus we search for roots between 1.5 and 2.5. (See Figure 5.3)

```
> options(digits=3)
> multiroot(fs, c(1.5,2.5))
$root
[1] 1.76 2.22

$f.root
[1] 1.79e-09 6.45e-07

$iter
[1] 5

$estim.precis
[1] 3.24e-07
```

This confirms the common estimate that about 7/8 of the height of an iceberg is under water.

Next we consider the set of two simultaneous equations

$$
\begin{aligned}
10x_1 + 3x_2^2 - 3 &= 0 \\
x_1^2 - e^{x_2} - 2 &= 0
\end{aligned}
\tag{5.5}
$$

We first use `multiroot` without an explicit Jacobian, so that the function does the Jacobian calculation internally.

```
> require(rootSolve)
> model = function(x) c(F1 = 10*x[1]+3*x[2]^2-3,
F2 = x[1]^2 -exp(x[2]) -2)
> (ss = multiroot(model,c(1,1)))
$root
[1] -1.445552 -2.412158
$f.root
          F1            F2
 5.117684e-12 -6.084022e-14
$iter [1] 10
$estim.precis
[1] 2.589262e-12
```

Providing an analytical Jacobian may provide a more quickly converging solution, but not always, as seen here.

```
> model = function(x) c(F1 = 10*x[1]+3*x[2]^2-3,
  F2 = x[1]^2 -exp(x[2]) -2)
> derivs = function(x) matrix(c(10,6*x[2],2*x[1],
  -exp(x[2])),nrow=2,byrow=T)
> (ssJ = multiroot(model,c(0,0),jacfunc = derivs))
$root
[1] -1.445552 -2.412158
$f.root
 1.166651e-09 -1.390243e-11
$iter [1] 29
$estim.precis
[1] 5.902766e-10
```

The help page explains how various convergence tolerances may be adjusted if the defaults are inadequate.

The `rootSolve` package has a variety of related functions, largely devoted to obtaining steady-state solutions to systems of ordinary and partial differential equations. We shall return to it later in this book. The package vignette at *http://cran.r-project.org/web/packages/rootSolve/vignettes/ rootSolve.pdf* is a valuable resource and should be consulted for more information.

5.9.2 `nleqslv`

Another nonlinear equation solver is `nleqslv` in the package of the same name. The package description states "Solve a system of non linear equations using a Broyden or a Newton method with a choice of global strategies such as linesearch and trust region. There are options for using a numerical or an analytical Jacobian and fixed or automatic scaling of parameters."

After installing the package with `install.packages("nleqslv")`, we load it and apply it to the same function we used with `multiroot`:

```
> install.packages("nleqslv")
> require(nleqslv)
> model = function(x) {
+   y = numeric(2)
+   y[1] = 10*x[1]+3*x[2]^2-3
+   y[2] = x[1]^2 -exp(x[2]) -2
+   y
+ }
> (ss = nleqslv(c(1,1), model))
$x
[1] -1.445552 -2.412158

$fvec
[1]  3.592504e-11 -1.544165e-11

$termcd
[1] 1

$message
[1] "Function criterion near zero"

$scalex
[1] 1 1

$nfcnt
[1] 22

$njcnt
[1] 1

$iter
[1] 18
```

Consult the help page for `nleqslv` to learn about its numerous options and to see more examples.

5.9.3 *`BBsolve()` in the BB package*

The third solver we shall discuss in this section is `BBsolve` in the BB package. According to the BB tutorial, accessed from R with `vignette("BB")`,

> " 'BB' is a package intended for two purposes: (1) for solving a nonlinear system of equations, and (2) for finding a local optimum (can be minimum or maximum) of a scalar, objective function. An attractive feature of the package

is that it has minimum memory requirements. Therefore, it is particularly well suited to solving high-dimensional problems with tens of thousands of parameters. However, BB can also be used to solve a single nonlinear equation or optimize a function with just one variable."

The vignette also includes an explanation of the underlying approach, with references.

In this chapter we shall deal with purpose (1), deferring purpose (2) to the Optimization chapter. BB has two basic functions for solving nonlinear systems of equations: `sane()` (spectral approach for nonlinear equations) and `dfsane()` (derivative-free spectral approach for nonlinear equations). `sane()` differs from `dfsane()` in requiring an approximation of a directional derivative (gradient) at every iteration of the merit function $F(x)^t F(x)$. The authors state that `dfsane()` tends to perform a bit better than `sane()`, which is a bit surprising since the gradient gives the direction of steepest descent to the minimum of the function. However, the reduced number of function evaluations in `dfsane()` apparently outweighs this advantage.

We first run the `dfsane()` function in BB on the same model function we've used for the other solvers.

```
> install.packages("BB")
> require(BB)
Loading required package: BB
Loading required package: quadprog
> model = function(x) c(F1 = 10*x[1]+3*x[2]^2-3,
+ F2 = x[1]^2 -exp(x[2]) -2)
> ans = dfsane(par=c(1,1), fn=model)
Iteration:  0  ||F(x0)||:  7.544058
iteration:  10  ||F(xn)|| =   2.564817
iteration:  20  ||F(xn)|| =   3.145361
iteration:  30  ||F(xn)|| =   2.421409
iteration:  40  ||F(xn)|| =   2.642886
iteration:  50  ||F(xn)|| =   2.115927
iteration:  60  ||F(xn)|| =   0.0463131
iteration:  70  ||F(xn)|| =   0.0001717358
> ans
$par
       F1        F2
-1.445552 -2.412158

$residual
[1] 2.15111e-08

$fn.reduction
[1] 10.66891

$feval
[1] 103
```

```
$iter
[1] 74

$convergence
[1] 0

$message
[1] "Successful convergence"
```

BBsolve() is a wrapper around dfsane() that automatically uses sequential strategies—detailed on its help page—in cases where there are difficulties with convergence. With the BBsolve() wrapper:

```
> ans = BBsolve(par=c(1,1), fn=model)
  Successful convergence.
> ans
$par
       F1        F2
-1.445552 -2.412158

$residual
[1] 7.036056e-08

$fn.reduction
[1] 0.0048782

$feval
[1] 174

$iter
[1] 60

$convergence
[1] 0

$message
[1] "Successful convergence"

$cpar
method      M     NM
     2     50      1
```

Here is an example where dfsane() doesn't converge but BBsolve() does, because it switches to a different method.

```
> froth = function(p){
+ f = rep(NA,length(p))
```

```
+ f[1] = -13 + p[1] + (p[2]*(5 - p[2]) - 2) * p[2]
+ f[2] = -29 + p[1] + (p[2]*(1 + p[2]) - 14) * p[2]
+ f
+ }
> p0 = c(3,2)
> BBsolve(par=p0, fn=froth)
  Successful convergence.
$par
[1] 5 4

$residual
[1] 3.659749e-10

$fn.reduction
[1] 0.001827326

$feval
[1] 100

$iter
[1] 10

$convergence
[1] 0

$message
[1] "Successful convergence"

$cpar
method      M     NM
     2     50      1
```

Compare this with

```
> dfsane(par=p0, fn=froth, control=list(trace=FALSE))
$par
[1] -9.822061 -1.875381

$residual
[1] 11.63811

$fn.reduction
[1] 25.58882

$feval
[1] 137
```

```
$iter
[1] 114

$convergence
[1] 5

$message
[1] "Lack of improvement in objective function"
```

Here is an example from the BB vignette in which 10,000 simultaneous equations are solved, demonstrating BB's impressive capability with large systems of equations.

```
> trigexp = function(x) {
+   n = length(x)
+   F = rep(NA, n)
+   F[1] = 3*x[1]^2 + 2*x[2] - 5 + sin(x[1] - x[2]) * sin(x[1] + x[2])
+   tn1 = 2:(n-1)
+   F[tn1] = -x[tn1-1] * exp(x[tn1-1] - x[tn1]) + x[tn1] *
+   ( 4 + 3*x[tn1]^2) + 2 * x[tn1 + 1] + sin(x[tn1] -
+   x[tn1 + 1]) * sin(x[tn1] + x[tn1 + 1]) - 8
+   F[n] = -x[n-1] * exp(x[n-1] - x[n]) + 4*x[n] - 3
+   F
+ }
>
> n = 10000
> p0 = runif(n) # n initial random starting guesses
> ans = dfsane(par=p0, fn=trigexp, control=list(trace=FALSE))
> ans$message
[1] "Successful convergence"
> ans$resid
[1] 9.829212e-08
> ans$par[1:10] # Just the first 10 out of 10,000 solution values
 [1] 1 1 1 1 1 1 1 1 1 1
```

The highest-order wrapper function in BB is `multiStart()`, which is useful if the system of equations has multiple roots or optima. `multiStart()` accepts a matrix of starting values, with as many columns as there are variables, and as many rows as there are trials. Here is an example taken from the BB vignette, with three variables and 300 trials, and with starting values taken from a uniform random distribution. Note that we did not set a random seed as in the example, so the number of converged trials `sum(ans$conv)` (294/300) is different from that in the example (287/300, but the 12 non-duplicated solutions are exactly the same, though in slightly different order. In the command `ans = multiStart(), action = "solve"` tells `multiStart` to solve rather than optimize. `quiet = T` suppresses the output of successes and failures for all 300 attempts, albeit at the cost of having the computer appear to do nothing while going through the attempts, which may take a minute or so.

```
> hdp = function(x) {
+   r = rep(NA, length(x))
+   r[1] = 5 * x[1]^9 - 6 * x[1]^5 * x[2]^2 + x[1] * x[2]^4 + 2 * x[1] * x[3]
+   r[2] = -2 * x[1]^6 * x[2] + 2 * x[1]^2 * x[2]^3 + 2 * x[2] * x[3]
+   r[3] = x[1]^2 + x[2]^2 - 0.265625
+   r
+ }
>
> p0 = matrix(runif(900), 300, 3)
> ans = multiStart(par = p0, fn = hdp, action = "solve", quiet=T)
> sum(ans$conv)
[1] 294
> pmat = ans$par[ans$conv, ]
> ord1 = order(pmat[, 1])
> ans = round(pmat[ord1, ], 4)
> ans[!duplicated(ans), ]
         [,1]    [,2]    [,3]
 [1,] -0.5154  0.0000 -0.0124
 [2,] -0.4670 -0.2181  0.0000
 [3,] -0.4670  0.2181  0.0000
 [4,] -0.2799  0.4328 -0.0142
 [5,] -0.2799 -0.4328 -0.0142
 [6,]  0.0000  0.5154  0.0000
 [7,]  0.0000 -0.5154  0.0000
 [8,]  0.2799  0.4328 -0.0142
 [9,]  0.2799 -0.4328 -0.0142
[10,]  0.4670 -0.2181  0.0000
[11,]  0.4670  0.2181  0.0000
[12,]  0.5154  0.0000 -0.0124
```

Tests reported on R-help show that BB appears to be considerably more efficient than `nleqslv`, as problems get larger, because of its low memory and storage requirements.

5.10 Case studies

5.10.1 Spectroscopic analysis of a mixture

As a practical example of solving a system of linear equations, we consider a calculation that arises frequently in chemistry and biochemistry: determining the concentrations of components in a mixture from their absorption spectra. The molar extinction coefficients of organic molecules are often well represented as Gaussian functions of wavelength x, with maximum at wavelength x_0, standard deviation sig, and integrated intensity I:

```
> gauss = function(I,x0,sig,x) {I/(sqrt(2*pi)*sig)*
  exp(-(x-x0)^2/(2*sig^2))}
```

We assign these parameters to each of four mixture components, choosing values typical of common biochemical molecules:

```
> A1 = function(x) gauss(6000,230,10,x)
```

```
> A2 = function(x) gauss(4500,260,15,x)
> A3 = function(x) gauss(3000,280,11,x)
> A4 = function(x) gauss(5700,320,20,x)
```

so that, for example, A1(x) will generate the spectrum of compound 1 as x is varied. Let the four compounds be present at concentrations of 7, 5, 8, and 2 millimolar, respectively.

At each wavelength, the optical density (OD) of the mixture is the sum of the extinction coefficient at that wavelength multiplied by the concentration of each component:

$$OD(x) = \sum_i A_i(x)C_i \tag{5.6}$$

In other words, OD(x) is the dot product of the vector A(x) with the vector C, which in R notation is written A%*%C.

If we know the concentrations (ultimately we will pretend we do not know them and will solve for them) we can calculate and plot the spectrum of the mixture as follows:

```
> x = 180:400 # Plot spectrum between 180 nm and 400 nm
> A = matrix(nrow = length(x), ncol = 4) # Initialize A matrix
> # Calculate Ais at each wavelength
> for (i in 1:length(x)) {
+   xi = x[i]
+   for (j in 1:4){
+    A[i,1] = A1(xi)
+    A[i,2] = A2(xi)
+    A[i,3] = A3(xi)
+    A[i,4] = A4(xi)
+    }
+ }
> conc = c(7,5,8,2)*1e-3   # Vector of concs (molar)
> OD = A%*%conc # Multiply A matrix into conc vector
> plot(x, OD, type="l")
```

Not knowing the concentrations and wishing to determine them, a chemist would choose at least four wavelengths at which to measure the OD. For example,

```
> x.meas = c(220,250,280,310)
```

We calculate the extinction coefficient matrix for the four compounds at the four wavelengths (Figure 5.4):

```
> A.meas = matrix(nrow = length(x.meas), ncol = 4)
> for (i in 1:length(x.meas)) {
+    A.meas[i,1] = A1(x.meas[i])
+    A.meas[i,2] = A2(x.meas[i])
+    A.meas[i,3] = A3(x.meas[i])
+    A.meas[i,4] = A4(x.meas[i])
+ }
```

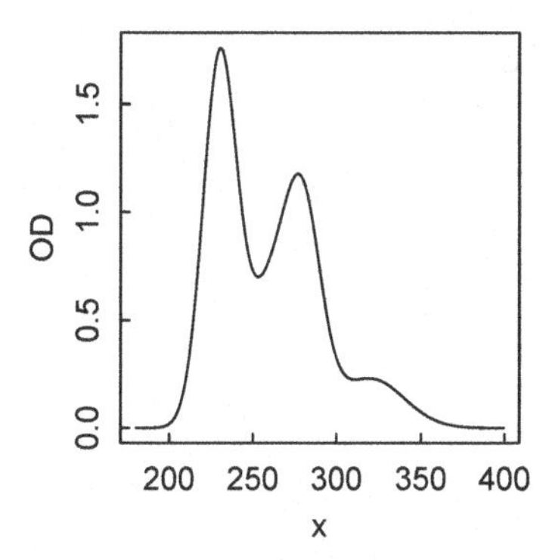

Figure 5.4: *Simulated spectrum of 4-component mixture.*

```
> conc = c(7,5,8,2)*1e-3
```

With these parameters, the measured ODs will be

```
> OD = A.meas %*% conc
> round(OD,3)
      [,1]
[1,] 1.033
[2,] 0.728
[3,] 1.147
[4,] 0.224
```

Then the concentrations (which we pretend we don't know) are calculated as

```
> solve(A.meas,OD)
      [,1]
[1,] 0.007
[2,] 0.005
[3,] 0.008
[4,] 0.002
```

recovering the input values.

The chemist would very likely measure the OD at more than four wavelengths. This would lead to an overdetermined system (see below), but still produce the correct results. For example, with measurements at six wavelengths, the A matrix is not square, so `solve()` will give an error; but `qr.solve()` will give what we want, returning a solution in the least-squares sense. For example,

```
> x.meas = c(220,250,265,280,300,310) # 6 measured wavelengths
> A.meas = matrix(nrow = length(x.meas), ncol = 4)
> for (i in 1:length(x.meas)) {
+   A.meas[i,1] = A1(x.meas[i])
+   A.meas[i,2] = A2(x.meas[i])
+   A.meas[i,3] = A3(x.meas[i])
+   A.meas[i,4] = A4(x.meas[i])
```

```
+ }
> OD = A.meas %*% conc
> round(OD,3)
      [,1]
[1,] 1.033
[2,] 0.728
[3,] 0.918
[4,] 1.147
[5,] 0.322
[6,] 0.224
> qr.solve(A.meas,OD)
      [,1]
[1,] 0.007
[2,] 0.005
[3,] 0.008
[4,] 0.002
```

again recovering the "unknown" concentrations.

Measurements at less than four wavelengths, however, will give an underdetermined system (e.g., three equations in four unknowns). `qr.solve()` will still return an answer, but one of the vector components will be zero, and the others will not be correct.

5.10.2 *van der Waals equation*

Probably every physical scientist has learned about the van der Waals equation

$$\left(P+\frac{n^s a}{V^2}\right)(V-nb)=nRT, \tag{5.7}$$

an equation of state for a gas that goes beyond the ideal gas law to take into account the intermolecular attractions and repulsions that occur in real gases. In this equation, P, V, and T are the pressure, volume, and Kelvin temperature, n is the number of moles of gas, a takes account of pairwise attractive interactions between the molecules that reduce the pressure, and b represents the excluded volume of a mole of molecules.

The van der Waals equation of state can be expressed in terms of reduced variables

$$P_r=\frac{P}{P_c},\ V_r=\frac{V}{V_c},\ T_r=\frac{T}{T_c} \tag{5.8}$$

where P_c is the critical pressure, T_c the critical temperature, and V_c the molar volume at the critical point (P_c,T_c):

$$P_c=\frac{a'}{27b'^2},\ V_c=3b',\ k_BT_c=\frac{8a'}{27b'} \tag{5.9}$$

where a' and b' are the molecular values of the molar parameters a and b, and k_B is the Boltzmann constant R/N_A where N_A is Avogadro's number.

The result of these substitutions is

$$\left(P_r + \frac{3}{V_r^2}\right)\left(V_r - \frac{1}{3}\right) = \frac{8}{3}T_r, \tag{5.10}$$

an equation that holds for all gases when expressed in terms of reduced variables.

With some algebraic manipulation we can write Equation 5.10 as a cubic equation in the reduced volume:

$$V_r^3 - \frac{1}{3}\left(1 + \frac{8T_r}{P_r}\right)V_r^2 + \frac{3}{P_r}V_r - \frac{1}{P_r} = 0 \tag{5.11}$$

We now use R's `polyroot()` function to solve for the real roots of this equation to construct a plot of V_r as a function of P_r at a given T_r that shows the special behavior of a van der Waals gas near its critical point. First we look at just a single point to understand the nature of the roots.

```
> Tr = 0.95
> Pr = 1.5
> # Write expressions for the coefficients in the cubic
> c0 = -1/Pr
> c1 = 3/Pr
> c2 = -1/3*(1+8*Tr/Pr)
> c3 = 1
> (prc = polyroot(c(c0,c1,c2,c3)))
[1] 0.5677520-0.0000000i 0.7272351+0.8033372i 0.7272351-0.8033372i
```

We see that there are three roots, as there should be for a cubic equation. It's the one with imaginary part equal to zero that we want, so our code has to have a way to pick that root. Since the roots are calculated numerically, the logical test`Im(prc) == 0` will likely fail. Also, it is found that the test `all.equal(Im(prc),0)` sometimes failed, suggesting that although the imaginary part of the root is displayed as 0 to seven decimal places, it may be larger than `{Machine\$double.eps ^ 0.5`, the test that `all.equal()` uses. Therefore, we use `abs(Im(prc)) <= 1e-12` as a heuristic test for a zero imaginary part, with the confidence that only one of the three roots would pass the test. These considerations lead to the following code.

```
> Tr = 0.95 # Temperature below the critical point
> pr = seq(0.5,3,by = 0.01) # From relatively dilute to compressed
> npr = length(pr)
> Vr = numeric(npr)
> for( i in 1:npr) {
+   Pr = pr[i]
+   c0 = -1/Pr
+   c1 = 3/Pr
+   c2 = -1/3*(1+8*Tr/Pr)
+   c3 = 1
+   prc = polyroot(c(c0,c1,c2,c3))
+   for (j in 1:3) if (abs(Im(prc[j])) <= 1e-12) Vr[i] = Re(prc[j])
+ }
```

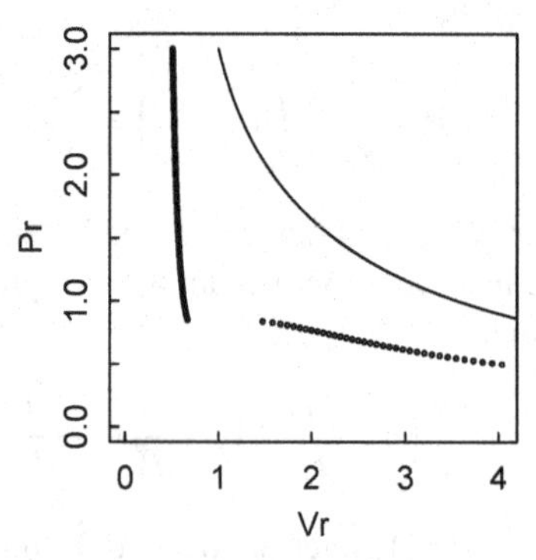

Figure 5.5: *Plots of reduced pressure vs. reduced volume below (points) and above (line) the critical temperature.*

It is conventional to plot pressure as a function of volume:

```
> plot(Vr,pr,xlim=c(0,max(Vr)),ylim=c(0,max(pr)), cex = 0.3,ylab="Pr")
```

For comparison, we do the same calculation for a reduced temperature well above critical, and add that (V,P) line to the plot (Figure 5.5).

```
> Tr = 1.5
> pr = seq(0.5,3,0.01)
> npr = length(pr)
> Vr = numeric(npr)
> for( i in 1:npr) {
+   Pr = pr[i]
+   c0 = -1/Pr
+   c1 = 3/Pr
+   c2 = -1/3*(1+8*Tr/Pr)
+   c3 = 1
+   prc = polyroot(c(c0,c1,c2,c3))
+   for (j in 1:3) if (abs(Im(prc[j])) <= 1e-12) Vr[i] = Re(prc[j])
+ }
> lines(Vr,pr)
```

We see that if the temperature is below critical, the PV curve of the van der Waals gas shows a discontinuity near the critical point, corresponding to the transition from gas to liquid. If the temperature is above critical, the familiar Boyle's law curve is observed.

5.10.3 *Chemical equilibrium*

Consider the system of chemical reactions that occur when 3 moles of hydrogen gas and 1 mole of carbon monoxide react in the presence of solid carbon to form methane, water, carbon dioxide, and ethane (Hanna and Sandall, 1995, p. 170).

1. $3H_2 + CO = CH_4 + H_2O$
2. $CO + H_2O = CO_2 + H_2$

3. $CO_2 + C = 2CO$
4. $5H_2 + 2CO = C_2H_6 + 2H_2O$

At one atm total pressure and 500 K, the equilibrium constants K_X on a mole fraction basis are 69.18, 4.68, 0.0056, and 0.141, respectively.

We wish to solve for the composition of the equilibrium mixture. There are six chemical species (plus solid carbon, which does not contribute to the gaseous mole fractions) and only four equilibrium constants, so some other constraints are needed. Hanna and Sandall use the "reaction coordinate" formulation of Smith and Van Ness (1987). The species H, CO, CH_4, H_2O, CO_2, and C_2H_6 are numbered 1–6, and the reaction coordinates are numbered r_1, r_2, r_3, r_4 for reactions 1–4. The equilibrium mole fractions X_i can then be expressed in terms of the reaction coordinates as

$$\begin{aligned} X_1 &= (3-3r_1+r_2-5r_4)/n_{tot} \\ X_2 &= (1-r_1-r_2+2r_3-2r_4)/n_{tot} \\ X_3 &= r_1/n_{tot} \\ X_4 &= (r_1-r_2+2r_4)/n_{tot} \\ X_5 &= (r_2-r_3)/n_{tot} \\ X_6 &= r_4/n_{tot} \end{aligned} \tag{5.12}$$

where n_{tot} is the total number of moles of gas at equilibrium,

$$n_{tot} = 4 - 2r_1 + r_3 - 4r_4. \tag{5.13}$$

Substituting these equations into the equilibrium constant expressions, we obtain

$$\begin{aligned} \frac{r_1(r_1-r_2+2r_4)n_{tot}^2}{(3-3r_1+r_2-5r_4)(1-r_1-r_2+2r_3-2r_4)} &= 69.18 \\ \frac{(r_2-r_3)(3-3r_1+r_2-5r_4)}{(1-r_1-r_2+2r_3-2r_4)(r_1-r_2+2r_4)} &= 4.68 \\ \frac{(1-r_1-r_2+2r_3-2r_4)^2}{(r_2-r_3)n_{tot}} &= 0.0056 \\ \frac{r_4(1-r_1-r_2+2r_3-2r_4)^2 n_{tot}^4}{(3-3r_1+r_2-5r_4)^5(1-r_1-r_2+2r_3-2r_4)} &= 0.141 \end{aligned} \tag{5.14}$$

We need to solve these equations for the r_i, and use the results in Equations 5.12 to calculate the equilibrium mole fractions. As they stand, Equations 5.14 are very difficult to solve numerically with any of the functions we have examined in this chapter, unless the starting guesses are unrealistically close to the true values. However, if the denominators are cleared, the difficulties largely disappear. Thus we use the following code to solve for the reaction coordinate values at equilibrium, choosing `nleqslv` as our solver.

```
> require(nleqslv)
> model = function(r) {
+ FX = numeric(4)
+ r1 = r[1]; r2 = r[2]; r3 = r[3]; r4 = r[4]
```

```
+ ntot = 4-2*r1+r3-4*r4
+ FX[1] = r1*(r1-r2+2*r4)*ntot^2-69.18*(3-3*r1+r2-5*r4)^3*
  (1-r1-r2+2*r3-2*r4)
+ FX[2] = (r2-r3)*(3-3*r1+r2-5*r4)-4.68*(1-r1-r2+2*r3-2*r4)*
  (r1-r2+2*r4)
+ FX[3] = (1-r1-r2+2*r3-2*r4)^2-0.0056*(r2-r3)*ntot
+ FX[4] = r4*(r1-r2+2*r4)^2*ntot^4-0.141*(3-3*r1+r2-5*r4)^5*
  (1-r1-r2+2*r3-2*r4)^2
+ FX
+ }
> # For initial guess, set all r equal
> (ss = nleqslv(c(.25,.25,.25,.25), model))
$x
[1]  6.816039e-01  1.589615e-02 -1.287031e-01  1.409549e-05
$fvec
[1]  1.015055e-08 -3.922078e-10  6.236144e-12 -9.467267e-09
$termcd
[1] 2
$message
[1] "x-values within tolerance 'xtol'"
$scalex
[1] 1 1 1 1
$nfcnt
[1] 78
$njcnt
[1] 2
$iter
[1] 63
```

Note that the equilibrium value for r_3 is negative, as it should be since solid carbon is consumed in the reaction.

We now set the r vector equal to the results ss$x and calculate the equilibrium mole fractions.

```
> r = ss$x
> ntot = 4-2*r[1]+r[3]-4*r[4]
> X = numeric(6)
> X[1] = (3-3*r[1]+r[2]-5*r[4])/ntot
> X[2] = (1-r[1]-r[2]+2*r[3]-2*r[4])/ntot
> X[3] = r[1]/ntot
> X[4] = (r[1] - r[2] + 2*r[4])/ntot
> X[5] = (r[2] - r[3])/ntot
> X[6] = r[4]/ntot
> X
[1] 3.871616e-01 1.796845e-02 2.717684e-01 2.654415e-01
[5] 5.765447e-02 5.620138e-06
```

Chapter 6

Numerical differentiation and integration

6.1 Numerical differentiation

6.1.1 Numerical differentiation using base R

6.1.1.1 Using the fundamental definition

Calculating numerical derivatives is straightforward using a finite difference version of the fundamental definition of a derivative:

$$\frac{df(x)}{dx} = \lim_{h \to 0} \frac{f(x+h) - f(x)}{h} \tag{6.1}$$

For example,

```
> f = function(x) x^3 * sin(x/3) * log(sqrt(x))
> x0 = 1; h = 1e-5
> (f(x0+h) - f(x0))/h
[1] 0.163603
```

while the true value of the derivative is $\frac{1}{2}\sin(\frac{1}{3}) = 0.163597348398076\ldots$

With h positive, this is called the *forward* derivative, otherwise it is the *backward* derivative. To take into account the slopes both before and after the point x, the central difference formula is chosen,

$$\frac{df(x)}{dx} = \lim_{h \to 0} \frac{f(x+h) - f(x-h)}{2h} \tag{6.2}$$

which is also numerically more accurate, with error $O(h^2)$ as h gets small:

```
> (f(x0+h) - f(x0-h))/(2*h)
[1] 0.1635973
```

It might be tempting, therefore, to make h as small as possible, e.g., as small as the machine accuracy `eps = .Machinedouble.eps` of about 2.2×10^{-16}, to minimize the error in the derivative. However, as the following numerical experiment shows, this strategy fails. Choose h to be 10^{-i} for i in $1 \ldots 16$ and plot the error as a function of i (Figure 6.1).

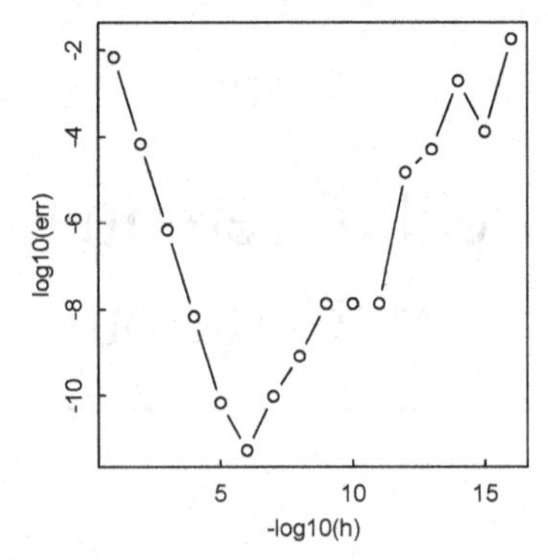

Figure 6.1: *Error in numerical differentiation of f as function of h.*

```
> f = function(x) x^3 * sin(x/3) * log(sqrt(x))
> x0 = 1
> err = numeric(16)
> for (i in 1:16) {
+ h = 10^-i
+ err[i] = abs( (f(x0+h)-f(x0-h))/(2*h) - 0.5*sin(1/3) )
+ }
> plot(log10(err), type="b", xlab="-log10(h)")
```

The difference between the numerical and the exact derivative is smallest for $h = 10^{-6}$ with error 10^{-11}, and increases again when h gets smaller. The reason is that, while the roundoff error between the computed derivative and its actual value gets smaller, the truncation error in the term $f(x+h) - f(x-h)$ increases with smaller h.

Theory says that the optimal step size is $\sqrt[3]{a}$ if a is the accuracy with which the function can be computed, and $\sqrt[3]{a^2}$ would be the accuracy of the computed derivative. Assuming that all basic math functions in R are calculated with accuracy `eps`, this corresponds quite well with the optimal size of h as found in the figure above.

6.1.1.2 diff()

`diff()` in base R is not a derivative function, but rather a function that returns lagged differences between entries in a vector. For central estimation use `lag = 2`. For example

```
> xfun = function(x0,h) seq(x0-h,x0+h,h)
> diff(f(xfun(x0,h)), lag = 2)/(2*h)
[1] 0.163597
```

6.1.2 *Numerical differentiation using the* numDeriv *package*

The standard R package for calculating numerical approximations to derivatives is `numDeriv`. It is used, for example, in all the standard optimization packages discussed in the next chapter. `numDeriv` contains functions to accurately calculate first derivatives (gradients and Jacobians) and second derivatives (Hessians). In searching for an optimum of a multivariate function, the gradient gives the direction of steepest descent (or ascent) and the Hessian gives the local curvature of the surface.

The usage for each of these functions is

```
dfun(func, x, method, method.args, ...)
```

where `dfun` is one of `grad`, `jacobian`, or `hessian`, `func` is the function to be differentiated, `method` is one of `"Richardson"` (the default), `"simple"` (not supported for `hessian`), or `"complex"`, indicating the method to be used for the approximation. `method.args` are the arguments—tolerances, number of repetitions, etc.—passed to the method (see the help page for `grad` for the most complete discussion of the details), and `...` stands for any additional arguments to be passed to `func`. In the examples following, we will use the defaults.

With `method="simple"`, these functions calculate forward derivatives with step size 10^{-4}, both choices that we already know are not optimal. Applying this method to the function above yields

```
> require(numDeriv)
> options(digits=16)
> grad(f, 1, method = simple)
[1] 0.1636540038633105              # error: 5.7e-5
```

With the default `method="Richardson"`, Richardson's extrapolation scheme is applied, a method for accelerating a sequence of related computations. The help page says: "This method should be used if accuracy, as opposed to speed, is important."

```
> grad(f, 1, method = "Richardson")
[1] 0.1635973483989158              # error: 8.4e-13
```

Method `"complex"` refers to the quite recent *complex-step derivative approach* and can be applied to complex-differentiable (i.e., analytic) functions that satisfy the conditions that x_0 and $f(x_0)$ are real. Then the complex step method computes derivatives to the same accuracy as the function itself. Almost all special functions available in R are complex-differentiable. Therefore, this method can be applied to the function above, returning the derivative to 16 digits, and with no loss in speed compared to method `"simple"`:

```
> grad(f, 1, method = "complex")
[1] 0.1635973483980761              # error: < 1e-15
```

One has to be careful with self-defined functions. Normally, the complex-step approach only works for functions composed of basic special functions defined in R.

6.1.2.1 *grad()*

To illustrate the use of grad() for multivariate functions, we consider the scalar function of three variables

$$f(x,y,z) = 2x + 3y^2 - \sin(z). \tag{6.3}$$

The gradient, as commonly defined in Cartesian coordinates, is the vector

$$\nabla f = \frac{\partial f}{\partial x}\mathbf{i} + \frac{\partial f}{\partial y}\mathbf{j} + \frac{\partial f}{\partial z}\mathbf{k} \tag{6.4}$$

where $\mathbf{i}, \mathbf{j}, \mathbf{k}$ are unit vectors along the x, y, z axes. For the function f defined in Equation 6.3, the gradient is therefore

$$\nabla f = 2\mathbf{i} + 6y\mathbf{j} - \cos(z)\mathbf{k} \tag{6.5}$$

We obtain the same result for particular numerical values of $x, y, z =$ c(1,1,0) using the grad() function as follows.

```
> require(numDeriv)
> f = function(u){
+   x = u[1]; y = u[2]; z = u[3]
+   return(2*x + 3*y^2 - sin(z))
+ }
> grad(f,c(1,1,0))
[1]  2  6 -1
```

6.1.2.2 *jacobian()*

The Jacobian matrix J of a vector function $\mathbf{F}(\mathbf{x})$ is the matrix of all first-order partial derivatives of $\mathbf{F}$ with respect to the components of $\mathbf{x}$. For a 2×2 system,

$$J = \begin{pmatrix} \frac{\partial F_1}{\partial x_1} & \frac{\partial F_1}{\partial x_2} \\ \frac{\partial F_2}{\partial x_1} & \frac{\partial F_2}{\partial x_2} \end{pmatrix} \tag{6.6}$$

With the function

$$F = x_1^2 + 2x_2^2 - 3, \cos(\pi x_1/2) - 5x_2^3, \tag{6.7}$$

we find that

$$J = \begin{pmatrix} 2x_1 & 4x_2 \\ -\frac{\pi}{2}\sin\frac{\pi x_1}{2} & -15x_2 \end{pmatrix} \tag{6.8}$$

Using the jacobian() function, we find numerical agreement with that result at the point c(2,1:

```
> require(numDeriv)
> F = function(x) c(x[1]^2 + 2*x[2]^2 - 3, cos(pi*x[1]/2) -5*x[2]^3)
> jacobian(F, c(2,1))
     [,1] [,2]
[1,]    4    4
[2,]    0  -15
```

6.1.2.3 hessian

The hessian is the matrix of second derivatives of a scalar function f with respect to coordinate components. It may be thought of as the jacobian of the gradient of the function. It gives the coefficients of the quadratic term of the Taylor series expansion of a function at the point in question. For a two-dimensional system,

$$H(f) = \begin{pmatrix} \frac{\partial^2 f}{\partial x_1^2} & \frac{\partial^2 f}{\partial x_1 \partial x_2} \\ \frac{\partial^2 f}{\partial x_2 \partial x_1} & \frac{\partial^2 f}{\partial x_2^2} \end{pmatrix} \tag{6.9}$$

For the function f defined in the subsection on grad above, we find

```
> hessian(f,c(1,1,0))
              [,1]          [,2]          [,3]
[1,]  0.000000e+00 -4.101521e-12  0.000000e+00
[2,] -4.101521e-12  6.000000e+00 -4.081342e-13
[3,]  0.000000e+00 -4.081342e-13  0.000000e+00

> zapsmall(hessian(f,c(1,1,0)))
     [,1] [,2] [,3]
[1,]    0    0    0
[2,]    0    6    0
[3,]    0    0    0
```

That is, as can be seen by inspection of Equation 6.5, all entries in the hessian matrix for f at the given point are 0 except for the [2,2] entry.

6.1.3 Numerical differentiation using the pracma package

The pracma package contains a variety of functions for both scalar and vector numerical differentiation. It has functions with the same names and roles as grad(), jacobian(), and hessian() in the numDeriv package, and in fact will mask those functions if it is loaded after numDeriv:

```
> require(numDeriv)
> require(pracma)
Loading required package: pracma
Attaching package: pracma
The following object(s) are masked from package:numDeriv:
    grad, hessian, jacobian
```

Which package one uses for these functions is largely a matter of choice, though those in numDeriv are probably more solid under a wider variety of circumstances. However, pracma is sometimes more accurate, as it uses the central difference formula plus an optimal step size. It also has some additional useful functions.

6.1.3.1 fderiv()

The fderiv() function enables numerical differentiation of functions from first to higher orders. Note that numerical derivatives get less accurate, the higher the order;

but derivatives up to the eighth order seem to be possible without problems. To obtain the nth derivative of a function f at a vector of points x, the usage with defaults is

```
fderiv(f, x, n = 1, h = 0, method="central", ...)
```

where `h` is the step size, set automatically if `h = 0`. Optimal step sizes for various orders of derivative are given in the help page. The central method should be used unless the function can be evaluated only on the right side (forward) or the left side (backward). As usual, ... stands for additional variables to be passed to `f`. An example of usage:

```
> require(pracma)
> f = function(x) x^3 * sin(x/3) * log(sqrt(x))
> x = 1:4
> fderiv(f,x)  # 1st derivative at 4 points
[1]  0.1635973  4.5347814 18.9378217 43.5914029
> fderiv(f,x,n=2,h=1e-5)  # 2nd derivative at 4 points
[1]  1.132972  8.699867 20.207551 27.569698
```

6.1.3.2 *numderiv() and numdiff()*

The `pracma` function `numderiv()` (not to be confused with the `numDeriv` package discussed above) implements Richardson's extrapolation method—a sequence acceleration method—to compute the numerical derivative at a single point, returning not only the value of the derivative, but also estimated absolute and relative errors and the number of iterations used.

```
> options(digits = 12)
> numderiv(f, x0=1, h=1/2)
$df
[1] 0.163597348398 # error: 1.859624e-15
$err
[1] 7.72992780895e-14
$relerr
[1] 4.72497133031e-13
$n
[1] 6
```

and we see that this returns two correct digits more than `grad` in the `numderiv` package. (Starting with the default $h = 1$ will lead to an error because the function f does not exist in $x_0 - h = 0$.)

`numderiv()` is not vectorized, i.e., x_0 must be a scalar, a single numerical value. To evaluate the derivative at a vector of points, use `numdiff()`, a function that simply wraps `numderiv()`. To evaluate the derivative at a vector of points, use `numdiff`.

```
> numdiff(f,x=2:4)
[1]  4.53478137145 18.93782173965 43.59140287422
```

6.1.3.3 *grad() and gradient()*

The pracma package has two functions for calculating gradients: grad() and gradient(). grad() calculates a numerical gradient at a single point x_0, given a function f of several variables, and an optimal step size h. In essence, grad() applies the central difference formula to each direction x_i.

For example, to calculate the electric field (the negative gradient of the potential) at x0 = (1,1,1) due to a unit charge at the origin, we proceed as follows.

```
> options(digits = 3)
> f = function(x) 1/sqrt(x[1]^2 + x[2]^2 + x[3]^2)
> x0 = c(1,1,1)
> -grad(f,x0)
[1] 0.192 0.192 0.192
```

The gradient() function takes as arguments a vector of function values or a matrix of values of a function of two variables, and x- and y-coordinates of grid points or values for the differences between grid points in the x and y directions, and returns the numerical gradient as a vector or matrix of discrete slopes in the x and y directions.

As an example of this useful capability of gradient(), we calculate and plot the two-dimensional electric field (the gradient of the potential) due to a dipole with unit positive charge at (-1,0) and unit negative charge at (1,0), on a square grid of points spaced 0.2 units apart (Figure 6.2)

```
> require(pracma)
> # Define the grid
> v = seq(-2, 2, by=0.2)
> X = meshgrid(v, v)$X
> Y = meshgrid(v, v)$Y
> # Define the potential Z
> Z = -(1/sqrt((X+1)^2 + Y^2) - 1/sqrt((X-1)^2 + Y^2))
> par(mar=c(4,4,1.5,1.5),mex=.8,mgp=c(2,.5,0),tcl=0.3)
> contour(v, v, t(Z), col="black",xlab="x",ylab="y")
> grid(col="white")
> # Calculate the gradient on the grid points
> grX = gradient(Z, v, v)$X
> grY = gradient(Z, v, v)$Y
> # Draw arrows representing the field strength at the grid
   points
> quiver(X, Y, grX, grY, scale = 0.2, col="black")
```

6.1.3.4 *jacobian()*

As noted in the numDeriv section, the Jacobian matrix is the matrix of first derivatives of the components of one vector x with respect to the components of another vector y: $\partial x_i/\partial y_j$. The determinant of this matrix is used as a multiplicative factor when changing variables from x to y when integrating a function over a region within

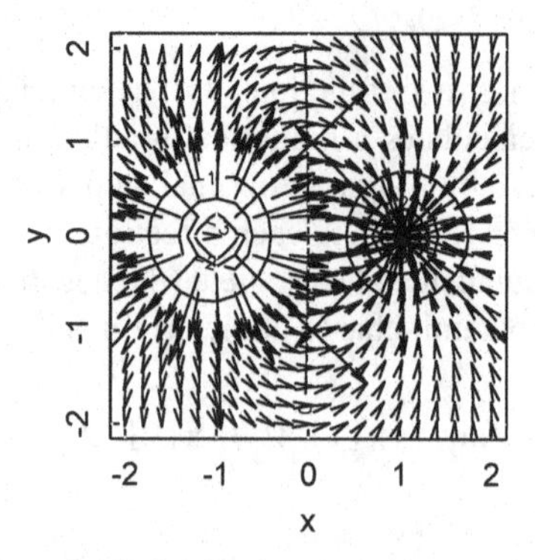

Figure 6.2: *Electric field of a dipole, plotted using the* `quiver` *function.*

its domain. Here is an example of the Jacobian in transforming from spherical polar to Cartesian coordinates:

```
> f = function(x) {
+   r = x[1]; theta = x[2]; phi = x[3];
+   return(c(r*sin(theta)*sin(phi), r*sin(theta)*cos(phi),
+   r*cos (theta)))
+   }
> x = c(2, 90*pi/180, 45*pi/180)
> options(digits=4)
> jacobian(f,x)
          [,1] [,2]   [,3]
[1,] 7.071e-01    0  1.414
[2,] 7.071e-01    0 -1.414
[3,] 6.123e-17   -2  0.000
```

This matrix accords with the analytical result.

6.1.3.5 `hessian`

The `hessian()` function in `pracma` behaves just as it does in `numDeriv`. Here is an example from the help page for `hessian()` in the `pracma` package.

```
> f = function(u) {
+   x = u[1]; y <- u[2]; z <- u[3]
+   return(x^3 + y^2 + z^2 +12*x*y + 2*z)
+   }
> x0 = c(1,1,1) # Point at which the hessian is calculated
> hessian(f, x0)
     [,1] [,2] [,3]
[1,]    6   12    0
[2,]   12    2    0
[3,]    0    0    2
```

`hessian()` functions are provided in many R packages. For example, one is included in the `rootSolve` package, where it is used in the context of solving differential equations. However, for stand-alone purposes, `numDeriv` or `pracma` are to be preferred.

6.1.3.6 `laplacian()`

The Laplacian is a differential operator given by the divergence of the gradient of a function, often denoted by ∇^2 or $\triangle$. In Cartesian coordinates, the Laplacian is given by the sum of second partial derivatives of the function with respect to x, y, and z.

$$\nabla^2 f = \frac{\partial^2 f}{\partial x^2} + \frac{\partial^2 f}{\partial y^2} + \frac{\partial^2 f}{\partial z^2}. \tag{6.10}$$

`pracma` numerically calculates this quantity, in as many dimensions as desired, with the `laplacian()`function. For example, in two dimensions:

```
> f = function(x) 2/x[1] - 1/x[2]^2
> laplacian(f, c(1,1))
[1] -2
```

6.2 Numerical integration

Numerical integration means the computation of an integral using numerical techniques. This numerical computation of a univariate integral is also called "quadrature" (and sometimes "cubature" to mean numerical computation of integrals in multidimensional space).

There is a wide range of approaches to numerical integration. Most scientists and engineers are probably familiar with the trapezoidal or Simpson's rules, which are based on dividing the integration interval into sections of equal width and simple shape (rectangle or trapezoid), calculating the area of each section, and summing the results. These are the 2-point and 3-point versions of the so-called *Newton–Cotes formulae*.

A more modern, and often more accurate, approach is some variant of *Gaussian quadrature*, which divides the integral into unequally spaced points, assigns weights to those points, and evaluates the integral as the product of the weight times the value of the function at each point, summed over the points. The points are chosen in such a way that the value of the integral will be exact for all polynomials up to a certain degree.

Both of these approaches can be used for *adaptive integration* where the value of the integral is approximated using one of these static rules on smaller and smaller subintervals of the integration domain. The process is stopped on subintervals for which an error estimate has fallen below a certain predefined tolerance.

Difficulties will arise with functions that have singularities in the integration domain or at the boundaries, domains that are unbounded (reach to infinity), or that involve multivariate functions. Especially useful for higher-dimensional functions are Monte Carlo integration and its variants.

We proceed to show how each of these approaches is implemented in R, and name packages that support numerical integration.

6.2.1 `integrate`: Basic integration in R

The main function for numerical integration is `integrate()` in base R. As an example we will integrate the function $f(x) = e^{-x}\cos(x)$ from 0 to π:

```
> f = function(x) exp(-x) * cos(x)
> ( q = integrate(f, 0, pi) )
0.521607 with absolute error < 7.6e-15

> str(q)
List of 5
 $ value       : num 0.522
 $ abs.error   : num 7.6e-15
 $ subdivisions: int 1
 $ message     : chr "OK"
 $ call        : language integrate(f = f, lower = 0, upper = pi)
 - attr(*, "class")= chr "integrate"
```

`integrate()` returns a list with entries `$value` for the approximate value of the integral, and `$abs.error` the estimated absolute error. Because the known exact value of the integral is $\frac{1}{2}(1+e^{-\pi})$ the true absolute error is:

```
> v = 0.5*(1+exp(-pi))
> abs(q$value - v)
[1] 1.110223e-16
```

The integrand function needs to be vectorized, otherwise one will get an error message, e.g., with the following nonnegative function:

```
> f1 = function(x) max(0, x)
> integrate(f1, -1, 1)
Error in integrate(f1, -1, 1) :
  evaluation of function gave a result of wrong length
```

The reason is that `f(c(x1, x2, ...))` is `max(0, x1, ...)` and not `c(max(0, x1), max(0, x2), ...)` as would be expected from a vectorized function. In this case, the behavior of the function can be remedied by using the `pmax()` function, which returns a vector of the maxima of the input values:

```
> f2 = function(x) pmax(0, x)
> integrate(f2, -1, 1)
0.5 with absolute error < 5.6e-15
```

In general, the help page suggests to vectorize the function by applying the `Vectorize()` function to it.

```
> f3 = Vectorize(f1)
> integrate(f3, -1, 1)
0.5 with absolute error < 5.6e-15
```

Sometimes, `integrate()` has difficulties with highly oscillating functions: one then sees a message like "maximum number of subdivisions reached." It may help to increase the number of subdivisions, but that is not guaranteed to solve the problem. It is sometimes recommended to set the number of subdivisions to 500 by default, anyway.

Note that the true absolute error will not always be smaller than the estimated one, there may be situations where the estimated absolute error will be misleadingly small. Consider for example the following function

$$f(x) = \frac{x^{1/3}}{1+x}, \tag{6.11}$$

which has ill-behaved derivatives at the origin.

```
> f = function(x) x^(1/3)/(1+x)
> curve(f,0,1)
> integrate(f,0,1)
0.4930535 with absolute error < 1.1e-09
```

Using `integrate()`, we get the same answer using x or the transformed variable $u = x^3$ in the integration, but with a considerably smaller estimated absolute error.

```
> # Now with transformed variable
> fu = function(u) 3*u^3/(1+u^3)
> integrate(fu,0,1)
0.4930535 with absolute error < 1.5e-13
```

`Example` — Consider the calculation of the mean-square radius of a sphere of radius R and constant density:

$$< R^2 >= \frac{1}{R^2}\frac{\int_0^R r^4 dr}{\int_0^R r^2 dr} \tag{6.12}$$

The integrals are trivial analytically, and lead to 3/5 as the answer. Numerically,

```
> f1 = function(r) r^2
> f2 = function(r) r^4
> f = function(R) integrate(f2,0,R)$value/
+     integrate(f1,0,R)$value/R^2
> f(1); f(10); f(100)
[1] 0.6
[1] 0.6
[1] 0.6
```

and we would get the same result for any value of R.

`Example` — The following exercise displays a combination of numerical differentiation and integration techniques.

Compute the surface area of rotating the curve $\sin(x)$ from 0 to 2π about the x-axis. The formula for an area of surface from a to b of revolving a curve f is

$$S_x = 2\pi \int_a^b f(x)\sqrt{1+f'(x)^2}\,\mathrm{d}x. \tag{6.13}$$

Assuming we do not know the derivative of $\sin(x)$ we have to apply a numerical gradient.

```
> library(numDeriv)
> fn = sin
> gr = function(x) grad(fn, x)
> F = function(x) fn(x) * sqrt(1 + gr(x)^2)
> ( I = integrate(F, 0, pi) )
2.295587 with absolute error < 2.1e-05
> S = 2*pi * I$value
> S
[1] 14.4236
```

with a theoretical value of $2\pi(\sqrt{2}+\mathrm{arcsinh}(1)) = 14.423599\ldots$ where arcsinh is the inverse of the hyperbolic sine function (available in package `pracma` as `asinh`).

6.2.2 *Integrating discretized functions*

A different situation that will often arise is when the function is not explicitly known, but is represented by a number of discrete points. Then one may imagine the function as linear between these known points and the classical “trapezoidal rule” could be applied. This rule is implemented in function `trapz()` in package `pracma`.

```
> require(pracma)
> f = function(x) exp(-x) * cos(x)
> xs = seq(0, pi, length.out = 101)
> ys = f(xs)
> trapz(xs, ys)
[1] 0.5216945
```

The help page reveals how this result can be slightly improved by correcting the end terms.

```
> h = pi/100
> ya = (ys[2] - ys[1])
> ye = (ys[101] - ys[100])
> trapz(xs, ys) - h/12 * (ye - ya)
[1] 0.521607
```

with an absolute error smaller than 0.5e-07, a good result when considering that a piecewise linear function between discrete points was assumed.

There is no straightforward implementation of Simpson’s rule available, but we can easily write our own discrete version:

```
> simpson = function(y, h) {
+   n = length(y)
+   if (n%%2 != 1) stop("Simpson's rule needs an uneven number
                             of points.")
+   i1 = seq(2, n-1, by=2)
+   i2 = seq(3, n-2, by=2)
+   h/3 * (y[1] + y[n] + 4*sum(y[i1]) + 2*sum(y[i2]))
+ }

> simpson(ys, h)
[1] 0.521607
```

One may attempt to reconstruct the original function through an approximation of the discrete points, for example a polynomial or spline approximation. `splinefun()` will generate such a function.

```
> fsp = splinefun(xs, ys)
> integrate(fsp, 0, pi)
0.521607 with absolute error < 6.7e-10
```

The absolute error concerns the spline approximation, not necessarily the error compared to the initial, unknown function from which the discrete points are derived.

Still another approach could be to approximate the points with a polynomial which has the advantage that polynomials can be integrated easily. With `pracma` we can do this as follows:

```
> require(pracma)
> p = polyfit(xs, ys, 6)              # fitting polynomial
> q = polyint(p)                      # anti-derivative
> polyval(q, pi) - polyval(q, 0)      # evaluate at endpoints
[1] 0.5216072
```

Which approach to use depends on the application, e.g., on possible oscillations or smoothness assumptions about the underlying function.

6.2.3 *Gaussian quadrature*

The `integrate()` function in the base R installation is an example of the modern approach to numerical integration, which emphasizes the high accuracy and efficiency of Gaussian integration methods. As stated above, Gaussian quadrature approximates the integral by a sum of the function values $f(x_i)$, multiplied by appropriate weights w_i, evaluated at a set of n points x_i:

$$\int_a^b f(x)dx \approx \sum_{i=1}^{n} w_i f(x_i). \tag{6.14}$$

It can be shown that the optimal abscissas x_i for a given n are the roots of the orthogonal polynomial for the same integral and weighting function. The resulting approximation to the integral is then exact for polynomials of degree $2n-1$ or less, and highly accurate for functions that are well approximated by polynomials. In some common cases, we have

$W(x)$	interval	polynomial
1	$(-1,1)$	Legendre $P_n(x)$
$(1-x^2)^{-1/2}$	$(-1,1)$	Chebyshev $T_n(x)$
$(1-x^2)^{1/2}$	$(-1,1)$	Chebyshev $U_n(x)$
e^{-x}	$(0,\infty)$	Laguerre $L_n(x)$
e^{-x^2}	$(-\infty,\infty)$	Hermite $H_n(x)$

If the interval in the first three cases is (a,b) rather than $(-1,1)$, the scaling transformation

$$\int_a^b f(x)dx = \frac{b-a}{2}\int_{-1}^{1} f(\frac{b-a}{2}x+\frac{b-a}{2})dx \tag{6.15}$$

accomplishes the change.

Package **gaussquad** encompasses a collection of functions for Gaussian quadrature. For example, function `legendre.quadrature.rules()` will return the nodes and weights for performing Gauss–Legendre quadrature on the interval $[-1,1]$.

```
> library(gaussquad)
> legendre.quadrature.rules(4)
[[1]]
  x w
1 0 2

[[2]]
           x w
1  0.5773503 1
2 -0.5773503 1

[[3]]
              x         w
1  7.745967e-01 0.5555556
2  7.771561e-16 0.8888889
3 -7.745967e-01 0.5555556

[[4]]
           x         w
1  0.8611363 0.3478548
2  0.3399810 0.6521452
3 -0.3399810 0.6521452
4 -0.8611363 0.3478548
```

Compute the integral of $f(x)=x^6$ on $[-1,1]$ with Legendre nodes and weights of order 4:

```
> f = function(x) x^6
> Lq = legendre.quadrature.rules(4)[[4]]   # Legendre of order 4
> xi = Lq$x; wi = Lq$w                    # nodes and weights
> sum(wi * f(xi))                           # quadrature
[1] 0.2857143
```

and this is exactly $2/7$, the value of integrating x^6 from -1 to 1. One can also directly calculate this integral with `legendre.quadrature()`:

```
> legendre.quadrature(f, Lq, lower = -1, upper = 1)
[1] 0.2857143
```

In `pracma` there is a `gaussLegendre()` function available. It takes as arguments the number of nodes and the limits of integration, and returns the positions and weights at the nodes. We illustrate with examples from the help page of the functions.

```
> f = function(x) sin(x+cos(10*exp(x))/3)
> curve(f, -1, 1)
```

Let us examine convergence with increasing number of nodes.

```
> nnodes = c(17,29,51,65)
> # Set up initial matrix of zeros for nodes and weights
> gLresult = matrix(rep(0, 2*length(nnodes)),ncol=2)
> for (i in 1:length(nnodes)) {
+   cc = gaussLegendre(nnodes[i],-1,1)
+   gLresult[i,1] = nnodes[i]
+   gLresult[i,2] = sum(cc$w * f(cc$x))
+   }
> gLresult
     [,1]       [,2]
[1,]   17 0.03164279
[2,]   29 0.03249163
[3,]   51 0.03250365
[4,]   65 0.03250365
> # Compare with integrate()
> integrate(f,-1,1)
0.03250365 with absolute error < 6.7e-07
```

We see that 51 nodes are enough to get a very precise result.

The `pracma` package has a number of other integration functions that implement Gaussian quadrature or some variants of it, most notably `quadgk()` for adaptive Gauss–Kronrod quadrature, and `quadgr()`, a Gaussian quadrature with Richardson extrapolation.

In Gauss–Kronrod quadrature the evaluation points are chosen so that an accurate approximation can be computed by reusing the information produced by the computation of a less accurate approximation. $n+1$ points are added to the n-point Gaussian rule to get a rule of order $2n+1$. The difference between these approximations leads to an estimate of the relative error.

The adaptive version applies this procedure recursively on refined subintervals of the integration interval, splitting the subinterval into smaller pieces if the relative error is greater than a tolerance level, and returning and adding up integral values on subintervals otherwise. Normally, Gauss–Kronrod works by comparing the $n = 7$ and $2n+1 = 15$ results.

Gauss–Kronrod quadrature is the basic step in `integrate` as well, combined with an adaptive interval subdivision and Wynn's "epsilon algorithm" for extrapolation.

quadgk, like all other functions in `pracma`, is written in R rather than, like `integrate`, in compiled C code. It therefore is slightly slower, but has the advantage of being more stable with oscillating functions while reaching a better level of accuracy. As an example, we will integrate the highly oscillating function $f(x) = sin(\frac{1}{x})$ on the intervall $[0, 1]$.

```
> require(pracma)
> f = function(x) sin(1/x)
> integrate(fun, 0, 1)
Error in integrate(fun, 0, 1) : maximum number of subdivisions reached
> integrate(fun, 0, 1, subdivisions=500)
0.5041151 with absolute error < 9.7e-05
> quadgk(fun, 0, 1)
[1] 0.5040670
```

with an absolute error of 1×10^{-7}. This accuracy will not be reached with `integrate()`. There are more complicated examples, where `integrate()` does not return a value while quadgk() does.

Therefore, the quadgk() function might be most efficient for high accuracies and oscillatory integrands. It can handle moderate singularities at the endpoints, but does not support infinite intervals.

6.2.4 More integration routines in *pracma*

There are some more integration routines in `pracma` that may be interesting to know about. quad() is an adaptive version of *Simpson's rule* that shows how much can be gained with a relatively simple formula through an adaptive approach.

```
> require(pracma)
> options(digits = 10)
> f = function(x) x * cos(0.1*exp(x)) * sin(0.1*pi*exp(x))
> curve(f, 0, 4); grid()

> quad(f, 0, 4)
[1] 1.282129075
```

quadl() uses adaptive *Lobatto quadrature*, which is similar to Gaussian quadrature, but includes the endpoints of the integration interval in the set of integration points. It is exact for polynomials up to degree $2n - 3$, where n is the number of integration points.

```
> quadl(f,0,1)
[1] 1.282129074
```

The quad() function might be more efficient for low accuracies with nonsmooth integrands, while the quadl() function might be more efficient than quad() at higher accuracies with smooth integrands.

Another advantage of quad() and quadl() is that the integrand does not need to be vectorized.

Function cotes() provides composite *Newton–Cotes formulas* of degrees 2 to 8. It takes as arguments the integrand, upper and lower limit, the number of subintervals to treat separately, and the number of nodes (the degree).

For the function above, because Newton–Cotes formulas are not adaptive, one needs a lot of intervals to get a good result.

```
> cotes(f, 0, 4, 500, 7)
[1] 1.282129074
```

No discussion of integration is complete without mentioning *Romberg integration*. Romberg's method approximates the integral with applying the trapezoidal rule (such as in trapz()) by doubling the number of subintervals in each step, and accelerates convergence by Richardson extrapolation.

```
> romberg(f, 0, 4, tol=1e-10)
$value
[1] 1.282129074
$iter
[1] 9
$rel.error
[1] 1.880781318e-13
```

The advantages of Romberg integration are the small number of calls to the integrand function compared to other integration methods—an advantage that will be relevant for difficult or costly to compute functions—and the quite high accuracy that can be reached. The functions should not have singularities and should not be oscillatory.

The last approach to mention is adaptive *Clenshaw–Curtis quadrature*, an integration routine that has gained popularity and is now considered to be a rival to Gauss–Kronrod. Clenshaw–Curtis quadrature is based on an expansion of the integrand in terms of Chebyshev polynomials. Unlike Gauss quadrature, which is exact for polynomials up to order $2n-1$, Clenshaw–Curtis quadrature is only exact for polynomials up to order n. However, since it uses the fast Fourier transform algorithm, the weights and nodes are computed in linear time. Its speed is further enhanced by the fact that the Chebyshev polynomial expansion of many functions converges rapidly. The function cannot have singularities.

```
> quadcc(f, 0, 4)
[1] 1.282129074
```

The implementation of `quadcc()` in `pracma` at the moment is iterative, not adaptive. That is, it will half all subintervals until the tolerance is reached. An adaptive version to come will be a strong competitor to `integrate()` and `quadgk()`.

`pracma` provides a function `integral()` that acts as a wrapper for some of the more important integration routines in this package. Some examples are given on the help page. Here we test it on the *dilogarithm function*

$$\int_0^1 \frac{\log(1-t)}{t}\, dx = \frac{\pi^2}{6} \tag{6.16}$$

```
> flog = function(t) log(1-t)/t
> val = pi^2/6
> for (m in c("Kron", "Rich", "Clen", "Simp", "Romb")) {
+     Q = integral(flog, 0, 1, reltol = 1e-12, method = m)
+     cat(m, Q, abs(Q-val), "\n")
+ }
Kron -1.644934067 9.858780459e-14    # Gauss-Kronrod
Rich -1.644934067 2.864375404e-14    # Gauss-Richardson
Clen -1.644934067 8.459899448e-14    # Clenshaw-Curtis
Simp -1.644934067 8.719469591e-12    # Simpson
Romb -1.645147726 0.0002136594219    # Romberg

> integrate(flog, 0, 1, rel.tol=1e-12)$value - val
[1] 0
```

Romberg does not come out well because the function has a pole at $x = 1$; and Gauss–Richardson is very accurate. But `integrate()` is certainly reliable and accurate in most cases.

6.2.5 *Functions with singularities*

If a function has one or more singularities (or discontinuities) within the integration domain (also called *improper* integrals), the result of a numerical integration can be strange or even unpredictable. For example, integrate the function $1/x^2$ from -1 to 1. In theory, the function is divergent, i.e., has no finite value.

```
> f = function(x) 1/x^2
> integrate(f, -1, 1)
Error in integrate(f, -1, 1) : non-finite function value

> integrate(f, -1, 1 - 1e10)
2753.484 with absolute error < 0

> integrate(f, -1, 1 - 1e-05)
Error in integrate(f, -1, 1 - 1e-05) :
the integral is probably divergent
```

The first error occurs because the integration tries to evaluate $f(0)$. If one of the boundary points is changed with a tiny value, `integrate` returns an answer that makes no sense; only the third call to `integrate` finds the correct answer.

If there are singularities (or discontinuities) in the integration interval, try to split the integral into a sum of integrals where singularities are on the boundary. As an example, we integrate the function $1/\sqrt{x}$ on $[0,1]$. The function is integrable; that is the integral has a finite value.

```
> f = function(x) 1/sqrt(x)
> integrate(f, 0, 1)
2 with absolute error < 5.8e-15
```

The result is exact because the antiderivative of f is $2\sqrt{x}$.

As another task, compute the following improper integral

$$\int_0^1 \frac{dx}{\sqrt{\sin x}} \tag{6.17}$$

whose exact, symbolic solution would require the hypergeometric series.

```
> f = function(x) 1/sqrt(sin(x)
> integrate(f, 0, 1)
2.034805 with absolute error < 9.1e-10
```

Note that since the quadrature rules will never use the value of the function on the boundary, the singularity at 0 will not disturb as $f(0)$ is never computed. For this reason, *removable* singularities such as $x = 0$ in $x \to \frac{\sin x}{x}$ pose no problem to integration routines based on quadrature rules.

But this approach of "ignoring the singularity" may not work if the integrand is oscillating, e.g.,

$$\int_0^1 \frac{1}{x} \sin\left(\frac{1}{x}\right) dx \tag{6.18}$$

which has a singularity in 0 that is approached in an oscillating manner.

```
> f = function(x) 1/x * sin(1/x)
> integrate(f, 0, 1)
Error in integrate(f, 0, 1) : the integral is probably divergent
```

But this "diagnosis" is *not* correct; in reality the integral converges. One can see this by transforming the variable with $u = 1/x$:

$$\int_0^1 \frac{1}{x} \sin(\frac{1}{x}) dx = \int_1^\infty \frac{\sin(u)}{u} du \tag{6.19}$$

and we will compute this integral in the next section. The example shows there is a kind of connection between improper and infinite integral—especially with oscillating functions—that often can be exploited with some background knowledge in mathematics.

6.2.6 *Infinite integration domains*

Functions that are integrated over infinite domains, such as $[-\infty,\infty]$ or $[0,\infty]$, will need to decrease sufficiently fast to 0 when approaching infinity. It is difficult for an integration routine to automatically recognize whether this is the case or not.

A well-behaved function is the Gauss error integral, well known in statistical applications and rapidly going to 0, defined as

$$\frac{1}{\sqrt{2\pi}} \int_{-\infty}^{\infty} e^{-\frac{1}{2}t^2} \mathrm{d}t \tag{6.20}$$

whose value must be 1. We define the function explicitly, though it is available in R as `pnorm()`.

```
> fgauss = function(t) exp(-t^2/2)
> ( q = integrate(fgauss, -Inf, Inf) )
2.506628 with absolute error < 0.00023
> q$value / sqrt(2*pi)
[1] 1
```

But if we put the peak far outside, `integrate()` has difficulties finding it there.

```
> mu = 1000
> fgauss = function(t) exp(-(t-mu)^2/2)
> integrate(fgauss, -Inf, Inf)
0 with absolute error < 0
```

For infinite domains it is recommended on the help page: "When integrating over infinite intervals do so explicitly, rather than just using a large number as the endpoint. This increases the chance of a correct answer."

And if using finite endpoints, try to put the "mass" of the integrand somewhere near the middle of the interval. (This may not be possible if the function is multi-modal with peaks far away from each other.)

```
> integrate(fgauss, 0, 2000)
2.506628 with absolute error < 5e-07
```

while `integrate(fgauss, 0, Inf)` will run into disaster again.

Not all integrable functions on infinite intervals are as rapidly decreasing as e^{-x^2} or e^{-x}. First, we will look at a critical example: $1/x$ from 1 to infinity.

```
> integrate(function(x) 1/x, 1, Inf)
Error in integrate(function(x) 1/x, 1, Inf) :
  maximum number of subdivisions reached
```

The function does not have a finite integral, and this error message is a typical—but not invariable—indication for integrals that do not converge.

There is a trick to cope with integrals on infinite domains by mapping the infinite range onto a finite interval, for instance with a transformation like $u = (1/x^2)f(1/x)$.

Function `integrate()` does this for us internally and thus can solve the following classical integrals almost exactly:

$$\int_0^\infty \sqrt{x}\,e^{-x}\,dx = \frac{\sqrt{\pi}}{2}, \qquad \int_0^\infty x e^{-x^2}\,dx = \frac{1}{2}, \qquad \int_{-\infty}^\infty \frac{1}{1+x^2}\,dx = \pi \tag{6.21}$$

```
> f = function(x) sqrt(x) * exp(-x)
> integrate(f, 0, Inf)
0.8862265 with absolute error < 2.5e-06

> f = function(x) x * exp(-x^2)
> integrate(f, 0, Inf)
0.5 with absolute error < 2.7e-06

> f = function(x) 1 / (1+x^2)
> integrate(f, -Inf, Inf)
3.141593 with absolute error < 5.2e-10
```

In the table of Section 6.2.3 Gauss–Laguerre and Gauss–Hermite quadrature were mentioned for integrals of the form

$$\int_0^\infty f(x) x^a e^{-x} dx \tag{6.22}$$

and

$$\int_{-\infty}^\infty f(x) e^{-x^2} dx \tag{6.23}$$

respectively, where function f does not increase too strongly.

Functions `gaussLaguerre()` and `gaussHermite()` in package `pracma` implement this approach. Applying them to the first function above results in

```
> require(pracma)
> cc = gaussLaguerre(4, 0.5)          # nodes and weights, a = 1/2
> sum(cc$w)                            # function f = 1
[1] 0.8862269

> cc = gaussHermite(8)                # nodes and weights
> sum(cc$w * cc$x^2)                   # function f(x) = x^2
[1] 0.8862269
```

The reader may verify that $\int_0^\infty \sqrt{x} e^{-x}\,dx = \int_{-\infty}^\infty x^2\,e^{-x^2}\,dx$ by applying the transformation $u = x^2$.

Example — There is still the task, left over from the last section, to compute the integral $\int_1^\infty \frac{sin(u)}{u} du$.

```
> f = function(u) sin(u)/u
> integrate(f, 1, Inf)
Error in integrate(f, 1, 10000): maximum number of subdivisions reached
```

But we know that this integral can be expressed as an alternating sum with smaller and smaller contributions, thus it *must* converge. Because the sign changes at every $n\pi$, we compute the integral to $10^6\pi$ and to $(10^6+1)\pi$,

```
> N = 10^6
> quadgk(f, 1, N*pi); quadgk(f, 1, (N+1)*pi) # takes some time
[1] 0.6247254
[1] 0.6247149
```

and the value of the integral will be 0.624720 ± 0.00001. (Do not use `integrate()` as it will declare "the integral is probably divergent" or say "maximum number of subdivisions reached.")

6.2.7 *Integrals in higher dimensions*

Multiple integrals, that is integrals of multivariate functions in higher dimensional space, are quite common in scientific applications. As an example we will try to compute the following integral

$$\int_0^1 \int_0^1 \frac{1}{1+x^2+y^2} dxdy \tag{6.24}$$

over the rectangular domain $[0,1]x[0,1]$. The first idea could be to solve this task as a twofold univariate integration by defining an intermediate function.

```
> fx = function(y) integrate(function(x) 1/(1+x^2+y^2), 0, 1)$value
> Fx = Vectorize(fx)
> ( q1 = integrate(Fx, 0, 1) )
0.6395104 with absolute error < 7.1e-15
```

This result will probably not be as accurate as the `abs.error` indicates because the inner function is itself an integral calculated with some error. There should be an easier and more accurate way to do this calculation.

For multidimensional integration two packages on CRAN, `cubature` and `R2Cuba`, provide this functionality on hyperrectangles using adaptive procedures internally. The integrand has to be a function of a vector, so in our case

```
> f = function(x) 1 / (1 + x[1]^2 + x[2]^2)
```

The integration function in `cubature` is called `adaptIntegrate`, so

```
> require(cubature)
> ( q2 = adaptIntegrate(f, c(0, 0), c(1, 1)) )
$integral
[1] 0.6395104
$error
[1] 4.5902e-06
$functionEvaluations
[1] 119
$returnCode
[1] 0
```

R2Cuba contains three different numerical integration routines—cuhre(), divonne(), and suave()—plus one Monte Carlo algorithm. The most commonly used one is cuhre(). The calling syntax is slightly more difficult than for adaptIntegrate(), and normally the accuracy of adaptIntegrate() is also a bit higher.

For two- and three-dimensional integrals there are two integration functions available in package pracma, integral2() and integral3(). Unfortunately, integral2 for 2-dimensional integrals needs a function definition using two variables explicitely.

```
> require(pracma)
> f = function(x, y) 1 / (1 + x^2 + y^2)
> ( q3 = integral2(f, 0, 1, 0, 1) )
$Q
[1] 0.6395104
$error
[1] 4.975372e-08
```

For each of these integration functions, a tolerance can be set in the call. With default tolerances, which of the three results is more accurate? This integral cannot be solved symbolically, still the true value up to 15 digits is $v = 0.6395103518703056\ldots$, thus

```
> print(q1$value, digits=16)    # 0.6395103518703110, abs error < 1e-14
> print(q2$integral,digits=16)  # 0.6395103518438505, abs error < 1e-10
> print(q3$Q, digits = 16)      # 0.6395103518702119, abs error < 1e-13
```

integral2() has some other nice and useful features:

- The endpoints of the integration interval of the inner integral can be (simple) functions of the value of the outer integration variable.
- integral2() can handle singularities at the endpoints (to a certain degree).
- The integrand can be integrated over domains characterized in polar coordinates.

The following example has been discussed on the R-help mailing list: Find the value of the integral

$$\frac{1}{2\pi}\int_0^5 \int_x^5 e^{-y/2} \sqrt{\frac{x}{y-x}} dy dx \tag{6.25}$$

The integrand is singular at the line $y = x$, and applying adaptIntegral() to it will not be successful. The lower endpoint of the integral is given through the value of x, thus

```
> require(pracma)
> f = function(x, y) 1/(2*pi) * exp(-y/2) * sqrt(x/(y-x))
> q = integral2(f, 0, 5, function(x) x, 5, singular = TRUE)
> q$Q
[1] 0.7127025
```

where `singular = TRUE` indicates to the `integral2` function that special care has to be taken along the boundaries.

As another example, we show how to compute the integral of the function $\ln(x^2 + y^2)$ in the ring defined by the two circles $x^2 + y^2 = 3$ and $x^2 + y^2 = 5$. To define the boundary of the integral as simple bounds on the variables x and y is not obvious; but in polar coordinates the region can be described through $\theta = 0 \ldots 2\pi$ and $r = 3 \ldots 5$. Thus, use `integral2` with `sector = TRUE`:

```
> require(pracma)
> f = function(x, y) log(x^2 + y^2)
> q = integral2(f, 0, 2*pi, 3, 5, sector = TRUE)
> q
$Q
[1] 140.4194
$error
[1] 2.271203e-07
```

There are many more two-dimensional integration routines in R, for instance in package `pracma`, `simpson2d()` for a 2D variant of Simpson's rule, or a 2-dimensional form of Gaussian quadrature in `quad2d()`. Readers are asked to look at the help pages and try out some examples for themselves.

6.2.8 Monte Carlo and sparse grid integration

For four- and higher-dimensional integrals the direct integration routines will become inaccurate and difficult to handle. This is where the Monte Carlo approach becomes most useful.

As a naive example, we try to compute the volume of the unit sphere in R^3. A set of N uniformly distributed points in $[0,1]^3$ is generated and the number of points is counted that lie in the volume of the sphere. Because the unit cube has volume one, the fraction of points falling into the sphere is also the volume of the sphere in $[0,1]^3$ or one eighth of the total volume.

```
> set.seed(4321)
> N = 10^6
> x = runif(N); y = runif(N); z = runif(N)
> V = 8 * sum(x^2 + y^2 + z^2 <= 1) / N
> V
[1] 4.195504
```

The formula for the volume of a sphere of radius r is $\frac{4}{3}\pi r^3$, the exact value being $V = 4.18879$ for radius 1. We see that even for a million points the result is not nearly exact. For good results one needs huge numbers of random points.

To improve the results, specialized techniques to perform Monte Carlo integration have been developed. In R the `R2Cuba` package provides the function `vegas()` that uses *importance sampling* to reduce the variance of the result.

Let f be the characteristic function of the sphere in three-dimensional space, i.e., f is 1 if $x^2+y^2+z^2 \leq 1$ and 0 otherwise. Function `vegas()` requests that the integrand is able to accept a second variable, the "weight," (that the user does not need to use in the function definition), as well as the dimension of space (3) and the number of components of the integrand (1).

```
> require(R2Cuba)
> f = function(u, w) {
>     x = u[1]; y = u[2]; z = u[3]
>     if (x^2 + y^2 + z^2 <= 1) 1 else 0
> }
> ndim = 3; ncomp = 1
> q = vegas(ndim, ncomp, f, lower = c(0,0,0), upper = c(1,1,1))
> ( V = 8 * q$value )
[1] 4.18501
```

Better than before, but still not very accurate. For these low-dimensional problems, standard integration procedures will probably work better in most cases.

As an example in dimension $D = 10$ we will compute the following integral

$$I = \int_0^1 \cdots \int_0^1 \prod_{d=1}^{D} (\frac{1}{2\pi} e^{-\frac{1}{2}x_d^2}) dx_D \ldots dx_1 \tag{6.26}$$

that in some form will often arise in statistical applications. Because the function is the product of one-dimensional functions, the integral can be calculated as the product of univariate integrals.

```
> f = function(x) prod(1/sqrt(2*pi)*exp(-x^2))
```

As this function is not vectorized(!) let's compute the one-dimensional integral with `quad()`, and then the 10th power will be a good approximation of the integral we are looking for.

```
> require(pracma)
> I1 = quad(f, 0, 1)
> I10 = I1^10
> I1; I10
[1] 0.2979397
[1] 5.511681e-06
```

`adaptIntegrate()` will not return a result for higher-dimensional integrals in an acceptable time frame. We test integration routines `cuhre()` and `vegas()` on this function in 10 dimensions:

```
> require(R2Cuba)
> ndim = 10; ncomp = 1
> cuhre(ndim, ncomp, f, lower=rep(0, 10), upper=rep(1, 10))
Iteration 1:  2605 integrand evaluations so far
[1] 5.51163e-06 +- 4.32043e-11      chisq 0 (0 df)
Iteration 2:  7815 integrand evaluations so far
[1] 5.51165e-06 +- 5.03113e-11      chisq 0.104658 (1 df)
integral: 5.511651e-06 (+-5e-11)
```

```
nregions: 2; number of evaluations: 7815; probability: 0.2536896

> vegas(ndim, ncomp, f, lower=rep(0, 10), upper=rep(1, 10))
Iteration 1:  1000 integrand evaluations so far
[1] 5.44824e-06 +- 1.64753e-07      chisq 0 (0 df)
...
Iteration 6:  13500 integrand evaluations so far
[1] 5.50905e-06 +- 4.75364e-09      chisq 1.17875 (5 df)
integral: 5.509047e-06 (+-4.8e-09)
number of evaluations: 13500; probability: 0.05310032
```

The results are quite good, though the error terms do not correctly indicate the true absolute error.

There is another routine for multiple integrals in higher dimensions in package `SparseGrid`. Applying it to our 10-D example is slightly more complicated, but the result is excellent.

First, a grid will be created with a certain accuracy level, where, e.g., $k = 2$ means the result will be exact for polynomials up to total order $2k-1$. Different types of quadrature rules are available.

```
> library(SparseGrid)
> ndim = 10
> k = 4
> spgrid = createSparseGrid(type = "KPU", dimension = ndim, k = k)
> n = length(spgrid$weights)
```

`spgrid` consists of nodes and weights. The integral will be calculated as the sum of weights times the values of the function at the nodes.

```
> I = 0
> for (i in 1:n)  I = I + f(spgrid$nodes[i, ])*spgrid$weights[i]
> I
[1] 5.507235e-06
```

The result is correct with an absolute error less than 0.005. For mid-sized dimensions a deterministic routine such as `cuhre` still seems better suited than a Monte Carlo or Sparse Grid approach.

6.2.9 Complex line integrals

In electrical engineering, complex line integrals are quite common. Most of the integration routines in R and its packages do not handle complex numbers and complex functions. We will look at an example and ways to compute line integrals.

The `trapz()` function in package `pracma` works with complex numbers. To compute the function $1/z$ in a circle of radius 1 around the origin, first generate points on the unit circle with the complex exponential, apply function $1/z$ and then `trapz()` on the generated points.

```
> require(pracma)
> N = 100
> s = seq(0, 1, length.out = N)
> z = exp(2*pi*1i * s)
> trapz(z, 1/z)
[1] 0+6.278968i
```

The exact result is $2\pi i$ because $\frac{1}{2\pi i}\int_C \frac{1}{z}dz = 1$ according to *Cauchy's integral theorem* for every simple closed curve C around the origin.

Another approach is to split the complex function into real and imaginary parts and integrate these functions separately as real functions. `cintegral()` in `pracma` does exactly this implicitly. The points along the integration curve are provided in the `waypoints` parameter.

```
> require(pracma)
> N = 100
> s = seq(0, 1, length.out = N)
> z = cos(2*pi*s) + 1i * sin(2*pi*s)
> f = function(z) 1/z
> cintegral(f, waypoints = z)
[1] 0+6.283185i
```

The result is much more accurate now as the two real functions representing real and imaginary parts are integrated utilizing a quadrature rule.

It is possible to integrate the function along a rectangle, e.g., with corners $(-1-1i, -1+1i, 1+1i, 1-1i, -1-1i)$ in this sequence.

```
> require(pracma)
> points = c(-1-1i, -1+1i, 1+1i, 1-1i, -1-1i)
> cintegral(function(z) 1/z, waypoints = points)
[1] 0+6.283185i
```

The result is the same as above because a complex line integral only depends on the residua of poles lying inside the closed curve. But the computation of the complex integrals along straight lines is in general faster and more accurate than along curved lines.

Of course, this function can also be used for *real* line integrals, that is, integrals of real functions in the plane along lines or curves.

Package `elliptic` (not currently available for Mac OS X) provides another routine for complex line integrals, here called "contour integrals," the function name being `integral.contour()`. The curve needs to be defined as a differentiable function, say u, the path runs from `u(0)` to `u(1)`, and the user has to supply the derivative function of u explicitly.

```
> install.packages("elliptic")
> require(elliptic)
> u = function(x) exp(2i*pi*x)
> uprime = function(x) 2i*pi*exp(2i*pi*x)
> integral.contour(f, u, uprime)
```

```
[1] 0+6.283185i
```

with the same accuracy as above. There is also a function `integral.segment()` that has a similar functionality as `cintegral()` with parameter `waypoints`.

6.3 Symbolic manipulations in R

R is not totally bereft of symbolic capabilities. Base R has two functions for returning symbolic derivatives: `D` and `deriv`. `D` is simpler, while `deriv` provides more information. According to the help page, "The internal code knows about the arithmetic operators +, -, *, / and ^, and the single-variable functions exp, log, sin, cos, tan, sinh, cosh, sqrt, pnorm, dnorm, asin, acos, atan, gamma, lgamma, digamma and trigamma, as well as psigamma for one or two arguments (but derivative only with respect to the first). (Note that only the standard normal distribution is considered.)"

6.3.1 D()

As an example of how to use `D()`, consider applying it to the function

$$f(x) = \sin(x)e^{-ax} \tag{6.27}$$

```
> # Define the expression and its function counterpart
> f = expression(sin(x)*exp(-a*x))
> ffun = function(x,a) sin(x)*exp(-a*x)
> # Take the first derivative
> (g = D(f,"x"))
cos(x) * exp(-a * x) - sin(x) * (exp(-a * x) * a)
> # Turn the result into a function
> gfun = function(x,a) eval(g)
> # Take the second derivative
> (g2 = D(g,"x"))
-(cos(x) * (exp(-a * x) * a) + sin(x) * exp(-a * x) + (cos(x) *
    (exp(-a * x) * a) - sin(x) * (exp(-a * x) * a * a)))
> # Turn the result into a function
> g2fun = function(x,a) eval(g2)
> # Plot the function and its derivatives, with a = 1
> curve(ffun(x,1),0,4, ylim = c(-1,1), ylab=c("f(x,1) and
+ derivatives"))
> curve(gfun(x,1), add=T, lty=2)
> curve(g2fun(x,1), add=T, lty=3)
> legend("topright", legend = c("f(x,1)", "df/dx", "d2f/dx2"),
+ lty=1:3, bty="n")
```

6.3.2 `deriv()`

An equivalent result is obtained with the `deriv()` function, albeit at the cost of greater complexity.

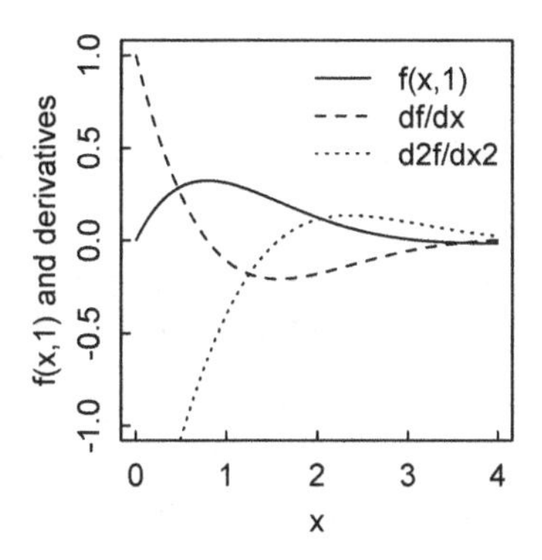

Figure 6.3: *Plot of the function defined by Equation 6.27 and its first and second derivatives.*

```
> (D1 = deriv(f,"x"))
expression({
    .expr1 <- sin(x)
    .expr4 <- exp(-a * x)
    .value <- .expr1 * .expr4
    .grad <- array(0, c(length(.value), 1L), list(NULL, c("x")))
    .grad[, "x"] <- cos(x) * .expr4 - .expr1 * (.expr4 * a)
    attr(.value, "gradient") <- .grad
    .value
})
> x = 0:3
> a = 1
> options(digits=3)
> eval(D1)
[1] 0.00000 0.30956 0.12306 0.00703
attr(,"gradient")
          x
[1,]  1.0000
[2,] -0.1108
[3,] -0.1794
[4,] -0.0563
```

To use the result of `deriv()` in a function, and to get the second derivative (the Hessian in this case) as well, proceed as follows

```
> (D1fun = deriv(f,"x", hessian = T, func=T))
function (x)
{
    .expr1 <- sin(x)
    .expr4 <- exp(-a * x)
    .expr5 <- .expr1 * .expr4
    .expr6 <- cos(x)
```

```
    .expr8 <- .expr4 * a
    .expr11 <- .expr6 * .expr8
    .value <- .expr5
    .grad <- array(0, c(length(.value), 1L), list(NULL, c("x")))
    .hessian <- array(0, c(length(.value), 1L, 1L), list(NULL,
        c("x"), c("x")))
    .grad[, "x"] <- .expr6 * .expr4 - .expr1 * .expr8
    .hessian[, "x", "x"] <- -(.expr11 + .expr5 + (.expr11 - .expr1 *
        (.expr8 * a)))
    attr(.value, "gradient") <- .grad
    attr(.value, "hessian") <- .hessian
    .value
}
> D1grad = function(x) attr(D1fun(x),"gradient")
> D1hess = function(x) attr(D1fun(x),"hessian")
```

The parameter a must be defined in this case as an external (global) variable

```
> a = 1
```

The following series of commands then reproduces the previous graph.

```
> curve(ffun(x,1), 0, 4, ylim = c(-1,1))
> curve(D1grad(x),  lty=2, add=T)
> curve(D1hess(x),  lty=3, add=T)
```

6.3.3 *Polynomial functions*

As noted in Chapter 4, the PolynomF package enables symbolic differentiation and integration of polynomial functions. An example of differentiation:

```
> require(PolynomF)
Loading required package: PolynomF
> (p <- poly.from.zeros(-2:5))
-240*x + 188*x^2 + 252*x^3 - 231*x^4 + 42*x^6 - 12*x^7 + x^8
> deriv(p)
-240 + 376*x + 756*x^2 - 924*x^3 + 252*x^5 - 84*x^6 + 8*x^7
```

The PolynomF package contains the integral() function for analytic indefinite integration of polynomials, and for numerical results with specified limits.

```
> require(PolynomF)
Loading required package: PolynomF
Attaching package: PolynomF
The following object is masked from package:pracma:
    integral
> x = polynom() # polynomial 'x'
> p = (x-1)^2 + 10*x^3 + 5*x^4
> p
1 - 2*x + x^2 + 10*x^3 + 5*x^4
> integral(p) # Note: no constant of integration
```

```
x - x^2 + 0.3333333*x^3 + 2.5*x^4 + x^5
> integral(p, limits = c(0,2))
[1] 72.66667
```

The `pracma` package contains the `polyder()` function to calculate the derivative of polynomials and products of polynomials. Remember that in `pracma`, polynomial coefficients are defined from highest to lowest order.

```
> require(pracma)
> p = c(3,2,1,1); q = c(4,5,6,0) # coefficients from high to low
> polyder(p)
[1] 9 4 1
> polyder(p,q)
[1]  72 115 128  63  22   6
```

6.3.4 *Interfaces to symbolic packages*

Beyond the limited (though still useful) symbolic capabilities discussed in this section, R has two packages that interface with broader symbolic mathematics systems. The package `Ryacas` provides an interface to `yacas` (yet another computer algebra system). And `rSymPy` provides access from within R to the `SymPy` computer algebra system running on Jython (java-hosted python). Detailed discussion of these packages is beyond the scope of this book; but for those interested, CRAN and various websites that can be located via Google will give pertinent information.

6.4 Case studies

6.4.1 *Circumference of an ellipse*

The area A of an ellipse with semi-axes (a,b) is well known to be the simple extension of the expression for a circle: $A = \pi ab$. However, the expression for the circumference C of the ellipse is a much more complicated issue. It can be shown that $C = 4aE(e^2)$ where E is the complete elliptic integral of the second kind, and e is the eccentricity $e = \sqrt{1-b^2/a^2}$. The `pracma` package calculates elliptic integrals with the function `ellipke()`, which returns a list with two components, k the value for an integral of the first kind, and e for the second kind. Thus an ellipse with $a = 1, b = 1/2$ has circumference

```
> require(pracma)
> a=1; b=1/2
> options(digits = 10)
> e = sqrt(1-b^2/a^2)
> E = ellipke(e^2)$e
> (C = 4*a*E)
[1] 4.84422411
```

A more intuitive way to do this calculation is to integrate along the arc length of the ellipse. `pracma` accomplishes this with the `arclength()` function, which applies

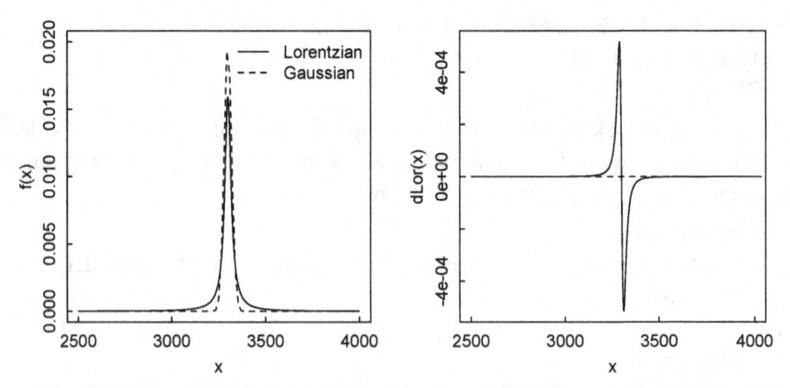

Figure 6.4: *(left) Plot of the function defined by Equation 6.28 compared with a Gaussian. (right) Derivative of the Lorentzian in the left panel.*

Richardson's extrapolation by refining polygon approximations to the parameterized curve.

```
> f = function(t) c(a*cos(t), b*sin(t))
> (C = arclength(f, 0, 2*pi, tol = 1e-10))
$length
[1] 4.84422411
$niter
[1] 10
$rel.err
[1] 2.703881563e-11
```

6.4.2 *Integration of a Lorentzian derivative spectrum*

The Lorentzian function

$$L(x) = \frac{1}{\pi} \frac{w}{(x - x_0)^2 + w^2} \tag{6.28}$$

describes the shape of some spectral lines, e.g., in electron paramagnetic resonance (EPR) spectroscopy. Here x_0 is the position of the maximum, and w is the half width at half height. The function is normalized to unity:

$$\int_{-\infty}^{\infty} L(x)\, dx = 1. \tag{6.29}$$

Compared with the Gaussian function with $\mu = x_0$ and $sd = w$, the Lorentzian is sharper near the maximum and decays more slowly away from the maximum, as can be seen in Figure 6.4 (left). The parameters (x_0, w) and the left and right limits are those typical of a free radical EPR spectrum, with the x-axis in magnetic field (Gauss) units.

```
> par(mfrow=c(1,2))
```

```
> Lor = function(x,x0=3300,w=20) 1/pi*w/((x-x0)^2 + w^2)
> Gau = function(x,x0=3300,w=20) 1/sqrt(2*pi*w^2)*exp(-(x-x0)^2/(2*w^2))
> curve(Lor,2500,4000,ylim = c(0,0.02), n=1000, lty=1,ylab="f(x)")
> curve(Gau,2500,4000,add=T, lty=2)
> legend("topright",legend=c("Lorentzian","Gaussian"),lty=1:2,bty="n")
```

Integration of the Lorentzian function between $\pm\infty$ yields the proper normalized value, but the function decays so slowly that integration between the experimental limits–a range of 25 halfwidths!–misses almost 2% of the total.

```
> integrate(Lor,-Inf,Inf)
1 with absolute error < 2.5e-05
> integrate(Lor,2500,4000)
0.9829518 with absolute error < 4.5e-07
```

Usually, EPR spectra are collected in derivative mode, which emphasizes the maximum and the width (Figure 6.4 (right)).

```
> require(pracma)
> dLor = function(x) numdiff(Lor,x)
> curve(dLor(x), 2500,4000, n=1000)
> abline(0,0,lty=2)
```

In a typical experiment, the derivative spectrum may be collected at 1000 equally spaced points. To determine the concentration of spins, the derivative spectrum must be integrated to get the "original" spectrum, then integrated again over the limits of observation to get the area under the curve. The `trapz()` and `cumtrapz()` functions in `pracma` can serve this purpose.

```
> xs = seq(2500,4000,len=1000)
> ys = Lor(xs)
> dys = dLor(xs)
> trapz(xs,ys) # Normalized to 1
[1] 0.9829518
```

As a check, we see that `trapz()` applied to the digitized spectrum gives the same result as `integrate()` applied to the Lorentzian function between the same limits.

We now apply `cumtrapz()` to recreate the digitized spectrum over the full range, and then `trapz()` to integrate the digitized spectrum over that range.

```
> intdys = cumtrapz(xs,dys)
> trapz(xs,intdys)
[1] 0.9680403
```

The integral is further decreased relative to the true value of 1, again due more to the finite range of integration rather than to inadequacy of the integration routine.

6.4.3 Volume of an ellipsoid

The volume of an ellipsoid with semi-axes A,B,C is $V = \frac{4}{3}\pi ABC$.

```
> A = 1; B = 2/3; C = 1/2
> (V = 4/3*pi*a*b*c)
[1] 1.396263
```

We use the `vegas()` function in the `R2Cuba` package to evaluate the volume using a Monte Carlo method.

```
> require(R2Cuba)
> f = function(u) {
+    x = u[1]; y=u[2]; z = u[3]
+    if (x^2/A^2 + y^2/B^2 +z^2/C^2 <=1) 1 else 0
+ }
> ndim=3; ncomp=1
> q = vegas(ndim,ncomp,f,lower=c(-A,-B,-C), upper=c(A,B,C))
Iteration 1:  1000 integrand evaluations so far
[1] 1.40533 +- 0.0421232   chisq 0 (0 df)
Iteration 2:  2500 integrand evaluations so far
[1] 1.38394 +- 0.0269662   chisq 0.182942 (1 df)
Iteration 3:  4500 integrand evaluations so far
...
Iteration 12:  45000 integrand evaluations so far
[1] 1.39231 +- 0.0116365   chisq 2.52888 (11 df)
Iteration 13:  52000 integrand evaluations so far
[1] 1.39815 +- 0.0112514   chisq 2.77392 (12 df)
> (V = q$value)
[1] 1.398148
```

Readers can judge whether this level of accuracy is sufficient for their purposes.

Chapter 7

Optimization

Scientists and engineers often have to solve for the maximum or minimum of a multi-dimensional function, sometimes with constraints on the values of some or all of the variables. This is known as optimization, and is a rich, highly developed, and often difficult problem. Generally the problem is phrased as a minimization, which shows its kinship to the least-squares data fitting procedures discussed in a subsequent chapter. If a maximum is desired, one simply solves for the minimum of the negative of the function. The greatest difficulties typically arise if the multi-dimensional surface has local minima in addition to the global minimum, because there is no way to show that the minimum is local except by trial and error.

R has three functions in the base installation: `optimize()` for one-dimensional problems, `optim()` for multi-dimensional problems, and `constrOptim()` for optimization with linear constraints. (Note that even "unconstrained" optimization normally is constrained by the limits on the search range.) We shall consider each of these with suitable examples, and introduce several add-on packages that expand the power of the basic functions. Optimization is a sufficiently large and important topic to deserve its own task view in R, at `http://cran.r-project.org/web/views/Optimization.html`.

In addition to the packages considered in this chapter, the interested reader should become acquainted with the `nloptr` package, which is considered one of the strongest and most comprehensive optimization packages in R. According to its synopsis,"nloptr is an R interface to NLopt. NLopt is a free/open-source library for nonlinear optimization, providing a common interface for a number of different free optimization routines available online as well as original implementations of various other algorithms."

7.1 One-dimensional optimization

The base R function for finding minima (the default) or maxima of functions of a single variable is `optimize()`. As a concrete example, given a 16×20 sheet of cardboard, find the size x of the squares to be cut from the corners that maximizes the volume of the open box formed when the sides are folded up. x must be in the range (0,8).

```
> optimize(function(x) x*(20-2*x)*(16-2*x), c(0,8), maximum=T)
```

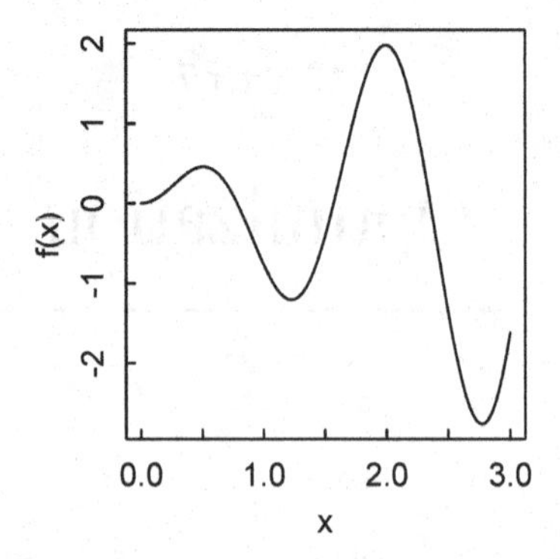

Figure 7.1: *Plot of function $f(x) = x\sin(4x)$ showing several maxima and minima.*

```
$maximum
[1] 2.944935

$objective
[1] 420.1104
```

Consider next the use of `optimize` with the function

```
> f = function(x) x*sin(4*x)
```

which plotted looks like this (Figure 7.1):

```
> curve(f,0,3)
```

It has two minima in the $x = 0-3$ range, with the global minimum near 2.8, and two maxima, with the global maximum near 2.0. Applying `optimize()` in the simplest way yields

```
> optimize(f,c(0,3))
$minimum
[1] 1.228297
$objective
[1] -1.203617
```

which gives the local minimum because it is the first minimum encountered by the search algorithm (Brent's method, which combines root bracketing, bisection, and inverse quadratic interpolation). Because we have a plot of the function, we can see that we must exclude the local minimum from the lower and upper endpoints of the search interval.

```
> optimize(f,c(1.5,3))
$minimum
[1] 2.771403
$objective
[1] -2.760177
```

To find the global maximum we enter

```
> optimize(f,c(1,3),maximum=TRUE)
$maximum
[1] 1.994684
$objective
[1] 1.979182
```

We could have obtained the same result by minimizing the negative of the function

```
> optimize(function(x) -f(x),c(1,3))
$minimum
[1] 1.994684
$objective
[1] -1.979182
```

which finds the maximum in the right place but, of course, yields the negative of the function value at the maximum.

If necessary, the desired accuracy can be adjusted with the `tol` option in the function call.

The `pracma` package contains the function `findmins()`, which finds the positions of all the minima in the search interval by dividing it n times (default $n = 100$) and applying `optimize` in each interval. To find the values at those minima, evaluate the function.

```
> require(pracma)
> f.mins = findmins(f,0,3)
> f.mins  # x values at the minima
[1] 1.228312 2.771382
> f(f.mins[1:2]) # function evaluated at the minima
[1] -1.203617 -2.760177
```

The Examples section of the help page for `optimize()` shows how to include a parameter in the function call. It also shows how, for a function with a very flat minimum, the wrong solution can be obtained if the search interval is not properly chosen. Unfortunately, there is no clear way to choose the search interval in such cases, so if the results are not as expected from inspecting the graph of the function, other intervals should be explored.

The function $f(x) = |x^2 - 8|$ yields some interesting behavior (Figure 7.2).

```
> f = function(x) abs(x^2-8)
> curve(f,-4,4)
```

Straightforward solving for the maximum over the entire range yields the result at $x = 0$.

```
> optimize(f,c(-4,4),maximum=T)
$maximum
[1] -1.110223e-16
$objective
[1] 8
```

Excluding the middle from the search interval finds the maxima at the extremes.

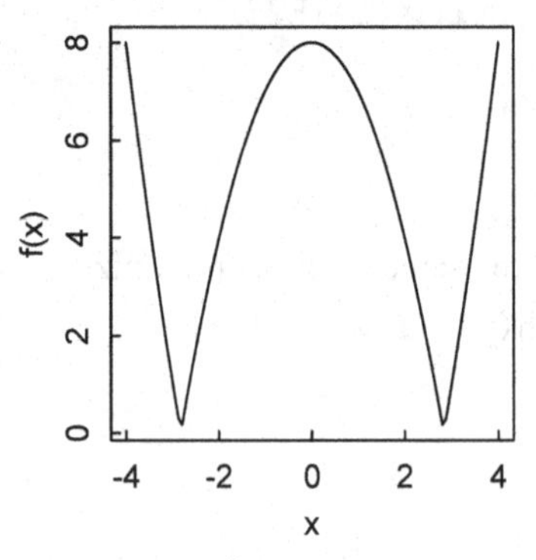

Figure 7.2: *Plot of function $f(x) = |x^2 - 8|$ showing several maxima and minima.*

```
> optimize(f,c(-4,-2),maximum=T)
$maximum
[1] -3.999959
$objective
[1] 7.999672
> optimize(f,c(2,4),maximum=T)
$maximum
[1] 3.999959
$objective
[1] 7.999672
```

However, "the endpoints of the interval will never be considered to be local minima" in `findmins`, because the function applies `optimize()` to *two* adjacent subintervals, and the endpoints have only one.

```
> findmins(function(x) -f(x),-4,4)
[1] -1.040834e-17
> findmins(function(x) -f(x),-4,-3)
NULL
```

7.2 Multi-dimensional optimization with `optim()`

Optimization in more than one dimension is harder to visualize and to compute. An example is a function arising in chemical engineering (Hanna and Sandall, p. 191).

$$f(x_1,x_2) = \frac{1}{x_1} + \frac{1}{x_2} + \frac{1-x_2}{x_2(1-x_1)} + \frac{1}{(1-x_1)(1-x_2)} \tag{7.1}$$

x_1 and x_2 are mole fractions, which must lie between 0 and 1. The surface defined by the function may be visualized by the `persp` function (Figure 7.3).

```
> x1 = x2 = seq(.1,.9,.02)
> z = outer(x1,x2,FUN=function(x1,x2) 1/x1 + 1/x2 +
```

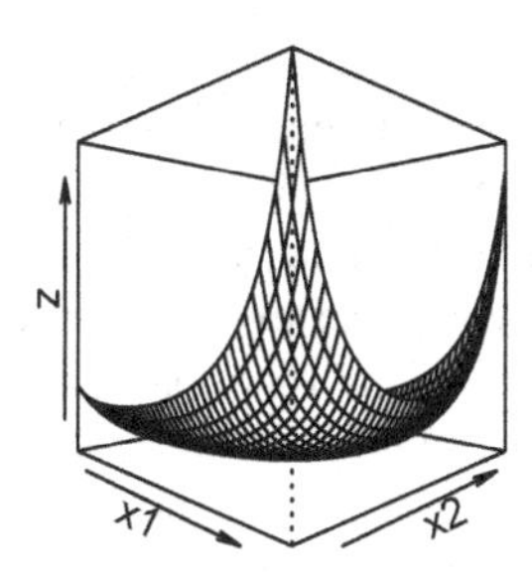

Figure 7.3: *Perspective plot of the function defined by Equation 7.1.*

```
+   (1-x2)/(x2*(1-x1)) + 1/((1-x1)*(1-x2)))
> persp(x1,x2,z,theta=45,phi=0)
```

The diversity and difficulty of optimization problems has led to the development of many packages and functions in R, each with its own strengths and weaknesses. It can therefore be somewhat bewildering to know which to try. According to Borchers (personal communication),

> Whenever one can reasonably assume that the objective function is smooth or at least differentiable, apply "BFGS" or "L-BFGS-B." If worried about memory requirements with high-dimensional problems, try also "CG." Apply "Nelder–Mead" only in other cases, and only for low-dimensional tasks. [All of these are methods of the optim() function.] If the objective function is truly non-smooth, none of these approaches may be successful.
>
> If you are looking for global optima, first try a global solver (GenSA, DEoptim, psoptim, CMAES, ...) or a kind of multi-start approach (in low dimensions). Try specialized solvers for least-squares problems as they are convex and therefore have only one global minimum.

In addition to these optimization solvers, and others which will be discussed below, there is also the package nloptwrap that is simply a wrapper for the nloptr package. This, in turn, is a wrapper for the free and powerful optimization library NLOPT. Consult the R package library for details.

7.2.1 optim() with "Nelder–Mead" default

The optim() function is the workhorse for multi-dimensional optimization in base R. By default, optim() performs minimization. Its calling usage is

```
optim(par, fn, gr = NULL, ...,
      method = c("Nelder-Mead", "BFGS", "CG", "L-BFGS-B",
         "SANN", "Brent"),
      lower = -Inf, upper = Inf,
      control = list(), hessian = FALSE)
```

with the Nelder–Mead method as the default. According to the `optim()` help page, Nelder–Mead "uses only function values and is robust but relatively slow. It will work reasonably well for non-differentiable functions." Nelder–Mead is a "downhill simplex method." In this context, a "simplex" is a figure with $n+1$ vertices in an n-dimensional space: a triangle in a plane, a tetrahedron in three dimensions, etc. In 2-D as a simple example, the "worst" vertex of the initial triangle is replaced by a better one, which lowers the value of the function enclosed by the simplex. The process continues, moving along the plane, until a minimum is reached within the desired tolerance. See Numerical Recipes, 3rd ed., pp 502–507, for an engaging explanation.

To calculate the position of the minimum of the function defined by Equation 7.1, we first define a function that takes a vector and returns a scalar, as required by `optim()`.

```
> f = function(x) {
+    x1 = x[1]
+    x2 = x[2]
+    return(1/x1 + 1/x2 + (1-x2)/(x2*(1-x1)) +
+    1/((1-x1)*(1-x2)))
+    }
```

To minimize f with respect to x1 and x2, we write

```
> optim(c(.5,.5),f)
$par
[1] 0.3636913 0.5612666
$value
[1] 9.341785
$counts
function gradient
      55       NA
$convergence
[1] 0
$message
NULL
```

A common test case for optimization routines in two dimensions is the Rosenbrock "Banana function,"

$$f(x_1,x_2) = 100(x_2 - x_1 x_2)^2 + (1 - x_1)^2 \tag{7.2}$$

used in the Examples section of the `optim()` help page (Figure 7.4).

```
> x1 = x2 = seq(-1.2,1,.1)
> z = outer(x1,x2,FUN=function(x1,x2) {100 *
    (x2 - x1 * x1)^2 + (1 -x1)^2})
> persp(x1,x2,z,theta=150)
```

It appears as if the minimum is somewhere near (1,1). Proceeding as in the previous example,

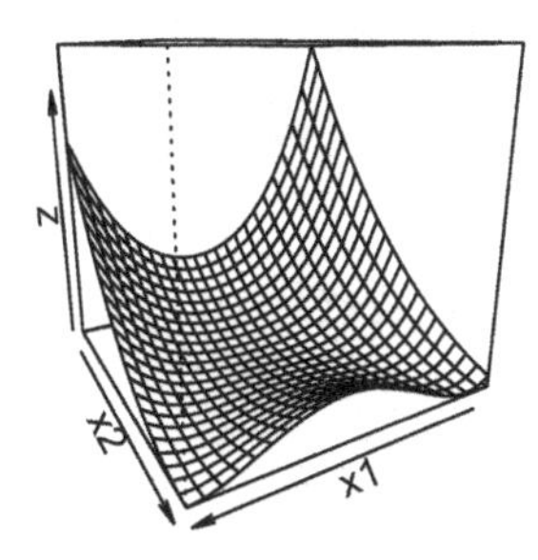

Figure 7.4: *Perspective plot of the Rosenbrock banana function defined by Equation 7.2.*

```
> fr = function(x) {  # Rosenbrock Banana function
+ x1 = x[1]
+ x2 = x[2]
+ 100 * (x2 - x1 * x1)^2 + (1 - x1)^2
+}
```

We then apply `optim()`, with the first argument being a starting guess for the vector specifying the set of parameters to be optimized over x1 and x2, and the second argument being the function to be minimized. Since we have not specified a method, `optim()` uses the Nelder–Mead default.

```
> optim(c(-1.2,1), fr)
$par
[1] 1.000260 1.000506
$value
[1] 8.825241e-08
$counts
function gradient
     195       NA
$convergence
[1] 0
$message
NULL
```

7.2.2 *`optim()` with "BFGS" method*

Sometimes a solution can be obtained more quickly and accurately if an analytical form of the gradient of the function is provided. This is the approach taken by the Broyden–Fletcher–Goldfarb–Shanno (BFGS) method, which uses an adaptation of Newton's method: $f(x)$ is approximated by a quadratic function around the current value of the x vector, and then a step is taken toward the minimum (or maximum) of that quadratic function. At the optimum, the gradient must be zero. The BFGS method is illustrated in the following example, which also uses the Rosenbrock banana function.

```
> grr  = function(x) { ## Gradient of fr
+   x1 = x[1]
+   x2 = x[2]
+   c(-400*x1*(x2-x1*x1)-2*(1-x1),
+      200 * (x2 - x1 * x1))
+ }
> optim(c(-1.2,1), fr, grr, method = "BFGS")
$par
[1] 1 1
$value
[1] 9.594956e-18
$counts
function gradient
     110       43
$convergence
[1] 0
$message
NULL
```

We see that in this case the `$par` components are found exactly, the value of `fr` at that point is essentially equal to zero, and the computation took 110 evaluations (plus 43 evaluations of the gradient) instead of 195 for Nelder–Mead.

If the Hessian (the matrix of second partial derivations of the function with respect to the coordinates) at the endpoint is desired, set `hessian = TRUE`.

```
> optim(c(-1.2,1), fr, grr, method = "BFGS", hessian=TRUE)
$par
[1] 1 1
$value
[1] 9.594956e-18
$counts
function gradient
     110       43
$convergence
[1] 0
$message
NULL
$hessian
          [,1] [,2]
[1,]  802.0004 -400
[2,] -400.0000  200
```

7.2.3 *optim() with "CG" method*

The "CG" (conjugate gradients) method of optim() fails for this function—it appears to converge, but to the wrong values—except if the Poliak–Ribiere updating (type = 2) is used. According to the help page, "Conjugate gradient methods will generally be more fragile than the BFGS method, but as they do not store a matrix they may be successful in much larger optimization problems."

```
> optim(c(2, .5), fn = fr, gr = grr, method="CG",
   control=list(type=1))
$par
[1] 0.9156605 0.8380146
$value
[1] 0.007103679
$counts
function gradient
     405      101
$convergence
[1] 1
$message
NULL
```

But control=list(type=2) gets it right:

```
> optim(c(2, .5), fn = fr, gr = grr, method="CG",
   control=list(type=2))
$par
[1] 1.000039 1.000078
$value
[1] 1.519142e-09
$counts
function gradient
     348      101
$convergence
[1] 1
$message
NULL
```

7.2.4 *optim() with "L-BFGS-B" method to find a local minimum*

If we want to find a minimum, even a local one, in a given region, we apply box constraints with method = "L-BFGS-B". L-BFGS is a limited memory form of BFGS, while L-BFGS-B applies box constraints to that method. For example, to find the minimum in the box with lower limits c(3.5,3.5) and upper limits c(5,5) we execute

```
> optim(fn = f, par=c(4,5), method="L-BFGS-B",
+ lower= c(3.5,3.5),upper=c(5,5))
$par
[1] 3.5 5.0
```

```
$value
[1] -0.09094568
$counts
function gradient
       3        3
$convergence
[1] 0
$message
[1] "CONVERGENCE: NORM OF PROJECTED GRADIENT <= PGTOL"
```

Note that the computed minimum value of the function in the box is at one corner, and it's not a local minimum of the function.

The material in this and the next section has strong ties to the discussion in Chapter 11 of least-squares fitting of data with nonlinear models, in which the sum of squared deviations between data and model is the function to be minimized. To give a preview, suppose we have data representing a functional connection between x and y:

```
> set.seed(237)
> x = seq(0, pi, length.out = 50)
> y = sin(x) + 0.1*rnorm(50)
> plot(x, y)
```

We try to find a smooth spline curve in $(0, \pi)$ that best approximates this curve in a least-squares sense, defined with ten equidistant nodes in this interval.

```
> xp = seq(0, pi, length.out = 12)
```

The y-coordinates yp of these points have to be determined by an optimization approach. If xp and yp are known a spline function through these points is defined by `f = splinefun(x, y)` and thus the sum-of-squares distance to the given points can be computed:

```
> F = function(p) {
+  fsp = splinefun(xp, c(0, p, 0))
+  sum((y - fsp(x))^2)
+ }
```

and the optimization procedure is

```
> opt = optim(rep(0.5, 10), F, method="L-BFGS-B",
+  lower = rep(0, 10), upper = rep(1, 10))
opt
$par
 [1] 0.2803714 0.4152705 0.8566495 0.9855610 0.9601668 0.9908711
 [7] 0.8670039 0.7114714 0.5768663 0.2435485
$value
[1] 0.2905246
$counts
function gradient
       9        9
```

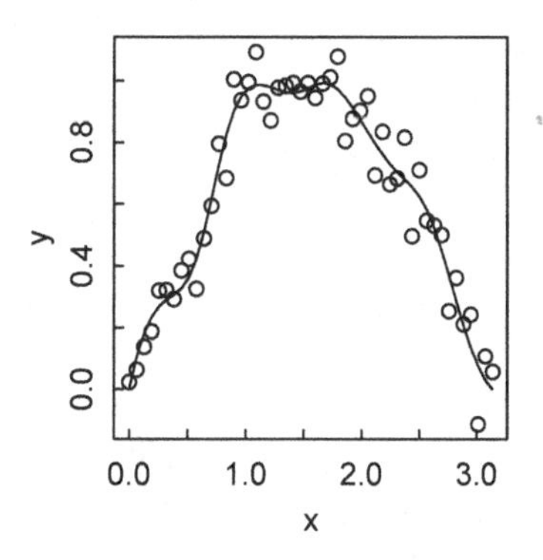

Figure 7.5: *Least squares fit of a spline function to data.*

```
$convergence
[1] 0
$message
[1] "CONVERGENCE: REL_REDUCTION_OF_F <= FACTR*EPSMCH"
```

Now plot the smoothed spline approximation (Figure 7.5):

```
> fsp = splinefun(xp, c(0, opt$par, 0))
> yy = fsp(x)
> lines(x, yy)
```

With fewer nodes a more sine-like shape will result.

In comparison, the default "Nelder-Mead" method finds a less satisfactory minimum:

```
> opt <- optim(rep(0.5, 10), F)
> opt
$par
[1] 0.2914511 0.3881710 0.8630475 0.9825521 0.9513082 1.0322379
[7] 0.8621980 0.7084492 0.5955802 0.2432722
$value
[1] 0.3030502
```

7.3 Other optimization packages

7.3.1 nlm()

The nlm() (nonlinear minimization) function in base R uses a Newton-type algorithm to find the minimum of a function. Here we apply it, with its defaults, to the Rosenbrock function:

```
> nlm(fr,c(-2,2))
$minimum
[1] 9.023082e-08
```

```
$estimate
[1] 0.9996997 0.9993989
$gradient
[1]  1.348420e-07 -6.012634e-08
$code
[1] 1
$iterations
[1] 49
```

The `nlm()` function can be applied without explicit derivatives (gradient) which are then calculated numerically, as in the example from the help page:

```
> f <- function(x, a) sum((x-a)^2)
> nlm(f, c(10,10), a = c(3,5))
$minimum
[1] 3.371781e-25
$estimate
[1] 3 5
$gradient
[1]  6.750156e-13 -9.450218e-13
$code
[1] 1
$iterations
[1] 2
```

or the gradient may be added as an attribute, which leads to somewhat better performance:

```
> f <- function(x, a)
+ {
+     res <- sum((x-a)^2)
+     attr(res, "gradient") <- 2*(x-a)
+     res
+ }
> nlm(f, c(10,10), a = c(3,5))
$minimum
[1] 0
$estimate
[1] 3 5
$gradient
[1] 0 0
$code
[1] 1
$iterations
[1] 1
```

`demo(nlm)` gives other instructive examples, including the use of derivatives.

7.3.2 *ucminf package*

The `ucminf()` function in the `ucminf` package employs a quasi-Newton type of algorithm for general-purpose unconstrained, nonlinear optimization. It uses the same updating of the inverse Hessian as the BFGS method in `optim()`, and has the same calling structure as `optim()`, so the two may be readily interchanged. As an example, let us find the position of the *maximum* of the function

$$f_1 = -(2x_1 - 5)^2 - (x_2 - 3)^2 - (5x_3 - 2)^2.$$

Since `ucminf()` finds minima, we apply it to the negative of f_1. We also note that, since each term is squared and thus always positive or zero, f_1 will be a maximum, or $-f_1$ a minimum, when each term equals zero, a condition that can be solved by inspection: $x = c(5/2, 3, 2/5)$. A calculation agrees within numerical precision:

```
> install.packages("ucminf")
> library(ucminf)
> f1 = function(x) (2*x[1]-5)^2 + (x[2]-3)^2 + (5*x[3]-2)^2
> ucminf(c(1,1,1), f1)
$par
[1] 2.4999992 2.9999994 0.4000001
$value
[1] 3.485199e-12
$convergence
[1] 4
$message
[1] "Stopped by zero step from line search"
$invhessian.lt
[1]  0.1257734298 -0.0001776904 -0.0001415078  0.5022988116
    -0.0001955878  0.0200489174
$info
 maxgradient     laststep      stepmax        neval
1.745566e-05 0.000000e+00 3.500000e-01 1.000000e+01
```

7.3.3 *BB package*

The BB package has the optimization functions `spg()`, `BBoptim`, and `multiStart` with `action = "optimize"`. These functions all have essentially the same calling structure, and seem to have about the same success as the various methods of `optim` in finding the global minimum of f, but may have various advantages of speed or alternative strategies depending on the problem. The reader is directed to the package vignette and help pages for details. As an example, let us find the values of (x_1, x_2) for which the function

$$f_2 = -\sin(x_1)\sin(x_2)\sin(x_1 + x_2).$$

is a minimum in the region around $(\pi/2, \pi/2)$.

```
> f2= function(x) -sin(x[1])*sin(x[2])*sin(x[1]+x[2])
> require(BB)
> spg(c(pi/2, pi/2), f2)
iter:  0  f-value:  -1.224647e-16  pgrad:  1
$par
[1] 1.047197 1.047197
$value
[1] -0.6495191
$gradient
[1] 1.89182e-06
$fn.reduction
[1] 0.6495191
$iter
[1] 6
$feval
[1] 7
$convergence
[1] 0
$message
[1] "Successful convergence"
```

7.3.4 *optimx() wrapper*

Given all of these—relatively similar—optimization routines, it has been deemed desirable to have a general-purpose "wrapper" function that calls other R optimization functions with a single calling usage. The function `optimx()` in the package of the same name serves that role. It calls all the methods discussed above in `optim()`, and also `spg()` from the BB package, `nlm()`, and `ucminf()`. The "Nelder–Mead" method is the default, and the "L-BFGS-B" method will be used automatically if upper and lower limits are supplied. For reasons discussed below, `optimx()` does not call the "SANN" simulated annealing method in `optim()`. It also does not call "Brent," because that is a one-dimensional optimization routine.

7.3.5 *Derivative-free optimization algorithms*

We conclude this section by noting two derivative-free optimization algorithms contained in the `dfoptim` package: `hk()` and `nmk()`. As stated in the Description of the package, "These algorithms do not require gradient information. ... They can also handle box constraints on parameters." In addition, `nmk()` and its bounded variant `nmkb()` can work with non-smooth functions. See the help pages for details and examples.

7.4 Optimization with constraints

It is often required to optimize an objective function (e.g., maximize yield or profit, or minimize cost) subject to constraints on the available resources. We shall briefly treat three aspects of this large topic in this chapter. Functions that may be nonlinear, but with linear constraints, are discussed in this section. Linear functions with linear constraints (linear programming) and quadratic functions with linear constraints (quadratic programming) are dealt with at the end of the chapter.

7.4.1 `constrOptim` to optimize functions with linear constraints

Base R has the function `constrOptim` to minimize a function with p unknown parameters subject to k linear inequality constraints. It uses `optim` to do most of the calculation, but adds a logarithmic barrier to enforce the constraints. `constrOptim` is called with the usage

```
constrOptim(theta, f, grad, ui, ci, mu = 1e-04, control = list(),
            method = if(is.null(grad)) "Nelder-Mead" else "BFGS",
            outer.iterations = 100, outer.eps = 1e-05, ...,
            hessian = FALSE)
```

where

- theta is a vector of the starting guesses for the p parameters;
- f is the function to be minimized;
- grad is either NULL or a function expressing the gradient of f;
- ui is the $k \times p$ constraint matrix;
- ci is the length k constraint vector;
- mu is a small multiplicative parameter that tunes the barrier term;
- the additional arguments are explained on the help page for `constrOptim`,

An example on the help page shows several ways to minimize the Rosenbrock banana function, which we have already encountered in our treatment of `optim`, but this time applying linear constraints.

```
## from optim
> fr = function(x) {
+     x1 = x[1]
+     x2 = x[2]
+     100 * (x2 - x1 * x1)^2 + (1 - x1)^2
+ }
> grr <- function(x) {  # gradient
+     x1 = x[1]
+     x2 = x[2]
+     c(-400 * x1 * (x2 - x1 * x1) - 2 * (1 - x1),
+        200 *       (x2 - x1 * x1))
+ }
```

We know that the minimum is at (1,1), so we first apply the constraints that both x1 and x2 must be ≤ 1 to test behavior of the function when the optimum is on the boundary. These constraints are expressed in matrix form as

$$\begin{pmatrix} 1 & 0 \\ 0 & 1 \end{pmatrix} \begin{pmatrix} x_1 \\ x_2 \end{pmatrix} \leq \begin{pmatrix} 1 \\ 1 \end{pmatrix} \tag{7.3}$$

The constraints must be of the form $lhs \geq 0$, so we rearrange the equation by moving the unit vector to the left-hand side of the equation, multiply both sides by -1, and thereby change the $\leq$ to a $\geq$, obtaining

$$\begin{pmatrix} -1 & 0 \\ 0 & -1 \end{pmatrix} \begin{pmatrix} x_1 \\ x_2 \end{pmatrix} - \begin{pmatrix} -1 \\ -1 \end{pmatrix} \geq \begin{pmatrix} 0 \\ 0 \end{pmatrix} \tag{7.4}$$

This gives the values for ui and ci in the function call below.

```
> constrOptim(c(-1.2,0.9), fr, grr, ui=rbind(c(-1,0),c(0,-1)),
   ci=c(-1,-1))
$par
[1] 0.9999761 0.9999521
$value
[1] 5.734115e-10
$counts
function gradient
     297       94
$convergence
[1] 0
$message
NULL
$outer.iterations
[1] 12
$barrier.value
[1] -0.0001999195
```

We can use constraints to find the optimum at locations away from the global minimum. For example, if we wish to find the minimum subject to $x_1 \leq 0.9$ and $x_2 - x_1 > 0.1$, a process similar to the above yields

$$\begin{pmatrix} -1 & 0 \\ 1 & -1 \end{pmatrix} \begin{pmatrix} x_1 \\ x_2 \end{pmatrix} - \begin{pmatrix} -0.9 \\ 0.1 \end{pmatrix} \geq \begin{pmatrix} 0 \\ 0 \end{pmatrix} \tag{7.5}$$

which leads to

```
> constrOptim(c(.5,0), fr, grr, ui=rbind(c(-1,0),c(1,-1)),
  ci=c(-0.9,0.1))
$par
[1] 0.8891335 0.7891335
$value
[1] 0.01249441
$counts
function gradient
```

```
      254        48
$convergence
[1] 0
$message
NULL
$outer.iterations
[1] 4
$barrier.value
[1] -7.399944e-05
```

7.4.2 External packages alabama and Rsolnp

R has two external packages, alabama and Rsolnp, that implement the augmented Lagrange multiplier method for general nonlinear optimization. Both are reliable and robust, and can handle equality and inequality constraints both linear and nonlinear. Their optimizer functions will take analytical expressions for the gradients and Hessians of the function to be minimized, and will run faster if these are provided; but we omit them, leaving the programs to employ numerical differentiation.

alabama has two functions for optimization, constrOptim.nl() and auglag(), which appear to be essentially identical. The important difference is that auglag() allows any initial vector of parameter values, even those that violate inequality constraints. constrOptim.nl(), on the other hand, requires that initial guesses be "feasible." Here we demonstrate constrOptim.nl() by seeking to minimize the function

$$\sin(x_1 * x_2 + x_3)$$

subject to the equality constraint

$$-x_1 x_2^3 + x_1^2 x_3^2 = 5$$

and the inequality constraints

$$x_1 \geq x_2 \geq x_3.$$

```
> f = function(x) sin(x[1]*x[2]+x[3])
> heq = function(x) -x[1]*x[2]^3 + x[1]^2*x[3]^2 -5
> hin = function(x) {
+   h = rep(NA,2)
+   h[1] = x[1]-x[2]
+   h[2] = x[2] -x[3]
+   h
+ }
```

The minimum value of the sine function is, of course, -1, but we pretend we don't know that; nor do we know the values of (x_1, x_2, x_3) that will result in that minimum. We make an initial guess and ask for the answer.

```
> p0 = c(3,2,1)
> require(alabama) # Also loads numDeriv package
```

```
> ans = constrOptim.nl(par=p0, fn = f, heq=heq, hin = hin)
Min(hin):  1 Max(abs(heq)):  20
Outer iteration:  1
Min(hin):  1 Max(abs(heq)):  20
par:  3 2 1
fval =   0.657

Outer iteration:  2
Min(hin):  0.1767693 Max(abs(heq)):  0.005111882
par:  3.21925 1.16012 0.983351
fval =   -1

...

Outer iteration:  6
Min(hin):  0.1758973 Max(abs(heq)):  1.515614e-06
par:  3.21885 1.15866 0.982768
fval =   -1

Outer iteration:  7
Min(hin):  0.1758972 Max(abs(heq)):  3.233966e-07
par:  3.21885 1.15866 0.982768
fval =   -1
```

If we had written p0 = c(1,2,3) we would have received an error message that "initial value violates inequality constraints." This message would not have appeared with auglag(). There are, of course, many (x_1, x_2, x_3) triplets that satisfy this minimization problem; which set is found depends on the initial guesses.

We now solve the same problem with the solnp() function in the Rsolnp package.

```
> f = function(x) -sin(x[1]*x[2]+x[3])
> heq = function(x) -x[1]*x[2]^3 + x[1]^2*x[3]^2 -5
> hin = function(x) {
+   h = rep(NA,2)
+   h[1] = x[1]-x[2]
+   h[2] = x[2] -x[3]
+   h
+ }
```

We load the Rsolnp package, which also loads the required truncnorm and parallel packages. solnp() asks for upper and lower bounds for the variables and the inequalities. It returns considerably more information about the solution than do the functions in alabama.

```
> require(Rsolnp)> upper = rep(5,3)
> lower = rep(0,3)
> p0 = c(3,2,1)
> ans = solnp(pars=p0, fun = f, eqfun=heq, ineqfun = hin, LB=lower,
```

```
+  UB=upper, ineqLB = c(0,0), ineqUB = c(5,5))

Iter: 1 fn: -1.0000  Pars:  3.14272 2.14793 1.10363
Iter: 2 fn: -1.0000  Pars:  3.14272 2.14793 1.10363
solnp--> Completed in 2 iterations
> ans
$pars
[1] 3.142718 2.147934 1.103631
$convergence
[1] 0
$values
[1] -0.6569866 -1.0000000 -1.0000000
$lagrange
              [,1]
[1,] -2.977098e-08
[2,]  4.690696e-08
$hessian
             [,1]         [,2]        [,3]        [,4]        [,5]
[1,]  0.998914483  0.003892931 -0.03114101 -0.05246969 -0.01155423
[2,]  0.003892931  0.964203357 -0.15306182 -0.17612761 -0.10200558
[3,] -0.031141014 -0.153061817  5.18027998  6.33407175  1.91315870
[4,] -0.052469692 -0.176127613  6.33407175 10.52298257  2.94535392
[5,] -0.011554233 -0.102005577  1.91315870  2.94535392  1.85308050
$ineqx0
[1] 0.9947835 1.0443029
$nfuneval
[1] 37
$outer.iter
[1] 2
$elapsed
Time difference of 0.04326391 secs
```

7.5 Global optimization with many local minima

If a function has many local minima, many starting values of `optim` may be required to find the global minimum, and success is not guaranteed. An example of a function with many local minima is the two-dimensional sum of two sinc ($\sin(\pi x)/\pi x$) functions (Figure 7.6).

```
> sinc = function(x) sin(pi*x)/(pi*x)
> x1 = x2 = seq(.1, 10, length=50)
> z = outer(x1,x2, FUN = function(x1,x2) sinc(x1)+sinc(x2))
> persp(x1,x2,z)
```

Actually, in this case we can find the minimum of sinc(x) along one axis:

```
> optimize(function (x) sinc(x), lower=.1, upper = 3)
$minimum
[1] 1.430279
```

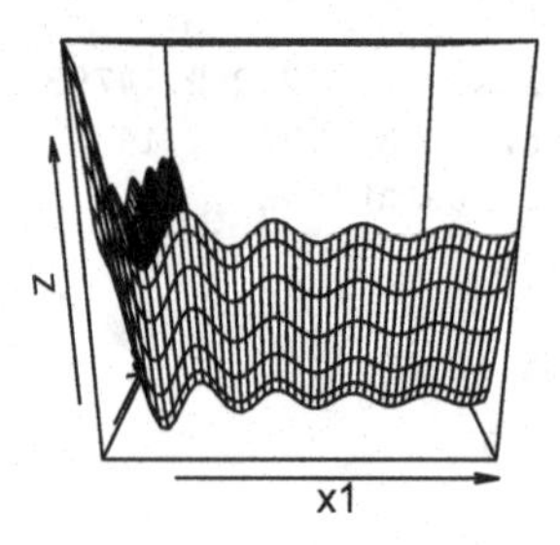

Figure 7.6: *Perspective plot of the sum of two sinc functions.*

```
$objective
[1] -0.2172336
```

so the minimum of the sum of two sinc functions should be near (1.43,1.43). Putting the function in vector form, pretending we don't know the answer, and using a plausible starting point with the default options of `optim()`, the function converges to a local minimum, but not the global minimum.

```
> f = function(x) sinc(x[1]) + sinc(x[2])
> optim(c(3,3),f)
$par
[1] 3.47089 3.47089
$value
[1] -0.1826504
$counts
function gradient
     107       NA
$convergence
[1] 0
$message
NULL
```

Two of the major approaches for optimizing functions with many local minima are simulated annealing and differential evolution.

7.5.1 Simulated annealing

The simulated annealing method is developed by analogy to annealing of a metal or alloy by slow cooling: at high T the material is in a highly disordered, high energy state. If cooled quickly, it will be trapped in that state, far from the orderly structure that gives it minimum free energy. The key is to cool slowly at first, so as to give the material enough energy to overcome local energy barriers as needed and proceed to the optimum without being trapped in local minima. As the temperature is lowered further, surmounting barriers becomes harder and excursions to higher energy states become rarer.

This behavior is encapsulated in the Boltzmann equation for the probability of a jump from a state of energy E_1 to a state of energy E_2, with change of energy $\triangle E = E_2 - E_1$:

$$P(E) = A\exp(-\triangle E/k_B T) \tag{7.6}$$

where k_B is the Boltzmann constant, T the temperature, and A a normalization constant. If $\triangle E > 0$, the system moves to a higher energy state with small but not zero probability. If $\triangle E < 0$, the system moves to a lower energy state with a higher probability. A jump to a higher energy state is harder if $\triangle E$ is larger or if T is smaller, but it is not impossible so long as $T > 0$. That is, annealing allows transitions to less probable states, which enables escaping from local minima.

This mechanism of thermal annealing is the basis for the simulated annealing algorithm for finding the optimum (minimum) state of some process or system. The system under examination may be either a continuous function or a finite set of states (e.g., how to minimize the total distance from point A to point B along a discrete set of city streets). We shall focus here on the continuous function problem. Starting at some point characterized by some quantity corresponding to energy E_1, the algorithm randomly chooses a nearby point corresponding to energy E_2. If $E_2 < E_1$, the energy is lowered by the move, which is therefore accepted. If $E_2 > E_1$, the probability $P(\triangle E)$ is calculated according to the Boltzmann equation and compared with a random number chosen from a uniform distribution between 0 and 1. If $P(\triangle E)$ is greater than that number the move is accepted; if less, it is rejected. This process requires a control parameter that is analogous to temperature, and generally starts at a very high T, such that $k_B T$ is much greater than the highest energy. As time (number of iterations) increases, T decreases, with the rate of decrease slowing with time.

`optim` contains the `''SANN''` method to implement simulated annealing, but it unfortunately is not adequate because it does not provide criteria for convergence; it simply prints `$convergence = 0` after executing `maxit` iterations. For this reason, `optimx` does not include `''SANN''` in its repertoire of optimization methods. Fortunately, however, the `GenSA` package contains an adequate implementation of the simulated annealing algorithm. For our two-dimensional sinc function, it yields the correct value and coordinates of the minimum:

```
> sinc = function(x) sin(pi*x)/(pi*x)
> sinc2D = function(x) sinc(x[1]) + sinc(x[2])
> library(GenSA)
> out = GenSA(c(1,1), sinc2D, lower = c(0,0), upper = c(10,10),
+   control = list(max.time=1))
> out[c("value", "par", "counts")]
$value
[1] -0.4344673
$par
[1] 1.430297 1.430297
$counts
```

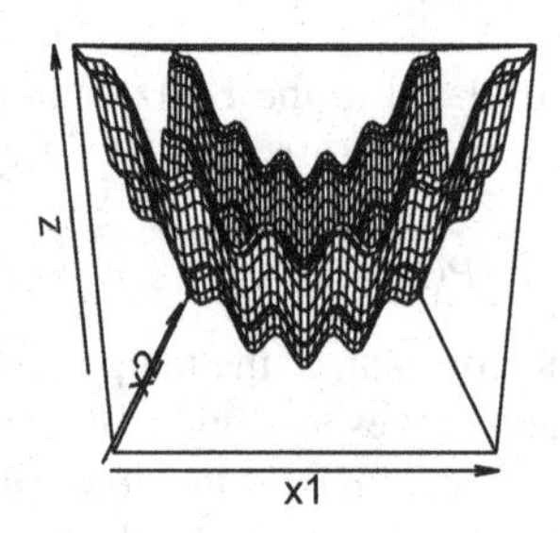

Figure 7.7: *Perspective plot of Equation 7.7.*

```
[1] 53285
```

The same results are obtained with max.time = 0.1, but with 10,257 counts.

As another example, consider the function (Figure 7.7)

$$\text{fsa}(x_1, x_2) = 0.2 + x_1^2 + x_2^2 - 0.1\cos(6\pi x_1) - 0.1\cos(6\pi x_2) \tag{7.7}$$

A perspective plot of this function

```
> fsa = function(x1,x2) 0.2 + x1^2 + x2^2 - 0.1*cos(6*pi*x1) -
+ 0.1*cos(6*pi*x2)
> x1 = x2 = seq(-1, 1, length=50)
> z = outer(x1,x2, fsa)
> par(mar=c(4,4,1.5,1.5),mex=.8,mgp=c(2,.5,0),tcl=0.3)
> persp(x1,x2,z)
```

yields the formidably bumpy plot shown in Figure 7.7.

Inspection of Equation 7.7 shows that the minimum is at (0,0) with value 0. `GenSA()` readily finds the correct result, though success with more complicated problems is not guaranteed.

```
> fsa = function(x) 0.2 + x[1]^2 + x[2]^2 - 0.1*cos(6*pi*x[1]) -
+ 0.1*cos(6*pi*x[2])
> library(GenSA)
> out = GenSA(c(1,1), fsa, lower = c(0,0), upper = c(10,10),
+   control = list(max.time=1))
> out[c("value", "par", "counts")]
$value
[1] 0
$par
[1] 0 0
$counts
[1] 31095
```

7.5.2 Genetic algorithms

Another approach that has proved successful for many complex, multi-dimensional optimization problems is one or another version of a genetic algorithm, which uses biological concepts such as mutation, crossover, and evolution to optimize fitness in a population. In the current context, "population" corresponds to a set of parameter vectors in a space whose dimension equals n, the number of parameters; "fitness" corresponds to the value of a scalar function that is to be minimized by choosing the best values of the parameters; "mutation" corresponds to random changes in the values of the parameters; "crossover" corresponds to exchanges of several components of a vector with those of another; and "evolution" corresponds to the gradual change of the population of parameter vectors as the function tends toward an optimum. In simple terms, the genetic algorithm approach mimics evolution to optimize a problem by maintaining a population of candidate solutions, creating new possible solutions by combining the existing ones according to some simple algorithm, and choosing that candidate solution which yields the optimum value of the function.

The genetic algorithm approach shares with simulated annealing the advantage that it can, by random trials, work its way out of local minima to find a global minimum. R has several useful contributed packages that implement some version of a genetic algorithm; three of these are `DEoptim`, `GA`, and `rgenoud`. We shall look at `DEoptim` in some detail, and briefly discuss the other two at the end of this section.

7.5.2.1 DEoptim

The "DE" in `DEoptim` stands for "differential evolution." The function, with the same name as the package, is called by the familiar set of arguments: `DEoptim(fn, lower, upper, control = DEoptim.control(), ..., fnMap=NULL)`. `fn` is the function to be minimized. Its first argument should be the vector of parameters to optimize, and it should return a real scalar result. `lower` and `upper` are vectors that specify the bounds on each parameter, with the ith component of each vector corresponding to the ith parameter. These vectors should encompass the full range of allowable values of the parameters.

`control` is a list of control parameters, discussed in the next paragraph. If `control` is not specified, its defaults are used. ... signifies further arguments, if any, to be passed to `fn`. `fnMap` is "an optional function that will be run after each population is created, but before the population is passed to the objective function. This allows the user to impose integer/cardinality constraints."

`DEoptim` will often run fine with the default control settings, but sometimes tweaks in those settings will lead to better results. Settings are modified with the `DEoptim.control` function, whose usage is

```
DEoptim.control(VTR = -Inf, strategy = 2, bs = FALSE, NP = NA,
  itermax = 200, CR = 0.5, F = 0.8, trace = TRUE, initialpop = NULL,
  storepopfrom = itermax + 1, storepopfreq = 1, p = 0.2, c = 0, reltol,
  steptol, parallelType = 0, packages = c(), parVar = c(),
  foreachArgs = list())
```

Some of the more commonly adjusted arguments of this function (see the help page for the full list and details) are:

- strategy: an integer between 1 and 6 (default = 2), specifying the way in which successive generations of candidate parameter vectors are chosen. We describe strategy = 2 in more detail below.
- NP: Number of population members. The default is 10×length(lower). Setting NP above 40 has been found empirically to not substantially improve convergence, regardless of the number of parameters.
- itermax: The maximum number of iterations allowed. The default is 200.
- CR: The crossover probability in the interval (0,1). Default is 0.5. However, it has been suggested that CR = 0.2 is a better choice if the parameters are substantially independent, while CR = 0.9 may be better if there is significant parameter dependence.
- F: Differential weighting factor (see strategy discussion below), with default = 0.8. However, according to the Differential Evolution Homepage (http://www1.icsi.berkeley.edu/ storn/code.html) "It has been found recently that selecting F from the interval [0.5, 1.0] randomly for each generation or for each difference vector, a technique called dither, improves convergence behaviour significantly, especially for noisy objective functions."

The default strategy = 2 replaces the classical random mutation strategy = 1 with the expression

$$v_{i,g} = old_{i,g} + (best_g - old_{i,g}) + x_{r0,g} + F * (x_{r1,g} - x_{r2,g})$$

where $old_{i,g}$ and $best_g$ are the ith member and best member, respectively, of the previous population. (i stands for the ith parameter, g for the gth generation.) See the Details section of the `DEoptim.control` help page for more information on available strategies.

As an example, we use the Rastrigin function, which is a common test function for multi-dimensional optimization. Inspection shows that its minimum is 0 at (0,0), with which the `DEoptim` calculation adequately agrees.

```
> fras = function(x) 10*length(x)+sum(x^2-10*cos(2*pi*x))
> require(DEoptim)
> optras = DEoptim(fras,lower=c(-5,-5),upper=c(5,5),
+   control=list(storepopfrom=1, trace=FALSE))
> optras$optim
$bestmem
         par1          par2
 1.161563e-06 -1.098620e-06


$bestval
[1] 5.071286e-10


$nfeval
```

```
[1] 402

$iter
[1] 200
```

7.5.2.2 *rgenoud*

The `genoud()` function in the `rgenoud` package combines genetic algorithm methods with a derivative-based, Newton-like method. The rationale is that over much of the landscape, a problem may be strongly nonlinear or even discontinuous: the sort of situation for which evolutionary methods were developed. On the other hand, near a solution many problems are regular and derivative information is useful. The approach and implementation are discussed at length, with examples, in the paper by Mebane and Sekhon (2011) and in the vignette[1] and help pages for the package. In this as in the other genetic algorithm methods, the population size and maximum number of generations are key variables in reaching an optimal solution, though of course there are trade-offs with computation time.

7.5.2.3 *GA*

The `GA` package developed by Scrucca (2012) is intended as "a flexible, general-purpose R package for implementing genetic algorithms search in both the continuous and discrete case, whether constrained or not." Scrucca continues, "Users can easily define their own objective function depending on the problem at hand. Several genetic operators are available and can be combined to explore the best settings for the current task. Furthermore, users can define new genetic operators and easily evaluate their performances." Interested readers should consult the package and its documentation[2] for details and examples.

7.6 Linear and quadratic programming

7.6.1 *Linear programming*

If the objective function is a linear function of the resources, and the constraints are linear, this is the domain of linear programming. There are several packages in R that can handle these kinds of problems very well, notably `lpSolve`, `Rglpk`,and `Rsymphony`. They all integrate existing free libraries into R that are well tested and can solve small to mid-sized problems, i.e., with several hundred variables, very efficiently.

As all three packages are comparable in power and have similar interfaces, package `lpSolve` will be used here to explain how to apply a linear programming solver to this problem class. The solver function `lp()` from `lpSolve` uses the Simplex algorithm of George B. Dantzig (1947) and will not only solve linear programs, but also *mixed-integer linear programs* (see next section).

[1] http://cran.r-project.org/web/packages/rgenoud/vignettes/rgenoud.pdf

[2] http://CRAN.R-project.org/package=GA

The syntax for calling function `lp()` in `lpSolve` is as follows:

```
lp(direction = "min", objective.in, const.mat, const.dir, const.rhs,
    transpose.constraints = TRUE, int.vec, presolve=0, compute.sens=0,
    binary.vec, all.int=FALSE, all.bin=FALSE, scale = 196, dense.const,
    num.bin.solns=1, use.rw=FALSE )
```

with

- `direction` is `"min"` or `"max"` for a minimization or maximization problem;
- `objective.in` is the linear objective function to be optimized, represented through a vector of coefficients;
- `const.mat` is the matrix of all coefficients of linear constraints, one row per inequality, one column per variable;
- `const.dir` a character vector giving the direction of the constraint, for each inequality should be one of "<", "<=","=","==", ">", or ">="
- `const.rhs` vector of constant values on the right-hand side of the inequalities.

These are the most important arguments. `int.vec` and `binary.vec` will be explained later. For the other, optional arguments see the help page of function `lp()`.

Suppose, for example, that a chemical engineer wants to maximize the revenue of a process by producing a mixture of products 1 and 2 from feedstocks A, B, and C. Product 1 yields 500 euros per unit, and product 2 yields 400 euros per unit. Feedstocks 1, 2, and 3 are available in quantities 100, 50, and 60, respectively. To make product 1 requires 20 units of A, 5 units of B, and 15 units of C; while the corresponding requirements for product 2 are 20, 30, and 7. How much of products 1 and 2 should be produced to maximize revenue, given the resource limitations and requirements?

We will translate the problem into a mathematical formulation. If m_1 and m_2 are the amounts produced of products 1 and 2, then the constraints of the problem are the amounts available of feedstocks 1, 2, and 3, that is

$$\begin{aligned} 20m_1 + 20m_2 &\leq 100 \\ 5m_1 + 30m_2 &\leq 50 \\ 15m_1 + 7m_2 &\leq 60 \end{aligned}$$

and the profit is $500m_1 + 400m_2$, where naturally $m_1, m_2 \geq 0$. For solving the problem, these equations and inequalities have to be defined in R as vectors and matrices.

```
> obj = c(500, 400)               # objective function 500*m1+400m2
> mat = matrix(c(20, 20,          # constraint matrix
+                  5, 30,
+                 15, 7), nrow=3, ncol=2, byrow=TRUE)
> dir = rep("<=", 3)              # direction of inequalities
> rhs = c(100, 50, 60)            # right hand side of inequalities
```

The fact that the variables have to be greater than 0 is always inplicitly assumed and does not need to be stated. Now we can easily solve this linear programming problem.

```
> require(lpSolve)
> soln = lp("max", obj, mat, dir, rhs)
> soln
Success: the objective function is 2180.723
> soln$solution
[1] 3.493976 1.084337
```

Of the first product about 3.494 units should be produced, and of the second about 1.084 units. To see that the constraints are respected, multiply the constraint matrix with the solution vector:

```
> mat %*% soln$solution
         [,1]
[1,] 91.56627
[2,] 50.00000
[3,] 60.00000
```

Therefore, feedstocks 2 and 3 have been completely used up.

For smaller linear programming problems there is another package, `linprog`, that is by far not as powerful as `lpSolve`, but returns more information about the problem. Applied to the problem above we see

```
> require(linprog)
> solveLP(obj, rhs, mat, maximum = TRUE)

Results of Linear Programming / Linear Optimization

Objective function (Maximum): 2180.72

Iterations in phase 1: 0
Iterations in phase 2: 2
Solution
      opt
1 3.49398
2 1.08434

Basic Variables
        opt
1   3.49398
2   1.08434
S 1 8.43373

Constraints
   actual dir bvec    free    dual dual.reg
1 91.5663  <=  100 8.43373  0.0000  8.43373
2 50.0000  <=   50 0.00000  6.0241 30.00000
3 60.0000  <=   60 0.00000 31.3253 48.33333
```

```
All Variables (including slack variables)
         opt cvec    min.c     max.c      marg marg.reg
1    3.49398  500  66.6667  857.1429        NA       NA
2    1.08434  400 233.3333 3000.0000        NA       NA
S 1 8.43373    0       NA   15.6250   0.0000       NA
S 2 0.00000    0     -Inf    6.0241  -6.0241  30.0000
S 3 0.00000    0     -Inf   31.3253 -31.3253  48.3333
```

Of interest here are the values of the dual variables (under "dual") for the constraints. These values indicate how much the objective will change if the constraints are changed by one unit. We see that increasing the stock in feedstock 3 is the most profitable action one can take:

```
> rhs = c(100, 50, 61)
> lp("max", obj, mat, dir, rhs)
Success: the objective function is 2212.048
```

and the increase 31.325 in profit is indeed exactly what is predicted by the dual variable shown in the output above.

7.6.2 *Quadratic programming*

A less frequently encountered—but still important—type of mathematical optimization problem is quadratic programming, which seeks to optimize a quadratic function of several variables subject to linear constraints on these variables. That is, we wish to minimize the function

$$f(x) = -d^T x + \frac{1}{2} x^T D x$$

subject to the constraints

$$A^T x \geq x_0.$$

where D is a matrix of quadratic coefficients, d a vector of linear coefficients, A a matrix of constraints, and x_0 a vector of constraint values. Quadratic programming can be handled in R by the function `solve.QP` of the package `quadprog`.

Consider, for example, the function

$$f(x) = \frac{1}{2} x_1^2 + x_2^2 - x_1 x_2 - 2x_1 - 6x_2$$

with the constraints

$$\begin{aligned} x_1 + x_2 &\leq 2 \\ -x_1 + 2x_2 &\leq 2 \\ 2x_1 + x_2 &\leq 3 \\ 0 \leq x_1, 0 &\leq x_2. \end{aligned}$$

These equations and inequalities lead to the following formulation and solution.

```
> require(quadprog)
> Dmat = matrix(c(1,-1,-1,2),2,2)
> dvec = c(2,6)
> Amat = matrix(c(-1,-1,1,-2,-2,-1),2,3)
> bvec = c(-2,-2,-3)
>
> solve.QP(Dmat, dvec, Amat, bvec)
$solution
[1] 0.6666667 1.3333333
$value
[1] -8.222222
$unconstrained.solution
[1] 10  8
$iterations
[1] 3 0
$Lagrangian
[1] 3.1111111 0.4444444 0.0000000
$iact
[1] 1 2
```

Quadratic programming plays an important role in *geometric optimization* problems, such as the "enclosing ball problem" and "polytope distance problem," or the "separating hyperplane problem" that is essential for the technique of support vector machines in machine learning.

Here we will look at finding a smallest circle in R^2 enclosing ten given points $p_1, \ldots, p_{10}$. (Figure 7.8) Let the points be $p_1 = (0.30, 0.21)$, etc., represented as columns in the following matrix.

```
> C = matrix(
+     c(0.30, 0.08, 0.30, 0.99, 0.31, 0.77, 0.23, 0.29, 0.92, 0.14,
+     0.21, 0.93, 0.48, 0.83, 0.69, 0.91, 0.35, 0.05, 0.03, 0.19),
+     nrow = 2, ncol = 10, byrow = TRUE)
```

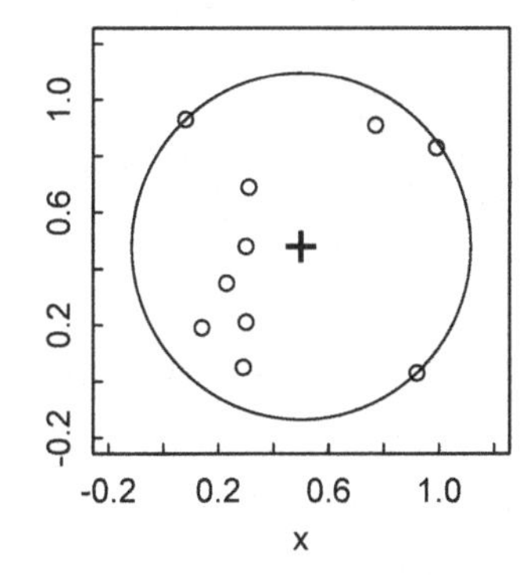

Figure 7.8: *Plot of smallest enclosing circle for ten points.*

The theory of geometric optimization says we need to solve the following quadratic programming problem: Minimize the quadratic form

$$x^T C^T Cx - \sum p_i^T p_i x_i$$

with the constraints $\sum x_i = 1$ and all $x_i \geq 0$. Then the point $p = \sum p_i x_i$ is the center of the ball, and the negative of the minimum value is the square of the radius.

Unfortunately, the matrix $C^T C$ is not positive definite. (Look at matrix `D` below and compute its eigenvalues with `eigen(D)`: eight of the ten eigenvalues are zero.) Thus function `solve.QP()` cannot be applied. We will instead turn to `ipop()` in package `kernlab`, an often used package and function.

For solving with `ipop()` we need to define `D`, `d`, `A`, and `b`.

```
> D = 2 * t(C) %*% C                        # D = 1/2 C' C
> d = apply(C^2, 2, sum)                    # d = (p1' p1, )
> A = matrix(rep(1, 10), 1, 10)             # sum xi = 1
> b = 1; r = 0                              # b <= A x <= b + r
> l = rep(0, 10); u = rep(1, 10)            # l <= x <= u
```

ipop requires explicit lower and upper bounds for the variables as well as for the inequalities. We can safely assume $x_i \leq 1$ because the x_i are positive and their sum is 1. Now everything is in place to compute a solution.

```
> require(kernlab)
> sol = ipop(-d, D, A, b, l, u, r)
> x = sol@primal
> sum(x)
[1] 1
```

The center of the ball will be at $p_0 = \sum x_i p_i$, that is

```
> p0 = C %*% x; p0
          [,1]
[1,] 0.4495151
[2,] 0.4029843
```

To see the distance of all points to the proposed center, do the following:

```
# Euclidean distance between p0 and all pi
> e = sqrt(colSums((C - c(p0))^2)); e
[1] 0.3360061 0.6155485 0.2000001 0.6021627 0.2831960
[5] 0.5077400 0.2996666 0.4785395 0.6155485 0.4622771
> r0 = max(e); r0
[1] 0.784569
```

To be convinced, we lay out the whole situation in a scatterplot.

```
> plot(C[1, ], C[2, ], xlim = c(-0.2, 1.2), ylim = c(-0.2, 1.2),
+       xlab = "x", ylab = "", asp = 1)
> grid()
# Draw the center of the circle
> points(p0[1], p0[2], pch = "+", cex = 2)
```

```
# Draw a circle with radius r0
> th = seq(0, 2*pi, length.out = 100)
> xc = p0[1] + r0 * cos(th)
> yc = p0[2] + r0 * sin(th)
> lines(xc, yc)
```

The circle cannot be made smaller because two points lie on the boundary on opposite sides.

7.7 Mixed-integer linear programming

7.7.1 *Mixed-integer problems*

Imagine we need to solve the following linear programming task: Maximize

$$y = 500x_1 + 450x_2$$

subject to

$$\begin{aligned} 6x_1 + 5x_2 &\leq 60 \\ 10x_1 + 20x_2 &\leq 150 \\ x_1 &\leq 8 \end{aligned}$$

and $x_1, x_2 \geq 0$, but the variables x_1, x_2 need to be integers, for example because we cannot break our product in smaller parts. If this problem is solved with `lpSolve`, the solution will be $x_1 = 6\frac{3}{7}, x_2 = 4\frac{2}{7}$ (with $y = 5142\frac{6}{7}$). Taking the integer parts, will $x_1 = 6, x_2 = 4$ (with $y = 4800$) be the optimal solution in integers?

Fortunately, `lpSolve`, as well as all the other LP solvers mentioned in the previous section, are capable of solving *mixed-integer linear programs* where some or all of the variables have to be binary (taking on only 0 or 1 as values) or integers. The user has to explicitly declare which of the variables fall into these classes; all others are still assumed to be positive reals.

```
> obj = c(500, 450)
> A = matrix(c( 6,  5,
+              10, 20,
+               1,  0), ncol = 2, byrow = TRUE)
> b = c(60, 150, 8)

# Declare which variables have to be integer (here all of them)
> int.vec = c(1, 2)

> soln = lp("max", obj, A, rep("<=", 3), b, int.vec = int.vec)
> soln; soln$solution
Success: the objective function is 4900
[1] 8 2
```

One can see that the solution $(8,2)$ is better than the nearest integer solution $(6,4)$, and is not even a direct "neighbor" in the integer grid. That means guessing from the unconstrained, real solution is not an appropriate step.

As another example, let us assume that x_1 and x_2 are allowed to be real, but the second variable is *semi-continuous*, that is either $x_2 = 0$ or $x_2 \geq 3$; it is not allowed to take on values between 0 and 3. This constraint cannot be expressed as a linear constraint. Instead, we introduce a new binary variable b that is assumed to be 0 if x_2 is zero, and 1 if not.

So $x_2 \geq 3b$ will guarantee that x_3, if not 0, is greater than 3. Of course, if $b = 0$ then x_2 shall be 0, too. This can be expressed with another constraint $x_2 \leq Mb$ where M is some large enough constant, e.g., we can assume $M = 20$. (In the literature on MILP, this is called the "big-M" trick.)

With variables x_1, x_2, b the set of constraints now look like

$$\begin{aligned}
6x_1 + 5x_2 &\leq 60 \\
10x_1 + 20x_2 &\leq 150 \\
x_1 &\leq 8 \\
-x_2 + 3b &\leq 0 \\
x_2 - 20b &\leq 0
\end{aligned}$$

and thus

```
> obj = c(500, 400, 0)
> A = matrix(c( 6,  5,   0,
+              10, 20,   0,
+               1,  0,   0,
+               0, -1,   3,
+               0,  1, -20), ncol = 3, byrow = TRUE)
> b = c(60, 150, 8, 0, 0)
> int.vec = c(3)
> soln = lp("max", obj, A, rep("<=", 5), b, int.vec = int.vec)
> soln; soln$solution
Success: the objective function is 4950
[1] 7.5 3.0 1.0
```

Semi-continuous variables and the "big-M" trick are often-used elements of modeling and solving linear programming tasks.

7.7.2 *Integer programming problems*

If all variables are assumed to be binary or integer, one speaks about *integer programming* problems. Here are some problems in discrete optimization that can be solved applying an integer programming approach.

7.7.2.1 *Knapsack problems*

Even the famous class of *knapsack problems* can be solved using integer programming. As an example, a peddler wants to fill a knapsack of weight capacity 105 with items that have the following values and weights, so as to maximize his profit:

```
> v = c(15, 100, 90, 60, 40, 15, 10,  1)     # value of items
> w = c( 2,  20, 20, 30, 40, 30, 60, 10)     # weight of items
> C = 105                                    # maximum capacity

> M = matrix(w, nrow=1)
> L = lp("max", v, M, "<=", 105, binary.vec = 1:8)
> L
Success: the objective function is 280

> (inds = which(L$solution == 1))     # solution indices
[1] 1 2 3 4 6
> sum(v[inds])                        # total value
[1] 280
> sum(w[inds])                        # used capacity
[1] 102
```

With `binary.vec = 1:8` all variables are forced to be binary. If the i-th variable is 1, it gets packed into the knapsack, otherwise not. `inds` represents all indices of items getting packed, so items 5, 7, and 8 will not be taken along.

7.7.2.2 *Transportation problems*

Consider a transportation problem with two origins and three destinations, for example three markets that have to be supplied from two production sites. The supplies at sites are 300 and 200 units, respectively, while at the markets the demands for this product are 150, 250, and 100, respectively. The transportation costs per unit are shown in the cost matrix C:

```
> C = matrix(c(10, 70, 40,
+              20, 30, 50), nrow = 2, byrow=TRUE)
```

where `C[i, j]` represents the cost to deliver one unit of this product from the i-th site to the j-th market.

We could attempt to model the problem with binary variables, but fortunately, `lpSolve` provides function `lp.transport()` to solve this problem for us. What is needed is a formulation of the constraints given above. For the production sites the sums per row (i.e., per site) are smaller or equal to 300 and 200,

```
> row.dir = rep("<=", 2)
> row.rhs = c(300, 200)
```

while the requests per column sum (i.e., per market) are greater or equal to 150, 250, and 100,

```
> col.dir = rep(">=", 3)
> col.rhs = c(150, 250, 100)
```

This is enough information for `lp.transport()` to get started:

```
> require(lpSolve)
> T = lp.transport(C, "min", row.dir, row.rhs, col.dir, col.rhs)
> T
Success: the objective function is 15000
> T$solution
     [,1] [,2] [,3]
[1,]  150   50  100
[2,]    0  200    0
```

The solution states that 150 units shall be brought from site 1 to market 1, 50 to market 2, and 100 to market 3, while 200 units shall be brought from site 2 to market 2. The total cost can be calculated explicitly with

```
> sum(C * T$solution)
[1] 15000
```

7.7.2.3 Assignment problems

Assume there are five machines to be assigned five jobs. The numbers in the following matrix M_{ij} indicate the costs for doing job j on machine i. How to assign jobs to machines to minimize costs? The matrix M may be given as

```
> M <- matrix(c(NA,  8,  6, 12,  1,
+               15, 12,  7, NA, 10,
+               10, NA,  5, 14, NA,
+               12, NA, 12, 16, 15,
+               18, 17, 14, NA, 13), 5, 5, byrow = TRUE)
```

where `NA` denotes that this job is not allowed on this machine. To make the matrix M fully numerical, replace all `NA`s by a large value: 100 should be enough.

```
> M[is.na(M)] = 100
```

Again, we could set up a model using binary variables to solve the problem. The function `lp.assign` in `lpSolve` does this for us automatically, thus

```
> A = lp.assign(M)
> A
Success: the objective function is 51
> A$solution
     [,1] [,2] [,3] [,4] [,5]
[1,]    0    0    0    0    1
[2,]    0    0    1    0    0
[3,]    0    0    0    1    0
[4,]    1    0    0    0    0
[5,]    0    1    0    0    0
```

Job 1, for instance, will be assigned to machine 4, job 2 to machine 5, etc., with a total cost of 51 for all five jobs together.

7.7.2.4 Subsetsum problems

The *subsetsum problem* can be formulated as follows: Given a set of (positive) numbers, is there a subset such that the sum of numbers in this subset equals a prescribed value? We will also require that the number of elements in this subset is fixed.

During a shopping trip, a man has bought 30 items with the following known prices (identified by price tags stuck to them) in, say, euros:

```
> p = c(99.28,  5.79, 63.31, 89.36,  7.63, 30.77, 23.54, 84.24,
        93.29, 53.47, 88.19, 91.49, 34.46, 52.13, 43.09, 76.40,
        21.42, 63.64, 28.79, 73.03, 8.29, 92.06, 26.69, 89.07,
        10.03, 10.24, 40.29, 81.76, 49.01,  3.85)
```

He inspects a receipt totaling 200.10 euros for four items, but not indicating which ones or for what single prices they were sold. Can he find out which items are covered by this receipt? (If a reader thinks this should be easy, he is invited to find out by hand.)

To avoid problems with comparing floating point numbers for equality, we will convert prices to cents, so all numbers are integers.

```
P = as.integer(100*p)
```

We introduce 30 binary variables $b = (b_i)$ indicating whether the i-th item belongs to the solution or not. There are two inequalities describing the problem:

$$\sum_{i=1}^{30} b_i = 4$$

$$\sum_{i=1}^{30} b_i P_i \leq 20010$$

the first one saying we want exactly four items, the second that the price for these four items shall be below 200.10 euros, but as close as possible. The reason is that a linear programming solver will maximize a linear function, not exactly reaching a certain value. But if the maximal value found by the solver is less than 20010, we know for sure that there are no four item prices exactly summing up to this value.

The objective function is also $\sum b_i P_i$; that is, we *use the same linear function as objective and as inequality*. Putting all the pieces together:

```
> obj = P
> M = rbind(rep(1, 30), P)
> dir = c("==", "<=")
> rhs = c(4, 20010)
> binary.vec = 1:40

> require(lpSolve)
> (L = lp("max", obj, M, dir, rhs, binary.vec = binary.vec))
Success: the objective function is 20010
```

This shows that there are four items whose prices add up to 200.10 euros, identifying their indices and single prices with

```
> inds = which(L$solution == 1)
> inds; P[inds]/100; sum(P[inds])/100
[1]  7  9 20 26
[1] 23.54 93.29 73.03 10.24
[1] 200.1
```

Note that there may be more than one combination of four prices that yield this total, because an optimization solver stops when it has found a solution, it does not try to produce all of the possible solutions. To make sure, remove one of those indices, for example 26, and try to find another combination of four prices with this same sum. In the following, we set `P[26]` to some high value, e.g., 21000, so it will not be a candidate for the sum. Then repeat the procedure from above.

```
> i = 26
> Q = P
> Q[i] = 21000
> N = rbind(rep(1, 30), Q)
> LL = lp("max", obj, N, dir, rhs, binary.vec = binary.vec)
> inds = which(LL$solution == 1)
> inds; Q[inds]/100; sum(Q[inds])/100
[1]  6 16 24 30
[1] 30.77 76.40 89.07  3.85
[1] 200.09
```

Without item 26 there is no subset of four prices adding up to 200.10. We try indices 20 or 9 (7 has not to be checked then), with the same negative result. Therefore we can safely conclude that exactly the goods with prices 23.54, 93.29, 73.03, and 10.24 are covered on this receipt!

7.8 Case study

7.8.1 *Monte Carlo simulation of the 2D Ising model*

The Ising model, named after the German physicist Ernst Ising, was developed as a statistical mechanical model of ferromagnetism. In its most common form it consists of discrete variables (usually ± 1) that represent spins arranged on a lattice. Each spin can interact with its neighbors with an interaction energy J (taken = 1 in the calculation that follows). The 2D Ising model on a square lattice is one of the simplest statistical mechanical models to display a phase transition. (The 1D model does not, though it still has many interesting properties.)

Lars Onsager developed an analytical solution to the 2D Ising model with periodic boundary conditions in 1942, an achievement acclaimed as one of the high points of theoretical physics in modern times. For our purposes in this chapter, the 2D Ising model provides a useful example of an optimization problem: i.e., finding the equilibrium position (minimum free energy) of an interacting spin system at a given temperature. It thereby has some similarity with the simulated annealing (SANN) method.

In developing R code to analyze the 2D Ising model, we follow the Fortran treatment of Larrimore[3] who in turn adapted the Basic code in Chandler (1987). The treatment implements importance sampling via the Metropolis algorithm. The procedure has three parts:

- Choose a spin randomly, and flip it
- Calculate the energy difference $\triangle E$
- If $\triangle E < 0$, the new state is favorable and is accepted. Otherwise a random number p between 0 and 1 is generated, and if $\exp(-\beta \triangle E) > p$, the new state is accepted. ($\beta = 1/k_B T$, where k_B is the Boltzmann constant and T the Kelvin temperature.) Equivalently, taking logs of both sides, the move is accepted if $-\beta \triangle E) > \log(p)$.

The computationally intensive part of this process comes from having to repeat these steps enough times that equilibrium should have been achieved, then repeat many more times to get adequate statistical sampling at the given T. For the rather small (12×12) lattice considered below, equilibration requires about 10^5 steps.

After a suitable number of equilibration and statistical averaging steps have been taken, we can calculate the average energy $\langle E \rangle$ and magnetization $\langle M \rangle$ directly by summing over the lattice. By calculating the mean-square energy $\left\langle E^2 \right\rangle$ and magnetization $\left\langle M^2 \right\rangle$, we use the thermodynamic relations

$$C_V = \frac{\beta}{T}\left[\left\langle E^2 \right\rangle - \langle E \rangle^2\right] \tag{7.8}$$

and

$$\chi = \beta\left[\left\langle M^2 \right\rangle - \langle M \rangle^2\right] \tag{7.9}$$

to calculate the heat capacity C_V and magnetic susceptibility χ from the variances of E and M.

We begin by setting the parameters of the simulation. Set up a square lattice (matrix) of spins **A** with `nr` rows and `nc` columns. The calculation runs rather slowly in R (for greater speed use Fortran or C) so we use a small 12×12 lattice. An even number of rows and columns assures consistency with the periodic boundary conditions. At each temperature we make `npass` random choices of a spin, flipping each and testing the energy of the resulting configuration against the Metropolis criterion. After the first `nequil` of these passes it is assumed that the system has reached equilibrium, so the remaining passes are used to obtain averages of the energy, magnetization, and their squares.

```
> nr = 12; nc = 12 # Number of rows and columns
> A = matrix(nrow = nr, ncol = nc)
> npass = 2e5 # Number of passes for each temperature
> nequil = 1e5 # Number of equilibration steps for each T
```

Set the upper and lower temperatures to be scanned, and the interval between them, thus determining the number of temperatures to be scanned. It is known that

[3] http://fraden.brandeis.edu/courses/phys39/simulations/Student%20Ising%20Swarthmore.pdf

the phase transition temperature (critical temperature) for this model is near $T_C \approx 2.3$, so we choose limits that bracket this value.

```
> T_hi = 3 # Temperature to start scan at
> T_lo = 1.5 # Temperature to finish scan at
> dT = 0.1 # Temperature scanning interval
> nscans = as.integer((T_hi - T_lo)/dT) + 1
```

Set up a table (matrix **M**) to accept the results at the end of each temperature scan.

```
> # Initialize results table
> M = matrix(nrow = nscans, ncol = 5, byrow=TRUE,
+   dimnames=list(rep("",nscans),c("T","E_av","Cv","Mag_av",
+   "Mag_sus")))
```

Construct a function `Ann(A,m,n)` that defines the nearest neighbors of the (m,n) spin in **A**, with special provision for edges to accommodate periodic boundary conditions.

```
> Ann = function(A, m, n) {
+       if (m == nr) Ann1 = A[1,n] else Ann1 = A[m+1,n] # bottom
+       if (n == 1) Ann2 = A[m,nc] else Ann2 = A[m,n-1] # left
+       if (m == 1) Ann3 = A[nr,n] else Ann3 = A[m-1,n] # top
+       if (n == nc) Ann4 = A[m,1] else Ann4 = A[m,n+1] # right
+    return(Ann1 + Ann2 + Ann3 + Ann4)
+    }
```

At each temperature we start the calculation anew: Initialize the variables in units chosen so that the thermal energy βT equals one, as does the spin interaction energy J (which therefore doesn't appear explicitly in the calculation). Set the energy and magnetization to zero.

```
> for (isc in 1:nscans) {  # T scan loop
+    temp = T_hi - dT*(isc - 1)
+    # Initialize variables
+    beta = 1/temp
+    oc = 0 # output count
+    E_av = 0
+    E2_av = 0
+    mag_av = 0
+    mag2_av = 0
```

Set up the lattice in a checkerboard configuration with alternating spins pointing up and down. (Other initial configurations are possible, but yield the same equilibrium results.)

```
+    # Set up initial checkerboard spin configuration
+    A[1,1] = 1
+    for (i in 1:(nr - 1)) A[i+1,1] = -A[i,1]
+    for (j in 1:(nc - 1)) A[,j+1] = -A[,j]
```

Begin passes at each temperature, using the Metropolis algorithm to accept or reject a trial. The first nequil passes are equilibration steps. The remainder are used to accumulate statistics on energy and magnetization.

```
+    for (ipass in 0:npass) { # Monte Carlo passes at T
+       if (ipass > nequil) {
+          oc = oc + 1 # output count
+          mag = sum(A)/(nr*nc)
+          mag_av = mag_av + mag
+          mag2_av = mag2_av + mag^2
+          E = 0
+          for (m in 1:nr) {
+             for (n in 1:nc) {
+                E = E - A[m,n]*Ann(A,m,n)
+             }
+          }
+          E = E/(2*nr*nc)
+          E_av = E_av + E
+          E2_av = E2_av + E^2
+       }
+     # Choose a random spin to change
+     m = sample(nr,1,replace=TRUE)
+     n = sample(nc,1,replace=TRUE)
+     ts = -A[m,n] # Flip sign of spin
+     dU = -2*ts*Ann(A,m,n)
+     log_eta = log(runif(1))
+     if(-beta*dU > log_eta) A[m,n] = ts
+    } # end MC passes at T
```

Fill a row with the temperature, energy, and magnetization results at that temperature.

```
+    M[isc,1] = temp
+    M[isc,2] = E_av/oc
+    M[isc,3] = beta^2*(E2_av/oc - (E_av/oc)^2)
+    M[isc,4] = abs(mag_av/oc)
+    M[isc,5] = beta*(mag2_av/oc - (mag_av/oc)^2)
+    cat(c(temp, mag_av,mag2_av,E_av,E2_av),"\n") # not shown
+ } # end T scans
```

Print and plot the results.

```
> M # print result (deleted from output)
> # plot results
> par(mar=c(4,4,1.5,1.5),mex=.8,mgp=c(2,.5,0),tcl=0.3)
> par(mfrow=c(2,2))
> plot(M[,1], M[,2], xlab="T", ylab="<E>")
> plot(M[,1], M[,3], xlab="T", ylab="<Cv>")
> plot(M[,1], M[,4], xlab="T", ylab="<M>")
```

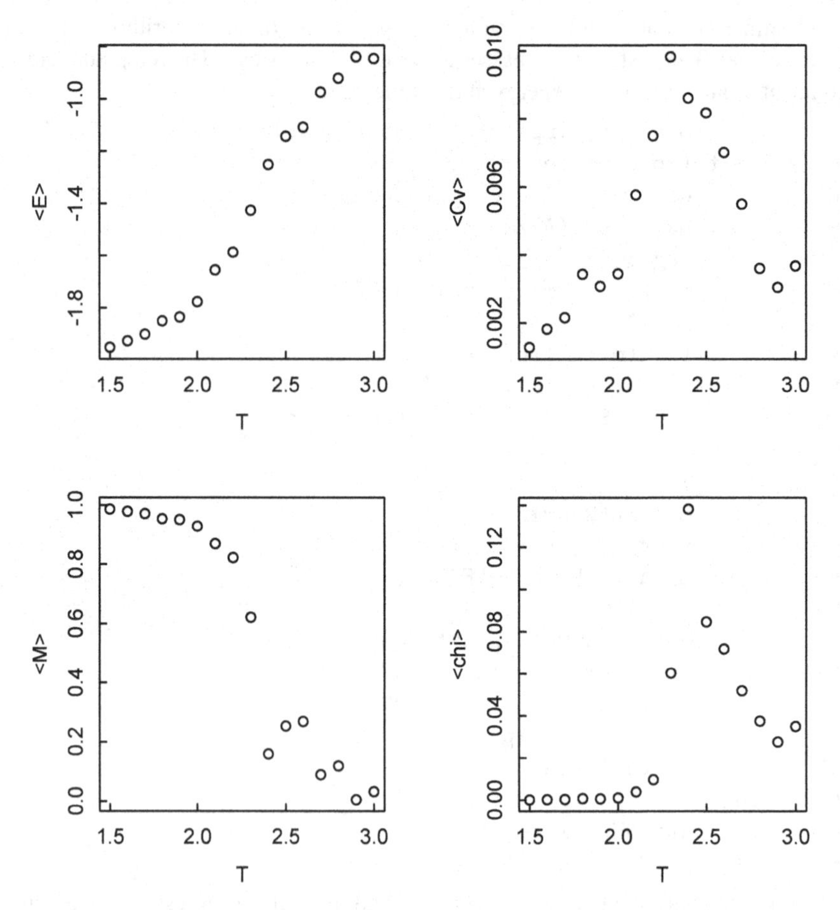

Figure 7.9: *Plots of thermodynamic and magnetic functions for 2D Ising model.*

```
> plot(M[,1], M[,5], xlab="T", ylab="<chi>")
```

It took nearly 38 minutes to complete the scan of 16 temperatures on a 12×12 array, with 2×10^5 passes at each T, of which 1×10^5 are equilibration steps. It takes about 10^5 steps to show the characteristic behavior of the thermodynamic functions without undue noise.

The results are as expected (Figure 7.9): The energy increases with T, while the magnetization decreases sharply near the critical temperature T_C and tends to zero above that temperature. The specific heat and magnetic susceptibility, both being related to fluctuations in the system, have peaks around T_C.

Chapter 8

Ordinary differential equations

Differential equations are ubiquitous in science and engineering, since they describe the rate of change of a system with time, position, or some other independent variable. It is conventional to classify differential equations according to certain characteristics.

Ordinary differential equations (ODEs) depend on a single independent variable, such as time; while partial differential equations (PDEs) depend on several independent variables, such as spatial coordinates as well as, perhaps, time. In this chapter we treat ODEs; PDEs are the subject of the next chapter.

First order differential equations involve only first derivatives of the dependent variables, while second and higher order differential equations involve second and higher order derivatives. In numerical solution of differential equations, all equations are reduced to first order by the expedient of defining, e.g., $dy/dt = y_1$ and then $d^2y/dt^2 = dy_1/dt$.

Initial value problems define the starting values of the variables, while boundary value problems specify the beginning and ending values of the variables.

Linear differential equations are linear in the dependent variables, while nonlinear differential equations involve higher or nonintegral powers. Most analytically soluble differential equations are linear, but numerical solutions can cope equally well with nonlinear equations.

R, through contributed packages, has powerful tools to numerically solve differential equations. The DifferentialEquations task view at `cran.r-project.org/web/views/DifferentialEquations.html` provides a useful overview. These are some of the most important packages:

- `deSolve` provides "functions that solve initial value problems of a system of first-order ordinary differential equations (ODE), of partial differential equations (PDE), of differential algebraic equations (DAE), and of delay differential equations. The functions provide an interface to the FORTRAN functions `lsoda`, `lsodar`, `lsode`, `lsodes` of the ODEPACK collection, to the FORTRAN functions `dvode` and `daspk` and a C-implementation of solvers of the Runge–Kutta family with fixed or variable time steps. The package contains routines designed for solving ODEs resulting from 1-D, 2-D and 3-D partial differential equations (PDE) that have been converted to ODEs by numerical differencing." The vignette "Package `deSolve`: Solving Initial Value Differential Equations in R" is available as a pdf file on the CRAN > Packages site, and should be consulted for orientation and examples. The help pages for the individual functions provide details and more examples.

- `bvpSolve` provides "functions that solve boundary value problems (BVP) of systems of ordinary differential equations (ODE). The functions provide an interface to the FORTRAN functions `twpbvpC` and `colnew` and an R-implementation of the shooting method."
- `ReacTran` provides "routines for developing models that describe reaction and advective-diffusive transport in one, two or three dimensions. Includes transport routines in porous media, in estuaries, and in bodies with variable shape."
- `rootSolve` contains "routines to find the root of nonlinear functions, and to perform steady-state and equilibrium analysis of ordinary differential equations (ODE). Includes routines that: (1) generate gradient and Jacobian matrices (full and banded), (2) find roots of non-linear equations by the Newton–Raphson method, (3) estimate steady-state conditions of a system of (differential) equations in full, banded or sparse form, using the Newton–Raphson method, or by dynamically running, (4) solve the steady-state conditions for uni-and multicomponent 1-D, 2-D, and 3-D partial differential equations, that have been converted to ODEs by numerical differencing (using the method-of-lines approach). Includes fortran code."
- `PBSddesolve` "solves systems of delay differential equations." (This capability also exists in `deSolve`.)
- `FME` (A Flexible Modelling Environment for Inverse Modelling, Sensitivity, Identifiability, Monte Carlo Analysis) "provides functions to help in fitting models to data, to perform Monte Carlo, sensitivity and identifiability analysis. It is intended to work with models written as a set of differential equations that are solved either by an integration routine from package `deSolve`, or a steady-state solver from package `rootSolve`. However, the methods can also be used with other types of functions."

These packages, and the functions in them, are explained with useful examples in their help pages and in the book *Solving Differential Equations in R* by Soetaert, Cash and Mazzia, Springer, 2012. Additional information can be found on the website of the special interest group about dynamic modeling with R, `https://stat.ethz.ch/pipermail/r-sig-dynamic-models/`. This site can also be reached from `r-sig-dynamic-models@r-project.org>Mailing lists`.

8.1 Euler method

We can numerically solve ODEs without using packages, by manually coding the steps. This is not as efficient as using a suitable package, nor are the results likely to be as good, but the exercise is instructive. We begin by illustrating the Euler method. The Euler method is the simplest method for numerically solving differential equations: the position at time $t+dt$ is computed from the position at t and the velocity at t. The error inherent in this approach is first order in dt, hence the method is often not very accurate except for very small dt, in which case numerical round-off introduces another kind of error. It often turns out, nonetheless, that the Euler method is good, or at least adequate, for some small problems.

8.1.1 Projectile motion

We look, for example, at projectile motion as exemplified by the path of a batted ball (After VanWyk, p. 24). This example also reminds us how to convert between units. The motion of the ball is determined by its velocity v_0 and angle θ_0 as it leaves the bat, its mass m with gravity (coefficent g) pulling it back to earth, and its frictional resistance in the air, which in turn is determined by its area A, the density of air ρ, and its drag coefficient C_{drag}. In Cartesian coordinates, the equations of motion are

$$m\frac{d^2x}{dt^2} = -\frac{1}{2}\rho AC_{drag}vv_x \tag{8.1}$$

$$m\frac{d^2y}{dt^2} = -mg - \frac{1}{2}\rho AC_{drag}vv_y \tag{8.2}$$

where the drag force is $\frac{1}{2}\rho AC_{drag}v^2$.

We establish base SI units and conversions from other systems of units.

```
> kg. = 1; s. = 1; m. = 1
> lb. = 0.4536*kg.; oz. = 1/16*lb.; hr. = 3600*s.
> ft. = 0.3048*m.; mile. = 5280*ft.; in. = 1/12*ft.;
> degree = pi/180
```

Then we convert various quantities pertinent to the calculation into SI units.

```
> v0.ball = 130*mile./hr. ; v0 = v0.ball/(m./s.)
> diam.ball = 2.9*in. ; diam = diam.ball/m.
> area = pi*(diam/2)^2
> air.density = 0.077*lb./ft.^3 ; dens = air.density/(kg./m.^3)
> mass.ball = 5.1*oz.; mass = mass.ball/kg.
> g = 9.806*m./s.^2
> Cdrag = 0.4
> C = 0.5*dens*area*Cdrag/mass
```

Set initial conditions ...

```
> x0=0 #homeplate
> y0 = 4*ft. # a high fastball
> theta0 = 30*degree # angle off the bat
> vx0 = v0*cos(theta0) # initial velocity in x-direction
> vy0 = v0*sin(theta0) # initial velocity in y-direction
```

... and initialize position vectors with the initial conditions.

```
> x = x0; y = y0; vx = vx0; vy= vy0; v = v0
```

Set the clock time to 0, and set the total time and the measurement interval.

```
> t0 = 0
> sec = t0
> dt = 0.1 # seconds
```

Now solve the equations of motion using the simple Euler method, stopping when the ball hits the ground ($y < 0$).

```
> i=1
> while (y[i] >= 0) {
+   dx = vx[i]*dt # increment in x
+   dy = vy[i]*dt # increment in y
+   v[i] = sqrt(vx[i]^2 + vy[i]^2) # total speed
+   dvx = -C*v[i]*vx[i]*dt # air resistance
+   dvy= (-g - C*v[i]*vy[i])*dt # gravity + air resistance
+   x[i+1] = x[i] + dx
+   y[i+1] = y[i] + dy
+   vx[i+1] = vx[i] + dvx
+   vy[i+1] = vy[i] + dvy
+   v[i+1]  = sqrt(vx[i+1]^2 + vy[i+1]^2)
+   sec[i+1] = sec[i] + dt
+   i=i+1
+ }
```

Tabulate results for the x and y coordinates and velocity as functions of time in the matrix mat, giving the columns appropriate names.

```
> x = round(x/ft.,0)
> y = round(y/ft.,0)
> v = round(v/(ft./s.),1)
> mat = cbind(sec, x,y,v)
> colnames(mat) = c("t/s", "x/ft", "y/ft", "v/(ft/s)")
```

Select and print the results for every fourth time step

```
> n = length(sec)
> show = seq(1,n,4)
> mat[show,]
      t/s x/ft y/ft v/(ft/s)
 [1,] 0.0    0    4    190.7
 [2,] 0.4   62   38    156.7
 [3,] 0.8  116   62    132.3
 [4,] 1.2  163   79    114.3
 [5,] 1.6  205   89    101.0
 [6,] 2.0  244   93     91.5
 [7,] 2.4  279   92     85.2
 [8,] 2.8  312   86     81.6
 [9,] 3.2  343   75     80.3
[10,] 3.6  371   60     80.8
[11,] 4.0  398   41     82.6
[12,] 4.4  422   18     85.2
```

If the fence is 20 feet high at 370 feet from home plate, the ball clears the fence with room to spare. This application of the Euler method converges satisfactorily, as can be verified by halving the time increment *dt* and re-running the simulation.

8.1.2 Orbital motion

Next we use the Euler method to calculate the dynamics of the orbit of the Earth around the Sun, assuming a circular orbit. (Adapted from S. VanWyk, "Computer Solutions in Physics," pp. 3–6.) We find that the calculated orbit is not stable, but drifts with time because the position and velocity get progressively out of phase with one another.

The equations of motion, combining Newton's second law with the law of gravity, can be expressed in Cartesian coordinates as

$$\frac{d^2x}{dt^2} = \frac{GM_{sun}x}{(x^2+y^2)^{3/2}}, \tag{8.3}$$

$$\frac{d^2y}{dt^2} = \frac{GM_{sun}y}{(x^2+y^2)^{3/2}}. \tag{8.4}$$

We set the physical parameters: gravitational constant `G` and Sun's mass `Msun`.

```
> G = 3600^2*6.673e-20
> Msun = 1.989e30
> GM = G*Msun
```

Next we specify the initial values of the x and y positions and velocities.

```
> x0 = 149.6e6; vx0 = 0
> y0 = 0; vy0 = 29.786*3600
```

We set the total elapsed time at 8800 hours, approximately 1 year, updating the calculation every hour.

```
> tmin = 0; tmax = 8800; dt = 1
> hrs = seq(tmin, tmax, dt)
> n = (tmax - tmin)/dt + 1 # Maximum number of updates
```

For the heart of the calculation, we set the first component of each vector to its initial value, then update the vectors according to the Euler scheme until the maximum time is reached.

```
> x = x0; vx = vx0; y = y0; vy = vy0
> for(i in 2:n) {
+    dx = vx[i-1]*dt
+    dvx = -GM*x[i-1]/(x[i-1]^2 + y[i-1]^2)^1.5*dt
+    dy = vy[i-1]*dt
+    dvy = -GM*y[i-1]/(x[i-1]^2 + y[i-1]^2)^1.5*dt
+    x[i] = x[i-1] + dx
+    vx[i] = vx[i-1] + dvx
+    y[i] = y[i-1] + dy
+    vy[i] = vy[i-1] + dvy
+    }
```

For presentation purposes, we round each value to three figures after the decimal point.

```
> r = round(sqrt(x^2 + y^2)*1e-8,3)
> v = round(sqrt(vx^2 + vy^2)/3600,3)
> x = round(x*1e-8,3)
> y = round(y*1e-8,3)
```

We select 23 of the 8800 time points, spaced 400 time units apart, and assign values of the hour, x and y positions, orbital radius, and velocity to the indicated vectors.

```
> show = seq(1,n,400)
>
> hrsdisp = hrs[show]
> xdisp = x[show]
> ydisp = y[show]
> rdisp = r[show]
> vdisp = v[show]
```

We then combine the column vectors into the matrix `mat`, label each column, and print the results using five digits to get three after the decimal point.

```
> options(digits=5)
> mat = cbind(hrsdisp,xdisp,ydisp,rdisp,vdisp)
> colnames(mat) = c("hrs", "x/1e8 km", "y/1e8 km",
   "r/1e8 km", "v km/ s")
> mat
       hrs x/1e8 km y/1e8 km r/1e8 km v km/ s
 [1,]    0    1.496    0.000    1.496  29.786
 [2,]  400    1.435    0.423    1.496  29.789
 [3,]  800    1.257    0.812    1.496  29.791
 [4,] 1200    0.976    1.134    1.497  29.792
 [5,] 1600    0.616    1.364    1.497  29.791
 [6,] 2000    0.205    1.483    1.497  29.787
 [7,] 2400   -0.222    1.482    1.498  29.781
 [8,] 2800   -0.632    1.359    1.499  29.772
 [9,] 3200   -0.990    1.127    1.500  29.761
[10,] 3600   -1.268    0.803    1.501  29.748
[11,] 4000   -1.444    0.414    1.502  29.734
[12,] 4400   -1.503   -0.009    1.503  29.719
[13,] 4800   -1.441   -0.431    1.504  29.704
[14,] 5200   -1.263   -0.818    1.505  29.690
[15,] 5600   -0.984   -1.140    1.506  29.677
[16,] 6000   -0.626   -1.370    1.507  29.666
[17,] 6400   -0.218   -1.492    1.507  29.657
[18,] 6800    0.208   -1.494    1.508  29.651
[19,] 7200    0.617   -1.377    1.508  29.648
[20,] 7600    0.977   -1.150    1.509  29.647
[21,] 8000    1.259   -0.832    1.509  29.648
```

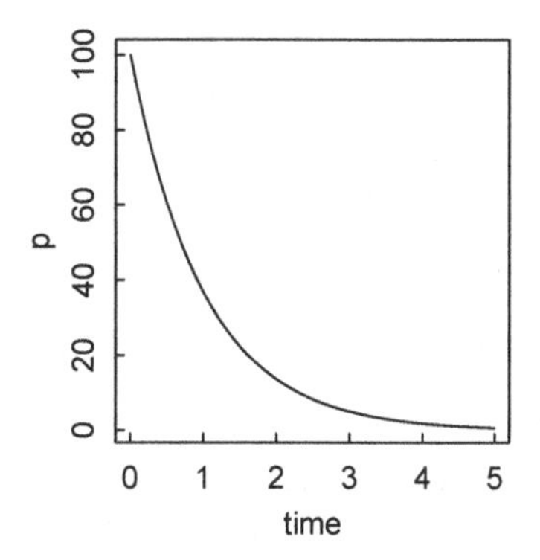

Figure 8.1: *Exponentially decaying population calculated by the improved Euler method.*

```
[22,] 8400     1.442    -0.447     1.509  29.650
[23,] 8800     1.509    -0.027     1.509  29.653
```

A proper solution would maintain a constant radius and velocity, but the Euler method does not, even though it uses 8800 very small time steps in this example.

8.2 Improved Euler method

The improved Euler method, does a better—but not perfect—job because the increment at each step is calculated as the mean of the increments at t and $d+dt$. We first illustrate the procedure with a simple exponential decay example (Figure 8.1).

```
> p0 = 100 # initial population value
> k = 1 # rate parameter
> tmin=0;tmax=5;dt=0.1 # Beginning and end times and increment
> time = seq(tmin,tmax,dt) # vector of times
> n = (tmax-tmin)/dt + 1 # number of evaluations of p
> p = p0 # initialize p vector
> for (i in 2:n) {
+     dpa = -k*p[i-1]*dt
+     pa = p[i-1] + dpa
+     dpb = -k*pa*dt
+     dp = (dpa + dpb)/2
+     p[i] = p[i-1] + dp
+ }
> plot(time, p, type = "l")
```

We now apply the improved Euler method to the orbital problem. Note that the calculation below uses only one tenth as many time steps as the simple Euler method, but achieves better results.

As before, we begin by defining parameters and initializing the variables.

```
> # Improved Euler
```

```
> G = 3600^2*6.673e-20
> Msun = 1.989e30
> GM = G*Msun
> x0 = 149.6e6; vx0 = 0
> y0 = 0; vy0 = 29.786*3600
```

With the improved Euler method, we use a time step corresponding to 10 hours.

```
> tmin = 0; tmax = 8800; dt = 10
> hrs = seq(tmin, tmax, dt)
> n = (tmax - tmin)/dt + 1
```

After starting the position and velocity vectors, we obtain first approximations to their values after each time step, by adding calculated increments to the previous values.

```
> x = x0; vx = vx0; y = y0; vy = vy0
> for(i in 2:n) {
+   dxa = vx[i-1]*dt
+   dvxa = -GM*x[i-1]/(x[i-1]^2 + y[i-1]^2)^1.5*dt
+   dya = vy[i-1]*dt
+   dvya = -GM*y[i-1]/(x[i-1]^2 + y[i-1]^2)^1.5*dt
+
+   xa = x[i-1] + dxa
+   vxa = vx[i-1] + dvxa
+   ya = y[i-1] + dya
+   vya = vy[i-1] + dvya
```

We then calculate new estimates of the increments, and average them with the previous estimates.

```
+   dxb = vxa*dt
+   dvxb = -GM*xa/(xa^2 + ya^2)^1.5*dt
+   dyb = vya*dt
+   dvyb = -GM*ya/(xa^2 + ya^2)^1.5*dt
+
+   dx = (dxa + dxb)/2
+   dvx = (dvxa + dvxb)/2
+   dy = (dya + dyb)/2
+   dvy = (dvya + dvyb)/2
```

The averaged increments are added to the previous values to get values at the next time points.

```
+   x[i] = x[i-1] + dx
+   vx[i] = vx[i-1] + dvx
+   y[i] = y[i-1] + dy
+   vy[i] = vy[i-1] + dvy
+   }
```

The remainder of the calculation proceeds as in the simple Euler example above.

```
> r = round(sqrt(x^2 + y^2)*1e-8,3)
> v = round(sqrt(vx^2 + vy^2)/3600,3)
> x = round(x*1e-8,3)
> y = round(y*1e-8,3)
> show = seq(1,n,40)
> hrsdisp = hrs[show]
> xdisp = x[show]
> ydisp = y[show]
> rdisp = r[show]
> vdisp = v[show]
> options(digits=5)
> mat = cbind(hrsdisp,xdisp,ydisp,rdisp,vdisp)
> colnames(mat) = c("hrs", "x/1e8 km", "y/1e8 km",
+     "r/1e8 km", "v km/ s")
> mat
       hrs x/1e8 km y/1e8 km r/1e8 km v km/ s
 [1,]    0    1.496    0.000    1.496  29.786
 [2,]  400    1.435    0.423    1.496  29.786
 [3,]  800    1.257    0.812    1.496  29.786
 [4,] 1200    0.976    1.134    1.496  29.786
 [5,] 1600    0.615    1.364    1.496  29.785
 [6,] 2000    0.205    1.482    1.496  29.785
 [7,] 2400   -0.223    1.479    1.496  29.785
 [8,] 2800   -0.632    1.356    1.496  29.784
 [9,] 3200   -0.990    1.122    1.496  29.784
[10,] 3600   -1.267    0.796    1.496  29.784
[11,] 4000   -1.440    0.406    1.496  29.784
[12,] 4400   -1.496   -0.018    1.496  29.784
[13,] 4800   -1.430   -0.440    1.496  29.784
[14,] 5200   -1.247   -0.827    1.496  29.784
[15,] 5600   -0.962   -1.145    1.496  29.784
[16,] 6000   -0.599   -1.371    1.496  29.784
[17,] 6400   -0.187   -1.484    1.496  29.785
[18,] 6800    0.240   -1.477    1.496  29.785
[19,] 7200    0.648   -1.348    1.496  29.785
[20,] 7600    1.003   -1.110    1.496  29.786
[21,] 8000    1.276   -0.781    1.496  29.786
[22,] 8400    1.445   -0.389    1.496  29.786
[23,] 8800    1.496    0.036    1.496  29.786
```

The orbital results are better with the improved Euler method, but the radius and velocity are still not quite constant.

8.3 deSolve package

There are many other more-or-less elementary algorithms to solve systems of ODEs, most of them elaborations of the improved Euler or Runge–Kutta approaches; but it is most useful to go directly to the more powerful methods employed in the `deSolve` package.

The central function in `deSolve` for solving a system of ODEs is `ode()`, which is a wrapper around 17(!) specific ODE solvers, each of which can be specified depending on the problem under consideration. `ode()` is called by

```
> ode(y, times, func, parms, method, ...)
```

where

- `y` is a vector of initial values for the dependent variables,
- `times` is the sequence of times at which output is desired (starting with the initial time),
- `func` is the function that computes the derivatives in the ODE system. The return value of `func` should be a list, whose first element is a vector containing the derivatives of `y` with respect to time, and whose next elements (if there are any) are global values that are required at each point in `times`. The derivatives should be specified in the same order as the state variables `y`.
- `parms` is the list of parameters passed to `func`.
- method is a string containing one of the integration methods: "`lsoda`", "`lsode`","`lsodes`", "`lsodar`", "`vode`", "`daspk`", "`euler`", "`rk4`", "`ode23`", "`ode45`", "`radau`", "texttbdf", "`bdf_d`", "`adams`", "`impAdams`", "`impAdams_d`", "`iteration`". `lsoda` is the default.
- ... represents additional arguments passed to the integrator or the method.

Solvers in the `deSolve` class yield a matrix with the number of rows equal to the number of elements in the times vector. The first column is the time value and the next columns are the values of the elements of the `y` vector at each time. If global values were returned in the second element of the list returned by `func`, their values will follow. If the initial values in the vector `y` are named, the names will label the columns of the output matrix.

Before delving deeper into the details of `ode` and its methods, let's demonstrate how it handles our problem of the Earth revolving in a circular orbit around the Sun. Note that we use a time increment `dt` of 400 hr, compared to 10 hr for Euler and 100 hr for improved Euler. However, this does not mean that the step is 40 times longer in `lsoda` than in Euler, since the step size in `lsoda` can vary as the calculation proceeds.

We load the `deSolve` package (first installing it if that has not already been done), then go through the initial steps of defining parameters, initializing position and velocity variables, and specifying the time sequence for the calculation.

```
> install.packages("deSolve")
> require(deSolve)
>
```

```
> # Compute parameter
> G = 3600^2*6.673e-20
> Msun = 1.989e30
> GM = G*Msun
> parms = GM
>
> # Initialize variables
> x0 = 149.6e6; vx0 = 0
> y0 = 0; vy0 = 29.786*3600
>
> # Set time sequence
> tmin = 0; tmax = 8800; dt = 400
> hrs = seq(tmin, tmax, dt)
```

We next define the function that computes the desired derivatives and returns them in a list.

```
> orbit = function(t,y,GM) {
+ dy1 = y[2]
+   dy2 = -GM*y[1]/(y[1]^2+y[3]^2)^1.5
+ dy3 = y[4]
+ dy4 = -GM*y[3]/(y[1]^2+y[3]^2)^1.5
+ return(list(c(dy1,dy2,dy3,dy4)))
+}
```

We then call the ode function, using the default `lsoda` method, to solve the system of differential equations. The arguments to the function are specified by position, and thus not explicitly named.

```
> out = ode(c(x0, vx0, y0, vy0), hrs, orbit, parms)
```

Finally, we display the results as before.

```
> options(digits=5)
> hrs = out[,1]; x = out[,2]; vx = out[,3]
> y = out[,4]; vy = out[,5]
> r = round(sqrt(x^2 + y^2)*1e-8,3)
> v = round(sqrt(vx^2 + vy^2)/3600,3)
> mat = cbind(hrs,x,y,r,v)
> colnames(mat) = c("hrs", "x km", "y km",
  "r/1e8 km", "v km/s")
> mat
      hrs      x km      y km r/1e8 km v km/s
 [1,]    0 149600000         0    1.496 29.786
 [2,]  400 143493179  42306604    1.496 29.786
 [3,]  800 125671404  81159285    1.496 29.786
 [4,] 1200  97589604 113386031    1.496 29.786
 [5,] 1600  61540456 136355691    1.496 29.786
 [6,] 2000  20467145 148193044    1.496 29.786
 [7,] 2400 -22277150 147931692    1.496 29.786
```

```
 [8,] 2800  -63202726  135592932     1.496 29.786
 [9,] 3200  -98968215  112184158     1.496 29.786
[10,] 3600 -126653781   79616413     1.496 29.786
[11,] 4000 -143999011   40548639     1.496 29.786
[12,] 4400 -149587807   -1829627     1.496 29.786
[13,] 4800 -142963858  -44058504     1.496 29.786
[14,] 5200 -124667908  -82690312     1.496 29.786
[15,] 5600  -96193662 -114571009     1.496 29.786
[16,] 6000  -59865843 -137097729     1.496 29.786
[17,] 6400  -18650363 -148431290     1.496 29.786
[18,] 6800   24087794 -147646368     1.496 29.786
[19,] 7200   64859326 -134807039     1.496 29.786
[20,] 7600  100335543 -110961522     1.496 29.786
[21,] 8000  127619992  -78056635     1.496 29.786
[22,] 8400  144484949  -38778991     1.496 29.786
[23,] 8800  149553679    3664689     1.496 29.786
```

With the `ode` function of `deSolve`, the radius and velocity of the orbit remain satisfactorily constant over the course of the year.

8.3.1 *`lsoda()` and `lsode()`*

The default integrator method of `ode` is `lsoda()`. It is able to switch automatically between stiff and non-stiff methods, depending on the problem. Other methods may be more efficient in particular cases, but `lsoda` will almost always get the job done. Here, for example, it numerically solves the Bessel differential equation of order ν when $\nu = 1$ (Figure 8.2).

$$x^2\frac{d^2J}{dx^2} + x\frac{dJ}{dx} + (x^2 - \nu^2)J = 0 \tag{8.5}$$

```
> require(deSolve)
>
```

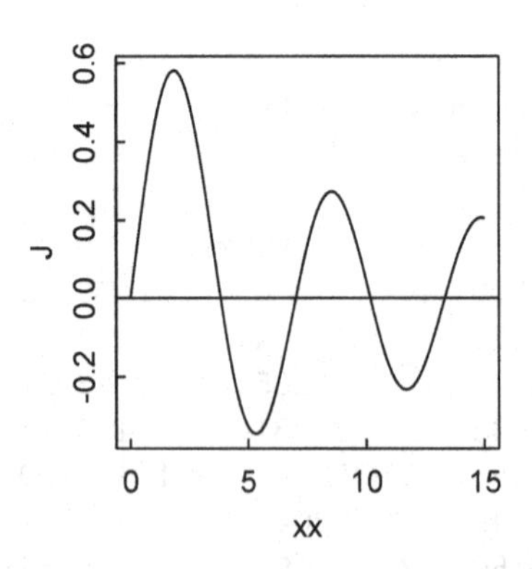

Figure 8.2: *Numerical solution of the Bessel equation of order 1.*

```
> # Function to feed to lsoda
> diffeqs = function(x,y,nu) {
+ J=y[1]
+ dJdx = y[2]
+   with(as.list(parms), {
+ dJ = dJdx
+     ddJdx = -1/x^2*(x*dJdx + (x^2-nu^2)*J)
+     res = c(dJ, ddJdx)
+ list(res)
+ })
+}
>
> # Abscissa steps
> xmin = 1e-15 # Don't start exactly at zero, to avoid infinity
> xmax = 15
> dx = 0.1
> xx = seq(xmin, xmax, dx)
>
> # Parameters
> parms = c(nu = 1) # Bessel equation of order 1
>
> # Initial values
> y0 = c(J = 0, dJdx = 1)
>
> # Solve with lsoda
> out = lsoda(y0, xx, diffeqs, parms)
>
> # Plot results and compare with built-in besselJ
> xx = out[,1]; J = out[,2]; dJdx = out[,3]
> plot(xx, J, type="l"); curve(besselJ(x,1),0,15,add=TRUE)
> abline(0,0)
```

The `lsode()` function is very similar to `lsoda()`, but requires that the user specify whether the problem is stiff. LSODE (Livermore Solver for Ordinary Differential Equations) is the basic solver of the ODEPACK collection on which `deSolve` is based. In turn, the solver `vode` is very similar to `lsode`, but uses a variable-coefficient method rather than the fixed-step interpolation methods in `lsode`. Also, in `vode` it is possible to choose whether or not a copy of the Jacobian is saved for reuse in the corrector iteration algorithm; in `lsode`, a copy is not kept. The solver `zvode` is like `vode`, but should be used when the dependent variables and derivatives, y and dy/dt, are complex. See the help pages for these solver functions for more details.

8.3.2 *"`adams`" and related methods*

For non-stiff ODEs, `lsoda` and `lsode` use the `adams` method (more generally, the Adams–Bashford–Moulton method) behind the scenes. This is a linear multistep

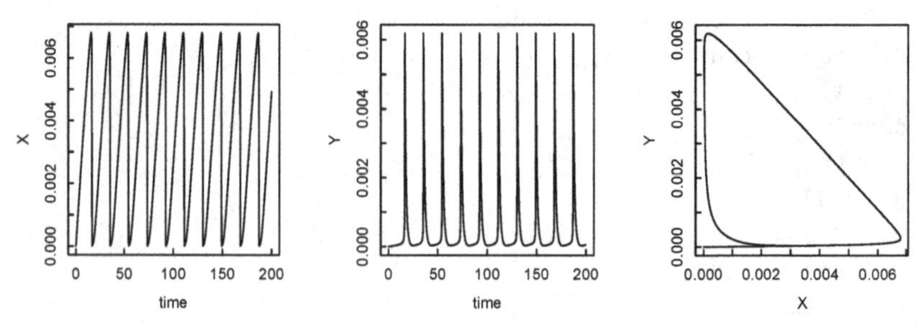

Figure 8.3: *Concentration changes with time in an oscillating chemical system (Equation 8.6).*

method for numerical solution of ODEs. To quote from Wikipedia: "Conceptually, a numerical method starts from an initial point and then takes a short step forward in time to find the next solution point. The process continues with subsequent steps to map out the solution. Single-step methods (such as Euler's method) refer to only one previous point and its derivative to determine the current value. Methods such as Runge–Kutta take some intermediate steps (for example, a half-step) to obtain a higher order method, but then discard all previous information before taking a second step. Multistep methods attempt to gain efficiency by keeping and using the information from previous steps rather than discarding it. Consequently, multistep methods refer to several previous points and derivative values. In the case of linear multistep methods, a linear combination of the previous points and derivative values is used." Variants of the `adams` method are `impAdams` and `impAdams_d` which use different methods of iteration and treatment of the Jacobian. See the `deSolve` help pages for details.

The `adams` method can be called directly as a method in `ode` (but not as a function `adams()`), as illustrated in this example of an oscillating chemical system of reactions (Equation 8.6) between X and Y, fed by reactant A which is maintained at a concentration of 0.05 units, while the concentrations of X and Y start at 0. Results are plotted in Figure 8.3.

$$\begin{aligned}\frac{dX}{dt} &= k_1A - k_2X - k_3XY^2 \\ \frac{dY}{dt} &= k_2X + k_3XY^2 - k_4Y\end{aligned} \tag{8.6}$$

```
> require(deSolve)
> # Reaction mechanism
> diffeqs = function(t,x,parms) {
+ X=x[1]
+ Y=x[2]
+ with(as.list(parms), {
+ dX = k1*A - k2*X - k3*X*Y^2
+ dY = k2*X + k3*X*Y^2 - k4*Y
```

```
+ list(c(dX, dY))
+ })}
>
> # Time steps
> tmin = 0; tmax = 200; dt = 0.01
> times = seq(tmin, tmax, dt)
>
> # Parameters: Rate constants and A concentration
> parms = c(k1 = 0.01, k2 = 0.01, k3 = 1e6, k4 = 1, A = 0.05)
>
> # Initial values
> x0 = c(X = 0, Y = 0)
>
> # Solve with adams method
> out = ode(x0, times, diffeqs, parms, method = "adams")
>
> # Plot results
> time = out[,1]; X = out[,2]; Y = out[,3]
>
> par(mfrow = c(1,3))
> plot(time, X, type="l") # Time variation of X
> plot(time, Y, type="l") # Time variation of Y
> plot(X,Y, type="l")  # Phase plot of X vs Y
> par(mfrow = c(1,1))
```

8.3.3 Stiff systems

A system of ODEs is called "stiff" if the dependent variables change according to two or more very different scales of the independent variable. `ode()` handles stiff systems with the backward differentiation formula (bdf) methods `bdf` or `bdf_d`. The help page, which explains the differences among these, also states that the method `daspk` may outperform method `bdf` for very stiff systems. The functions `lsoda()` and `lsode()` invoke these methods when faced with stiff equations.

For example, a set of equations that arise in a chemical engineering context (Hanna and Sandall, pp. 302-3)

$$\begin{aligned}\frac{dy_1}{dt} &= -0.1y_1 - 49.9y_2 \\ \frac{dy_2}{dt} &= -50y_2 \\ \frac{dy_3}{dt} &= 70y_2 - 120y_3\end{aligned} \tag{8.7}$$

with $y_1(0) = 2, y_2(0) = 1, y_3(0) = 2$ can be solved with `bdf` as follows:

```
> require(deSolve)
> # model equations
```

```
> diffeqs = function(x,y,p) {
+     y1=y[1];y2=y[2];y3=y[3]
+     dy1 = -0.1*y1 - 49.9*y2
+     dy2 = -50*y2
+     dy3 = 70*y2 - 120*y3
+     list(c(dy1, dy2, dy3))
+     }
>
> # Time steps
> xmin=0;xmax=2;dx=0.1
> x = seq(xmin, xmax, dx)
>
> # Initial values
> y0 = c(2,1,2)
>
> # Solve with bdf
> out = ode(y0, x, diffeqs, parms = NULL, method = "bdf")
> mat = cbind(out[,1], out[,2], out[,3], out[,4])
> mat[nrow(mat),]
[1]  2.000000e+00  8.187308e-01 -1.475708e-08 -1.698035e-08
```

The analytical solution for y[1] at time 2.0 is 0.8187.

To solve with `lsoda()`, we need change only the invocation of the particular solver:

```
> out = lsoda(y0, x, diffeqs, parms = NULL)
```

As stated in the `deSolve` package vignette (p. 22) "The solvers from ODEPACK can be fine-tuned if it is known whether the problem is stiff or non-stiff, or if the structure of the Jacobian is sparse." See the documentation for an illustration of how to use the Jacobian to enhance solution of stiff systems.

8.4 Matrix exponential solution for sets of linear ODEs

Since the above set of ODEs is linear, it can also be solved by a matrix exponential technique. The concentrations of each of the dependent variables can be shown to decay as a weighted sum of exponentials exp($-\lambda t$). The decay constants λ are given by the eigenvectors of the rate coefficient matrix **m**, and the weights w are obtained by solving the equation $mathbfEvw = c_0$, where **Ev** is the matrix of eigenvectors and c_0 is the vector of initial concentrations. The procedure is illustrated below, and leads to the same result as with the ode method.

```
> m = matrix(c(-0.1,-49.9,0,0,-50,0,0,70,-120), nrow=3, byrow=TRUE)
> lam = c(); expt = c(); A = 0
> time = 2
> Ev = eigen(m)$vectors
> c0 = c(2,1,2) # initial concentrations
> w = solve(Ev, c0)
> for (i in 1:nrow(m)) {
```

```
+    lam[i] = eigen(m)$values[i]
+    A = A + w[i]*Ev[1,i]*exp(lam[i]*time)
+    }
> A
[1] 0.8187308
```

8.5 Events and roots

The solvers in ode can handle events, instantaneous changes in the variables that affect the model. If the variable is external, so that it is essential to the model but not calculated by the model, it is called a forcing function. The deSolve vignette (p.37) presents an example of a forcing—the measured flux of sediment oxygen consumption—that is imposed on the model in the form of a data series. Here we present another example: the voltage V across an RC circuit subject to a periodic rectangular voltage pulse built with a combination of the rep() and approxfun() functions. The differential equation for the change in voltage with time is

$$\frac{dV}{dt} = ForcingFunction - \frac{V}{RC}. \tag{8.8}$$

We begin in the more-or-less standard way, loading deSolve and setting up the time sequence over which the solution will be calculated.

```
> require(deSolve)
> # Time sequence
> tmin=0; tmax=4; dt=.01 #millisec
> times = seq(tmin, tmax, dt)
```

We next define the forcing voltage: a sequence of (1,0,1,0) pulses, each extending for $1/dt$ time steps. The time and voltage pulses are bound together as a matrix representing a rectangular wave.

```
> # Forcing behavior
> pulse = c(rep(1,1/dt), rep(0,1/dt), rep(1,1/dt), rep(0,1/dt+1))
> sqw = cbind(times, pulse)
```

The pulse is now converted into a function with approxfun(). The interpolation method is "linear" and rule = 2 avoids NaNs while interpolating.

```
> SqWave = approxfun(x = sqw[,1], y = sqw[,2],
+   method = "linear", rule = 2)
```

We write the RC circuit equation for dV/dt as a function to be fed to lsoda()

```
> voltage = function( t, V, RC) list (c(SqWave(t) - V/RC))
```

and define the RC value as a parameter, and the initial value of V

```
> parms = c(RC = 0.6) # millisec
> V0 = 0 # Initial condition
```

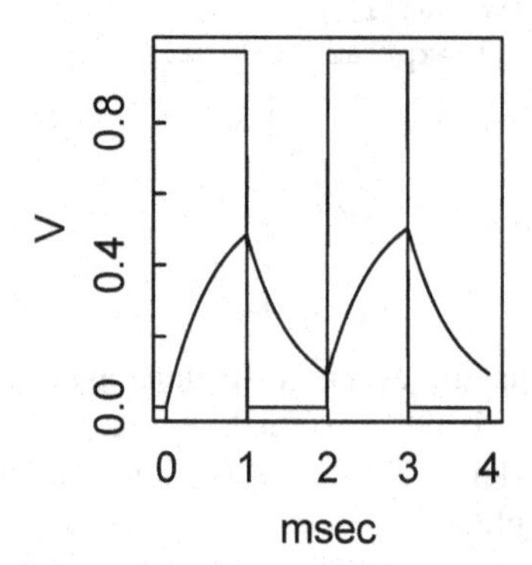

Figure 8.4: *RC circuit with periodic pulse as example of an event-driven ODE.*

The solution is now generated simply by a single line of code:

```
> Out = lsoda(y = V0, times = times, func = voltage,
+ parms = parms)
```

The first column of the solution matrix Out is the time, the second column is the voltage. These are plotted against each other, with the result in Figure 8.4.

```
> plot(Out[,1], Out[,2], type = "l", ylim = c(0,1), xlab = "msec",
+ ylab = "V")
> lines(times, pulse)
```

Events may be triggered by roots, i.e., when a certain condition equals 0. Here is an example from a hypothetical drug delivery protocol. The administered drug is A, which is then metabolized to the active form B in a reversible process with rate constants k_{f1} and k_{r1}. B is also irreversibly removed from the system with rate constant k_{f2}. The kinetic rate equations are

$$\frac{dA}{dt} = -k_{f1}A + k_{r1}B$$
$$\frac{dB}{dt} = k_{f1}A - k_{r1}B - k_{f2}B \tag{8.9}$$

These equations are translated into a function returning the time derivatives:

```
> # Drug metabolism kinetic equations
> diffeqs = function(t,y,parms) {
+ #A = y[1]
+ #B = y[2]
+ with(as.list(parms), {
+ dy1 = - kf1*y[1] + kr1*y[2]
+ dy2 = kf1*y[1] - kr1*y[2] - kf2*y[2]
+ return(list(c(dy1,dy2)))
+ })
+ }
```

We want to keep B at a concentration 1 or higher, so whenever the concentration of B falls to 1, we add enough A to bring its concentration to 2. The code for the root and event functions is:

```
> # event triggered if B = 1
> rootfun <- function (t, y, parms) y[2] -1
>
> # sets A = 2
> eventfun = function(t, y, parms) {
+   y[1] = 2
+   return(y)
+   }
```

Setting up to solve the equations proceeds in the standard way:

```
> # Time steps
> tmin = 0; tmax = 10; dt = 0.1
> times = seq(tmin, tmax, dt)
>
> # Parameters: values of rate constants
> parms = c(kf1 = 1, kr1 = 0.1, kf2 = 1)
>
> # Initial values of A and B
> y0 = c(3,2)
```

We invoke `ode()` as the wrapper to solve the set of equations, while feeding `events` and `rootfun` as arguments. This automatically calls the `lsodar` solver. See the help page for details.

```
> # Solve with lsodar
> out = ode(times = times, y = y0, func = diffeqs, parms = parms,
+   events = list(func = eventfun, root = TRUE),
+   rootfun = rootfun)
```

The results may now be plotted in the standard way (Figure 8.5).

```
> # Plot results
> # par(mfrow = c(1,2))
> time = out[,1]; A = out[,2]; B = out[,3]
> plot(time, A, type="l", ylim = c(0,3), ylab = "conc")
> lines(time, B, lty = 2)
> legend("topright", legend = c("A","B"), lty = 1:2, bty="n")
```

For another example, we consider the Lotka–Volterra model (Figure 8.6) for the population dynamics of a predator–prey system. The prey population grows according to first-order kinetics (`k2*prey`) and is reduced by second-order interactions with predators (`-k1*pred*prey`). The predator population grows by second-order interactions with prey (`k3*pred*prey`, where `k3` is a conversion efficiency coefficient),

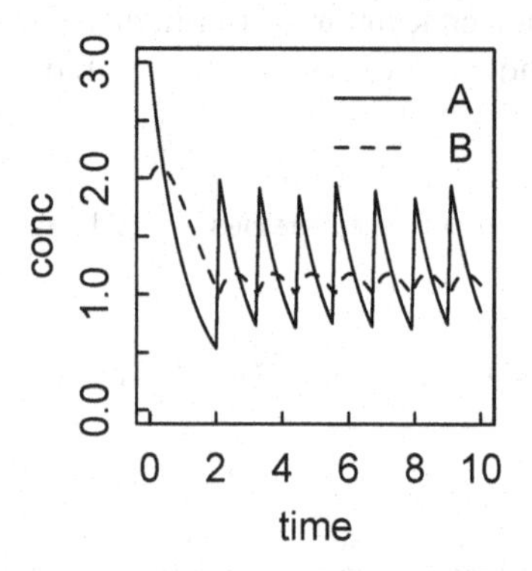

Figure 8.5: *Drug delivery protocol illustrating root-triggered event: when B falls below 1, A is added to bring it to 2.*

and is reduced by a first-order death rate (`-k4*pred`).

$$\frac{dprey}{dt} = -k_1 pred \times prey + k_2 prey$$
$$\frac{dpred}{dt} = k_3 * pred \times prey - k_4 pred \tag{8.10}$$

We first run the model without any events, showing that the two populations vary in a regular cyclic manner.

```
> require(deSolve)
>
> # Population dynamics
> diffeqs = function(t,x,parms) {
+   prey = x[1]
+   pred = x[2]
```

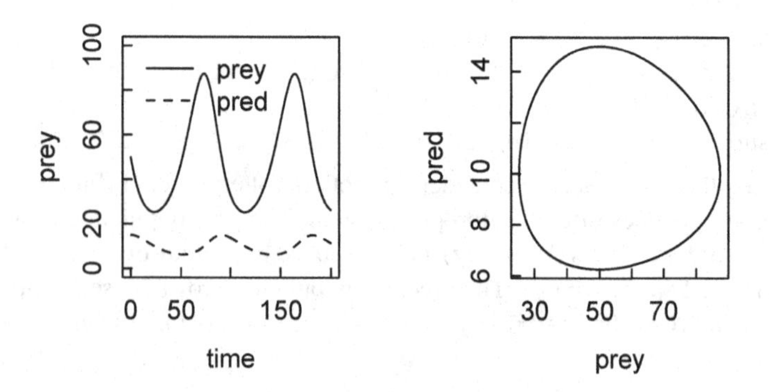

Figure 8.6: *Lotka–Volterra predator–prey simulation.*

```
+     with(as.list(parms), {
+       dprey = -k1*pred*prey + k2*prey
+       dpred = k3*pred*prey - k4*pred
+       res = c(dprey, dpred)
+   list(res)
+   })
+ }
>
> # Time steps
> tmin = 0; tmax = 200; dt = 1
> times = seq(tmin, tmax, dt)
>
> # Parameters
> parms = c(k1 = 0.01, k2 = 0.1, k3 = 0.001, k4 = 0.05)
> # Initial values
> x0 = c(prey = 50, pred = 15)
>
> # Solve with lsoda
> out = lsoda(x0, times, diffeqs, parms, rtol = 1e-6,
+ atol = 1e-6)
>
> # Plot results
> par(mfrow = c(1,2))
> time = out[,1]; prey = out[,2]; pred = out[,3]
> plot(time, prey, type="l", ylim = c(0,100))
> lines(time, pred, lty = 2)
> legend("topleft", legend = c("prey","pred"), lty = 1:2,
+ bty="n")
> plot(prey, pred, type = "l")
> par(mfrow = c(1,1))
```

Now we repeat the simulation, but perturb the system (impose an event) by adding two predators at time 50 with the function `eventdat`.

```
> eventdat = data.frame(var = "pred", time = 50, value = 2, method = "add")
```

The calculation proceeds just as above, except that the call to `lsoda()` is modified by addition of the `events` argument, with results as shown in Figure 8.7.

```
> out = lsoda(x0, times, diffeqs, parms, events = list(data =
+ eventdat))
```

Surprisingly, the populations of predators and prey settle into a new steady state in which both are reduced! Essentially, the extra predators kill off too many prey, thereby reducing their food supply and thereby their own population.

We have demonstrated the root-finding syntax for the `lsoda` solver; `lsodar`, `lsode`, and `lsodes` work the same way. The `radau` solver works slightly differently; one or another of these may be more efficient depending on the problem. See the help pages for details.

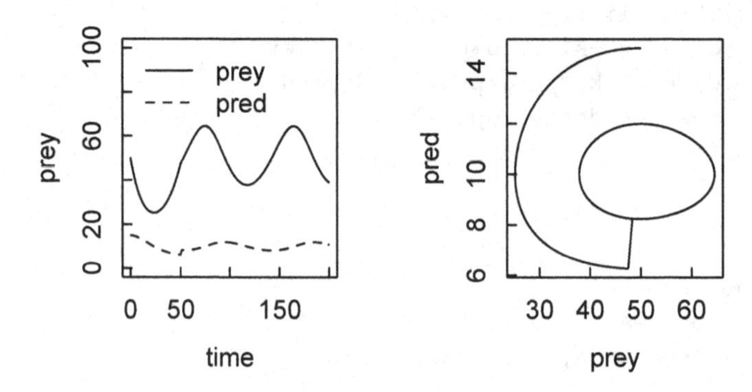

Figure 8.7: *Lotka–Volterra predator–prey simulation with added event at t = 50.*

8.6 Difference equations

The "iteration" method in ode() enables the numerical solution of difference equations. The model function (Population in the example below) returns the new values, rather than the rates of change of the variables. Consider a model of a population divided into three age groups: child (0–12 years), childbearing (13–40 years), and aged (41 years or more). The 0–12 group increases by births in the 13–40 group, and decreases by deaths and by passage into the 13–40 group. The 13–40 group increases by gains from the 0–12 group, and decreases by deaths and by passage into the >= 41 group. The >= 41 group increases by gains from the 13–40 group, and decreases by death. The parameters are chosen to represent fairly high birth and death rates as found in many developing societies.

We begin by writing a function, Population, whose variables are the populations in the three age groups, and giving equations for the growth and decline in each group. The function returns a list with the new populations after a unit time increment.

```
> Population = function(t,y,param) {
+   y1 = y[1] # 0-12 group population
+   y2 = y[2] # 13-40 group population
+   y3 = y[3] # 41 and older population
+   y1.new = b*y2 + 11/12*y1*(1 - d1)
+   y2.new = 1/12*y1*(1 - d1) + 26/27*y2*(1 - d2)
+   y3.new = 1/27*y2*(1 - d2) + y3*(1 - d3)
+   return(list(c(y1.new, y2.new, y3.new)))
+   }
```

We specify the birth and death rate parameters, the initial populations, and the time span over which the population is simulated.

```
> b = 0.5 # Birth rate in 13-40 group
> d1 = 0.1 # Death rate of 0-12 group
> d2 = 0.1 # Death rate of 13-40 group
```

```
> d3 = 0.25 # Death rate of 41 and older
>
> y = c(200, 400, 400) # Initial populations in each group
>
> times = 0:50 # Time span
```

We then call ode with the iteration method to calculate the population over time.

```
> out = ode(func = Population, y = y, times = times, parms =
+ c(b,d1,d2,d3), method = "iteration")
```

The plot command, when applied to a deSolve object, automatically produces a separate plot for each variable. To put all three populations on the same plot, we use matplot as indicated (Figure 8.8).

```
> plot(out) # Three separate plots
>
> # Now alll three age groups on a single plot
> matplot(times,out[,2:4], type = "l", lty=1:3, col = rep(1,3),
+ ylab = "Population")
> legend("topleft", legend = c("0-12","13-40","40"),
+ bty="n",lty=1:3)
> par(mfrow=c(1,1)) # Return to 1 row, 1 column
```

8.7 Delay differential equations

Delay differential equations (DDEs) differ from ODEs in that they contain derivatives that depend on the values of the variables at previous times. The initial conditions for DDEs must specify not only the values at time = 0, but also at times < 0 back to the time of the longest lag in the problem. DDEs also present difficulties in that they often have discontinuities in low-order derivatives; e.g., a constant value up to $t = 0$, followed by a non-zero slope for $t > 0$. Therefore, numerical methods developed for ODEs must be modified to deal with DDEs. A useful discussion of DDEs and their solutions in the context of MATLAB® is Shampine and Thompson 2000 (http://www.radford.edu/thompson/webddes/tutorial.html).

In R, DDEs may be solved using the dede function in deSolve. The PBSddesolve package contains the function dde which operates similarly, albeit with slight differences in the names of some functions. In the discussion that follows we use dede. We show just two relatively simple examples; more examples of increasing complexity can be found in the reference manuals *deSolve.pdf* and *PBS-ddesolve.pdf* (including demos), the vignette *Package deSolve: Solving Initial Value Differential Equations in R* (all available on the CRAN Packages website), and the help pages for dede and lagvalue in the deSolve package. The above-referenced paper by Shampine and Thompson has further examples, whose translation from MATLAB to R would be a useful exercise for the reader.

The approach, function calls, and syntax for DDEs are similar to those for ODEs, with a few additions. The time lags—of which there may be more than one—are

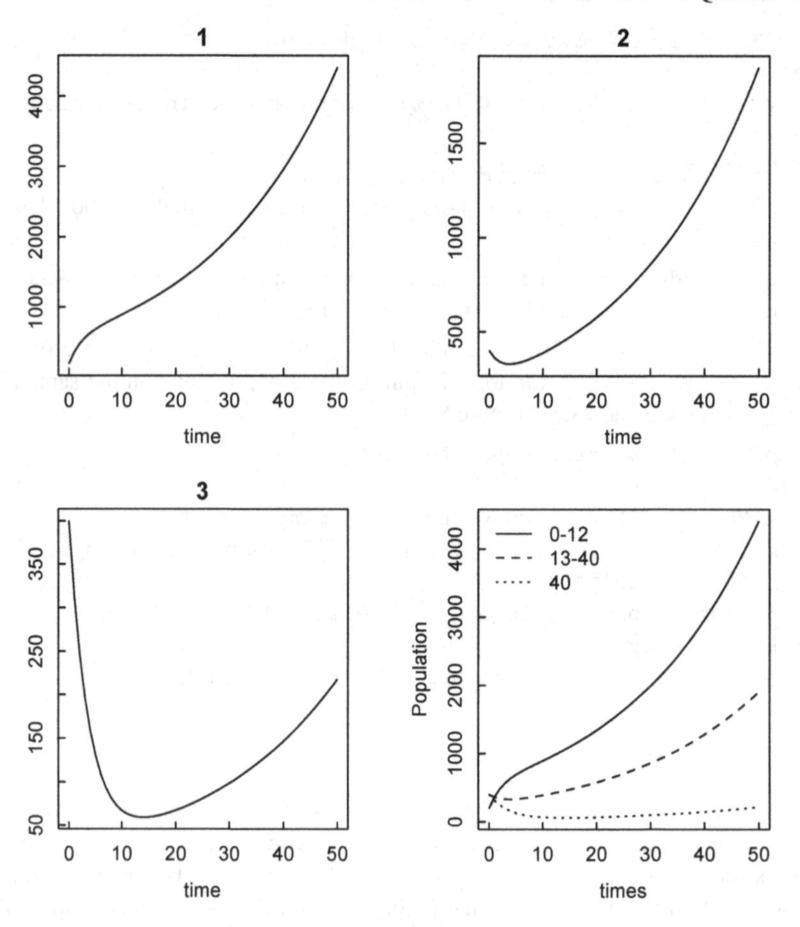

Figure 8.8: *Graphs of three population groups (1: 0–12, 2: 13–40, 3: greater than 40).*

generally (though not necessarily) called `tau`. The value of the dependent variable(s) at the lagged time is defined by the function `lagvalue`. If there is more than one variable, `lagvalue` becomes a vector with variables specified by indices [1], [2], etc. If necessary, the lagged derivative is defined by the function `lagderiv`.

Our first example is the Hutchinson equation of population dynamics

$$\frac{dy}{dt} = ry(1 - \frac{y(t-\tau)}{K}) \tag{8.11}$$

as presented by Y. Kuang in `math.la.asu.edu/~kuang/paper/STE034KuangDDEs.pdf`.

The derivative function for the Hutchinson model, which is to be fed to `dede`, is

```
> func = function(t, y, parms) {
+   tlag=t-tau
+   if (tlag < 0) dy = 1 else dy = r*y*(1 - lagvalue(tlag)/K)
+   return(list(c(dy))) }
```

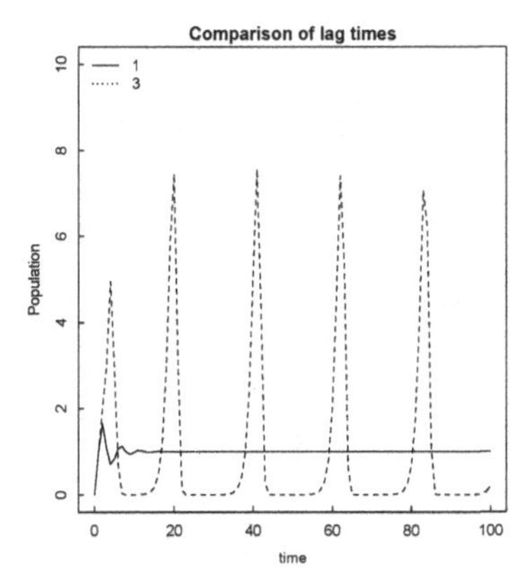

Figure 8.9: *Solutions to Hutchinson Equation 8.11 using* `dede` *with time lag* τ = *1 and 3).*

We enter the initial value of the population, the desired time sequence, and the parameters of the model:

```
> yinit = 0
> times = 0:100
> r = 1; K = 1; tau = 1
```

To get a solution, we load `deSolve` and put the arguments into `dede`.

```
> require(deSolve)
> yout1 = dede(y = yinit, times = times, func = func,
+ parms = c(r,K, tau))
```

We test the sensitivity of the model to the delay time by choosing another value of τ, running the calculation again ...

```
> tau = 3
> yout2 = dede(y = yinit, times = times, func = func,
+ parms = c(r,K, tau))
```

... and plotting both results on the same plot (Figure 8.9).

```
> plot(yout1,yout2, type = "l", main="Comparison of lag times",
+ col = rep(1,2), ylim = c(0,10), ylab = "Population")
> legend("topleft", legend = c("1","3"), lty = c(1,3), bty="n")
```

As Huang points out, τ has no clear physical meaning, and the fact that modest variation in τ leads to such different results leads one to be skeptical of the utility of the equation. However, it does serve as a useful example of how a delay differential equation can be formulated and solved.

The second example, adapted from the `PBSddesolve` reference, shows how to treat a system of DDEs with two dependent variables (Figure 8.10).

```
> require(deSolve)
>
> # Create a function to return derivatives
```

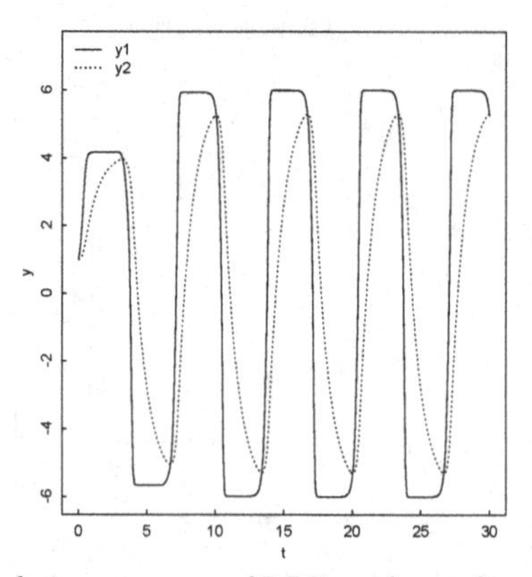

Figure 8.10: *Solution to system of DDEs with two dependent variables.*

```
> derivs = function(t,y,parms) {
+  if (t < tau) lag = yinit else lag = lagvalue(t - tau)
+  dy1 = a * y[1] - (y[1]^3/3) + m*(lag[1] - y[1])
+  dy2 = y[1] - y[2]
+  return(list(c(dy1,dy2))) }
>
> # Define initial values, parameters, and time sequence
> yinit = c(1,1)
> tau = 3; a = 2; m = -5
> times = seq(0,30,0.1)
>
> # Solve the dede system
> yout = dede(y=yinit,times=times,func=derivs,parms=c(tau,a,m))
>
> # Plot the results
> plot(yout[,1], yout[,2], type="l", xlab="t", ylab="y",
+  ylim=c(min(yout[,2], yout[,3]), 1.2*max(yout[,2], yout[,3])))
> lines(yout[,1], yout[,3], lty=3)
> legend("topleft", legend = c("y1","y2"),lty = c(1,3),bty="n")
```

8.8 Differential algebraic equations

Differential algebraic equations (DAEs) contain a combination of ODEs, which are responsible for the evolution of the system, and algebraic equations, which impose constraints on the solution. Common examples arise in chemical reaction kinetics, where differential equations describe the time variation of the concentrations of the chemical species, while equilibrium or conservation equations constrain the allowable values of the concentrations.

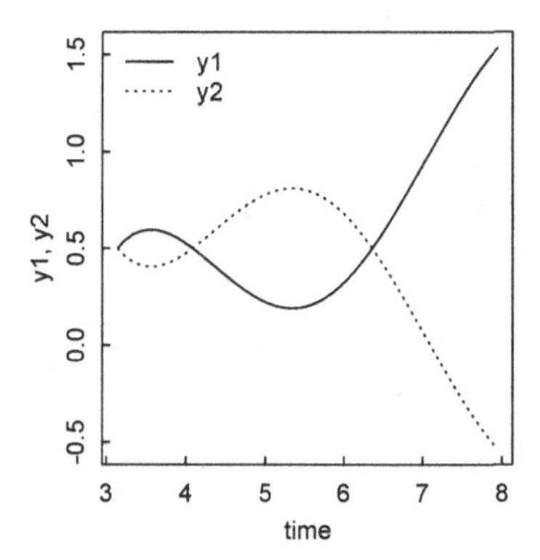

Figure 8.11: *Solution to system of differential algebraic Equations 8.12.*

As Wikipedia (en.wikipedia.org/wiki/Differential_algebraic_equation#Examples) explains, "The solution of a DAE consists of two parts, first the search for consistent initial values and second the computation of a trajectory. To find consistent initial values it is often necessary to consider the derivatives of some of the component functions of the DAE. The highest order of a derivative that is necessary in this process is called the differentiation index. The equations derived in computing the index and consistent initial values may also be of use in the computation of the trajectory."

The `deSolve` package in R contains two DAE solvers, `daspk` and `radau`. `daspk` solves problems with index 0 or 1, and is useful for non-stiff systems. `radau` solves problems with index <= 3, and may be used for stiff systems.

As an example, consider the system of equations

$$\begin{aligned} x(t) - y(t) &= \sin(t) \\ x(t) + y(t) &= 1 \end{aligned} \tag{8.12}$$

with initial conditions $x(\pi) = 1/2$. `daspk` requires that the equations be written in residual form, as in `eq1` and `eq2` in the code below with results shown in Figure 8.11.

```
> require(deSolve)
>
> # Function defining the system
> Res_DAE = function (t, y, dy, pars){
+   y1=y[1]; y2=y[2]; dy1=dy[1]; dy2=dy[2]
+   eq1 = dy1 - y2 - sin(t)
+   eq2 = y1 + y2 - 1
+   return(list(c(eq1, eq2), c(y1,y2)))
+ }
>
> # Time sequence and initial values
> times = seq(pi,8,.1)
```

```
> y = c(y1 = 0.5, y2 = 0.5)
> dy = c(dy1 = 0.5, dy2 = -0.5)
>
> # Solution with daspk
> DAE = daspk(y = y, dy = dy, times = times,
+ res = Res_DAE, parms = NULL, atol = 1e-10, rtol = 1e-10)
>
> # Output and plotting
> time = DAE[,1]; y1 = DAE[,2]; y2 = DAE[,3]
> matplot(time, cbind(y1, y2), xlab = "time", ylab = "y1, y2",
+ type = "l", lty=c(1,3), col = 1)
>
> legend("topleft", legend = c("y1", "y2"), lty = c(1,3),
+ col = 1, bty = "n")
>
```

As another example, this time with a second derivative, we consider the system of equations

$$\begin{aligned} \frac{d^2x}{dt^2} &= y(t) \\ x(t)+4y(t) &= \sin(t) \end{aligned} \tag{8.13}$$

with initial conditions $x(\pi) = 1, \frac{dx}{dt}(\pi) = 0$.

```
> require(deSolve)
>
> Res_DAE = function (t, y, dy, pars){
+   y1 = y[1]; y2 = y[2]; y3 = y[3]
+   dy1 = dy[1]; dy2 = dy[2]; dy3 = dy[3]
+   eq1 = dy1 - 1/4*(cos(t) - dy2)
+   eq2 = dy2 - y3
+   eq3 = dy3-y1
+   return(list(c(eq1, eq2, eq3), c(y1,y2,y3)))
+ }
>
> # times and initial values
> times = seq(pi,7,.1)
> y = c(y1 = -0.25, y2 = 1, y3 = 0)
> dy = c(dy1 = -0.25, dy2 = 0, dy3 = -0.25)
>
> DAE <- daspk(y = y, dy = dy, times = times,
+ res = Res_DAE, parms = NULL, atol = 1e-10, rtol = 1e-10)
>
> time = DAE[,1]; y = DAE[,2]; x = DAE[,3]
>
> matplot(time, cbind(y, x), xlab = "time", ylab = "y,
```

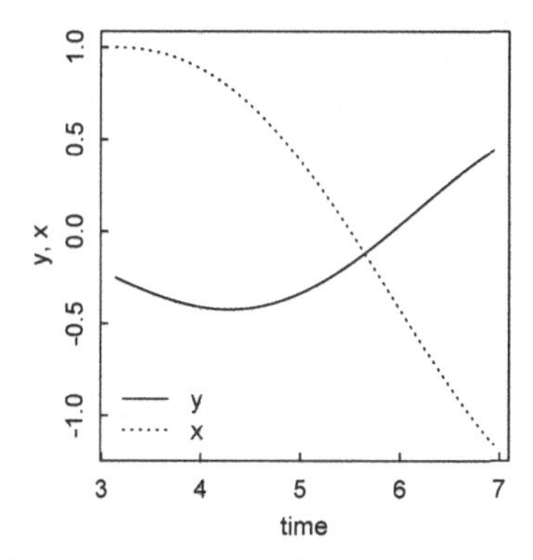

Figure 8.12: *Solution to system of differential algebraic Equations 8.13.*

```
+ x", type = "l", lty=c(1,3), col = 1)
> legend("bottomleft", legend = c("y", "x"), lty = c(1,3),
+ col = 1, bty = "n")
```

The `radau` function in the `deSolve` package is able to solve DAEs with indices up to 3, and can solve stiff systems as well. The equations must be written in the form $\mathbf{M}dy/dt = f(t,y)$ where $\mathbf{M}$ is the "mass matrix." The help page for `radau` explains its many options and gives several instructive examples, including classic pendulum and stiff chemical kinetic examples.

8.9 `rootSolve` for steady state solutions of systems of ODEs

The time evolution of dynamic systems often leads to a steady state. The `rootSolve` package provides the function `steady()`, with several variants depending on the Jacobian of the system, to calculate steady-state values of the system variables. See the help page for a useful example. For another example, consider a solution containing an enzyme obeying reversible Michaelis–Menten kinetics, in which the net velocity v is

$$v = \frac{\frac{V_f}{K_f}S - \frac{V_r}{K_r}P}{1 + \frac{S}{K_f} + \frac{P}{K_r}} \tag{8.14}$$

where V_f and V_r are maximum velocities in the forward and reverse directions and K_f and K_r are the Michaelis constants (dissociation constants of the enzyme–substrate or enzyme–product complexes). A flux F_{in} of S is supplied to the system, and a flux F_{out} of P is removed from the system. The kinetic behavior is modeled by the user-defined function `enzyme()`. Solving the system of differential equations for the concentrations of S and P for a particular set of parameters gives the following behavior.

```
> require(deSolve)
> enzyme = function(t, state, pars) {
+ with (as.list(c(state,pars)), {
```

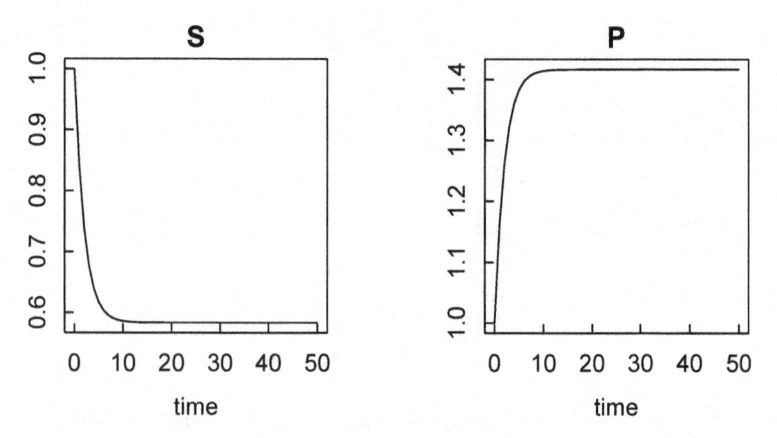

Figure 8.13: *Decrease in substrate S and increase in product P according to Michaelis–Menten Equation 8.14.*

```
+   dS = Fin - (Vf/Kf*S - Vr/Kr*P)/(1 + S/Kf + P/Kr)
+   dP = (Vf/Kf*S - Vr/Kr*P)/(1 + S/Kf + P/Kr) - Fout
+   return(list(c(dS,dP)))})
+ }
```

Note that in the code above, the variable `state` corresponds to the set of variable concentrations, used to calculate dS and dP, that are passed between function and solver.

```
> pars = list(Fin = 0.1, Fout = 0.1, Vf = 1, Vr = 0.5,
+ Kf = 1, Kr = 2)
>
> out = ode (y = c(S = 1, P = 1), times = 0:50,
+ func = enzyme, parms = pars)
> plot(out)
```

The concentrations appear to reach steady-state levels in a relatively short period of time. To calculate those levels, and the time required to reach them, we use the "`runsteady`" method of the `steady()` function.

```
> require (rootSolve)
> ysteady = steady(y = c(S = 1, P = 1), time = c(0,100), func = enzyme,
+ parms = pars, method = "runsteady")
> ysteady$y
        S         P
0.5833333 1.4166667
```

The function `steady()` returns a list. The steady-state values of S and P are given in ysteady$y. Other attributes of the calculation can be accessed by asking for the *structure* of ysteady.

```
> str(ysteady)
List of 1
```

```
 $ y: Named num [1:2] 0.583 1.417
  ..- attr(*, "names")= chr [1:2] "S" "P"
 - attr(*, "istate")= int [1:23] 2 1 1 0 0 5 100000 0 0 0 ...
 - attr(*, "rstate")= num [1:5] 3.88 3.88 66.81 0 0
 - attr(*, "precis")= num 8.08e-09
 - attr(*, "steady")= logi TRUE
 - attr(*, "time")= num 66.8
 - attr(*, "steps")= int 75
 - attr(*, "class")= chr [1:3] "steady" "rootSolve" "list"
 - attr(*, "nspec")= int 2
 - attr(*, "ynames")= chr [1:2] "S" "P"
```

The attributes that are probably of greatest interest are "precis", the precision to which a steady state is reached; "steady", which indicates whether a steady state has been reached (TRUE or FALSE); "time", the time in problem units at which the steady state is reached; and "steps", the number of steps required to reach the steady state.

On the other hand, although we should be able to use steady() with the default method "stode", we get

```
> steady(y = c(S = 1, P = 1), time = 100, func = enzyme, parms = pars)
diagonal element is zero
[1] 2
$y
S P
1 1

attr(,"precis")
[1] 0.2
attr(,"steady")
[1] FALSE
attr(,"class")
[1] "steady"    "rootSolve" "list"
attr(,"nspec")
[1] 2
attr(,"ynames")
[1] "S" "P"
Warning messages:
1: In stode(y, time, func, parms = parms, ...) :
  error during factorisation of matrix (dgefa);  singular matrix
2: In stode(y, time, func, parms = parms, ...) : steady-state not
    reached
```

Apparently this difficulty arises because the system has several steady states depending on initial conditions, and the Newton–Raphson solution method flips unpredictably between them. This example serves as a warning that numerical methods, in R or any other language, are not foolproof.

8.10 `bvpSolve` package for boundary value ODE problems

Boundary value problems (BVPs) are systems of ODEs with values and derivatives specified at more than one point—commonly two points, at the boundaries. The `bvpSolve` package contains three functions for solving boundary value problems: `bvpshoot()`, `bvptwpl()`, and `bvpcol()`.

8.10.1 *bvpshoot()*

Perhaps the most common approach to the solution of one-dimensional BVPs is the shooting method, which gains its name from the analogy with the method of training artillery on a distant target: guess an initial direction and velocity for the projectile, try to improve the initial guesses, and repeat until the target is hit. The code for methods such as `bvpshoot()` contains systematic procedures (usually Newtonian iteration) for using the results of previous trials to iterate subsequent guesses. The method can fail if the ODE is highly nonlinear or unstable, because the guesses may need to be unrealistically close to the true value.

For an example of a case where the shooting method does work, we consider the equation (in reduced units) for the height y of a liquid droplet on a flat surface as a function of surface distance x (from Higham and Higham, p. 163):

$$\frac{d^2y}{dx^2} + (1-y)\left[1 + \left(\frac{dy}{dx}\right)^2\right]^{3/2} = 0, \tag{8.15}$$

along with the boundary conditions at the edges of the drop:

$$y(-1) = y(1) = 0. \tag{8.16}$$

We load the `bvpSolve` package (which must of course be installed first), and it causes the loading of two other packages on which it depends:

```
> require(bvpSolve)
Loading required package: bvpSolve
Loading required package: rootSolve
Loading required package: deSolve
Attaching package: bvpSolve
The following object(s) are masked from package:stats:
    approx
```

We define the function, according to Equation 8.15, that returns the first and second derivatives of y

```
> fun = function(t, y, parms)
+  { dy1 = y[2]
+  dy2 = -(1-y[1])*(1 + dy1^2)^(3/2)
+  return(list(c(dy1,
+  dy2))) }
```

and specify the boundary conditions according to Equation 8.16. The first derivatives at the boundaries are unknown, so they are given as `NA`.

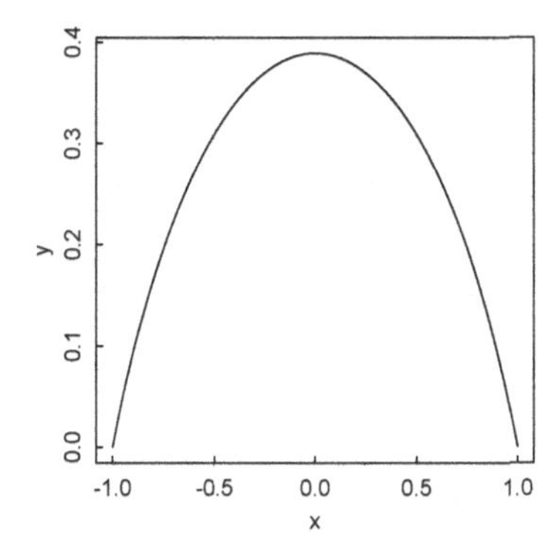

Figure 8.14: *Solution to Equation 8.15 for the shape of a liquid drop on a flat surface, by the shooting method.*

```
> init = c(y = 0, dy = NA)
> end =c(y=0,dy=NA)
```

Now we solve the boundary value problem by the shooting method, providing the initial (`yini`) and final (`end`) values for the ODE system, the sequence over which the independent variable ranges, the function that computes the derivatives, the parameters (none in this case), and the `guess` for the unknown values of the initial conditions. There are other possible inputs to `bvpshoot` depending on the problem; see the help page for details.

```
> sol = bvpshoot(yini = init, x = seq(-1,1,0.01),
+  func = fun, yend = end, parms = NULL, guess = 1)
```

The solution provides x and y vectors, which we use to plot the results shown in Figure 8.14.

```
> x = sol[,1]
> y = sol[,2]
> plot(x,y, type = "l")
```

8.10.2 bvptwp()

We next consider an example of a strongly nonlinear problem with which `bvpshoot()` does not deal well, but which the function `bvptwp()` (twp stands for two-point) handles nicely. The differential equation is

$$\frac{d^2y}{dx^2} = 100y^2 \tag{8.17}$$

with the boundary conditions

$$\begin{aligned} y(0) &= 1 \\ \left(\frac{dy}{dx}\right)_{x=1} &= 0. \end{aligned} \tag{8.18}$$

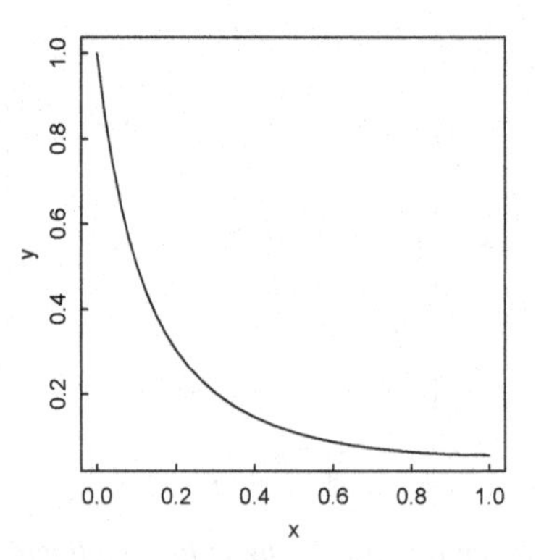

Figure 8.15: *Solution to Equation 8.17 by the two-point method.*

As above, we define the function that returns the derivatives,

```
> fun = function(t, y, p)
+   { dy1 = y[2]
+     dy2 = p*y[1]^2
+     return(list(c(dy1,
+                       dy2))) }
```

define the parameter,

```
> p = 100
```

and specify the initial and final conditions, setting the unknown ones to NA.

```
> init = c(y = 1, dy = NA)
> end =c(y=NA,dy=0)
```

We then solve and plot the solution as before (Figure 8.15).

```
> # Solve bvp
> sol =  bvptwp(yini = init, x = seq(0,1,0.1),
+              func = fun, yend = end, parms = p)
>
> x = sol[,1]
> y = sol[,2]
> plot(x,y, type = "l")
```

8.10.3 bvpcol()

The third function in the bvpSolve package, bvpcol(), is based on FORTRAN code developed for solving multi-point boundary value problems of mixed order. col stands for collocation. The idea of the collocation method "is to choose a finite-dimensional space of candidate solutions (usually, polynomials [often splines] up to a certain degree) and a number of points in the domain (called collocation points), and to select that solution which satisfies the given equation at the collocation points" (Wikipedia).

Here is a simple example from Acton, *Numerical Methods that Work*, p. 157

$$y'' + y = 0 \tag{8.19}$$

subject to the boundary conditions

$$\begin{aligned} y(0) &= 1 \\ y(1) &= 2 \end{aligned} \tag{8.20}$$

for which the analytical solution is $y(t) = 1.7347\sin(t) + \cos(t)$.

Following our by now familiar process, we solve the problem numerically and plot the result.

```
> fun = function(t, y, p)
+   { dy1 = y[2]
+     dy2 = -y[1]
+     return(list(c(dy1,
+   dy2))) }
>
> # initial and final condition; second conditions unknown
> init = c(y = 1, dy = NA)
> end =c(y=2,dy=NA)
>
> # Solve bvp
> sol <- bvpcol(yini = init, x = seq(0,1,0.01),
+   func = fun, yend = end, parms = NULL)
>
> x = sol[,1]
> y = sol[,2]
> plot(x,y, type = "l")
>
> # Verify boundary conditions
> y[1]
[1] 1
> y[length(y)]
[1] 2
```

8.11 Stochastic differential equations: GillespieSSA package

Ordinary differential equations generally assume that the variables are continuous functions. In biological systems, among others, this is not necessarily the case. There may be only a few molecules in a given region of a cell, or a small number of members of a population subject to dynamic processes. Such situations are more properly modeled by stochastic equations, in which processes occur by jumps rather than continuously. An algorithm that generates a possible solution of a set of stochastic equations, obeying the properties (proved by

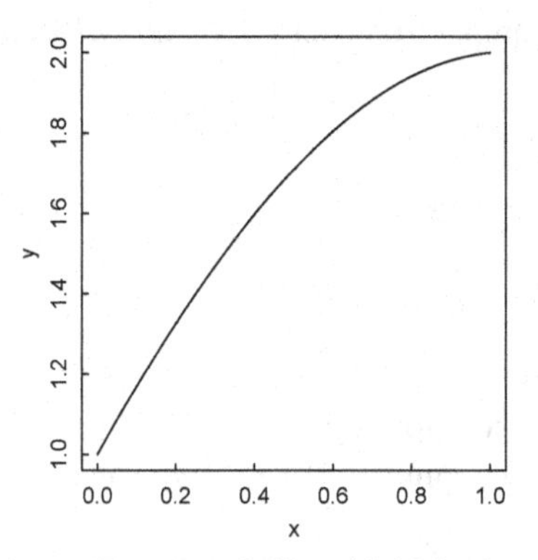

Figure 8.16: *Solution to Equations 8.19 and 8.20 by the collocation method.*

William Feller), that "the time-to-the-next-jump is exponentially distributed and the probability of the next event is proportional to the rate," was developed and popularized by Daniel Gillespie (Gillespie, Daniel T. (1977). "Exact Stochastic Simulation of Coupled Chemical Reactions." *The Journal of Physical Chemistry* **81** (25): 23402361; `http://en.wikipedia.org/wiki/Gillespie_algorithm`). The `GillespieSSA` package provides the function `ssa()` that implements the exact Gillespie algorithm and several approximate "tau-leaping" methods.

As an example, consider a region within a cell containing a few binding sites S to which one of several copies of a protein P may bind, forming a complex SP. The rate constants for formation and dissociation of the complex are k_f and k_r, respectively. The fractional occupancy of the site, $S/(S+SP)$, regulates some further process in the cell. We first consider the system of ODEs, in which the concentrations of the components are treated as continuous variables.

First, load `deSolve` and define the rate equations through the `binding` function.

```
> require(deSolve)
> binding = function(Time, State, Pars) {
+ with(as.list(c(State, Pars)), {
+   rate = kf*S*P - kr*SP
+   dS = -rate
+   dP = -rate
+   dSP = rate
+   return(list(c(dS, dP, dSP)))
+   })
```

Specify the parameter values and the time sequence.

```
> pars = c(kf = 0.005, kr = 0.15)
> yini = c(S = 10, P = 120, SP = 0)
> times = seq(0, 10, by = 0.1)
```

Solve the system of equations and plot the results (Figure 8.17).

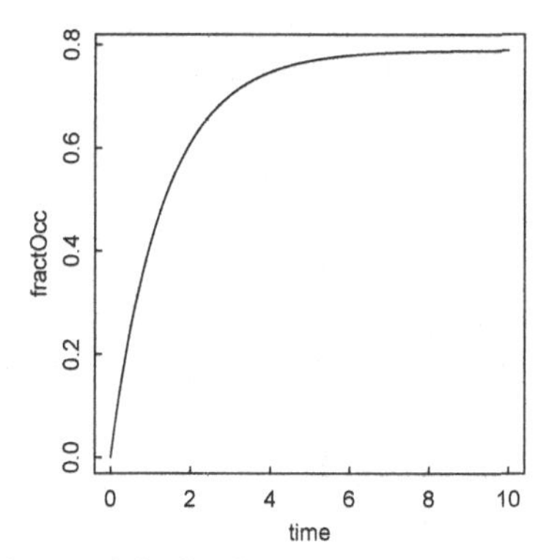

Figure 8.17: *Time dependence of the binding reaction $S + P = SP$ treated as a continuous process.*

```
> out = ode(yini, times, binding, pars)
> time = out[,1]
> fractOcc = out[,4]/(out[,2] + out[,4])
> plot(time, fractOcc, type = "l")
```

Now consider the same reversible binding reaction, treated by the Gillespie algorithm. There are three species and two reaction channels, since the forward `S + P --kf--> SP` and reverse `SP --kr--> S + P` reactions are treated as separate channels in the Gillespie formalism. Load the `GillespieSSA` package (installing it first if that has not been done), then set up the problem, beginning with the parameter values `pars` and the initial state vector `yini`.

```
> require(GillespieSSA)
> pars = c(kf = 0.005, kr = 0.15)
> yini = c(S = 10, P = 120, SP = 0)
```

Formulate the state-change matrix `nu`, with species in rows and reactions in columns:

```
> nu = matrix(c(-1, +1,
               -1, +1,
                1, -1),
                ncol = 2, byrow = TRUE)
```

and the "propensity vector" `a`, the rate for each channel:

```
> a = c("kf*S*P", "kr*SP")
```

Specify the final time `tf` to which the simulation is to be carried out, and the name of the simulation.

```
> tf = 10
> simName = "Reversible Binding Reaction"
```

Start with the direct method `D` in `ssa()`, with verbose output = TRUE giving wall clock time, computed time, and variable values at each `consoleinterval`. (If verbose = FALSE the reaction progress is plotted, but none of the numerical results are printed. However, this is the default because the calculation runs much faster.)

```
> # Direct method
> set.seed(1)
> out = ssa(yini,a,nu,pars,tf,method="D",
simName,verbose=TRUE,consoleInterval=1)
Running D method with console output every 1 time step
Start wall time: 2012-10-31 10:39:47...
t=0 : 10,120,0
(0.003s) t=1.230886 : 6,116,4
(0.005s) t=2.217471 : 5,115,5
(0.006s) t=3.08914 : 3,113,7
(0.007s) t=4.172439 : 2,112,8
(0.009s) t=5.043667 : 4,114,6
(0.01s) t=6.125707 : 5,115,5
(0.012s) t=7.476262 : 3,113,7
(0.013s) t=8.317041 : 2,112,8
(0.014s) t=9.31306 : 5,115,5
t=10.04292 : 3,113,7
tf: 10.04292
TerminationStatus: finalTime
Duration: 0.015 seconds
Method: D
Nr of steps: 33
Mean step size: 0.3043309+/-0.2187604
End wall time: 2012-10-31 10:39:47

> # Plot the result
> ssa.plot(out)
```

The function ssa.plot() provides an easy way to visualize the time course of the set of reactions, along with some useful characteristics of the simulation. Note the clearly fluctuating concentrations (Figure 8.18).

The output of ssa(), out in this case, is a list with three different kinds of information: a matrix out$data about the time dependence of the processes, a list out$stats about the course of the calculation, and a list out$args summarizing the inputs to the calculation.

```
> str(out)  # Structure of out
List of 3
 $ data : num [1:35, 1:4] 0 0.165 0.182 0.204 0.295 ...
  ..- attr(*, "dimnames")=List of 2
  .. ..$ : chr [1:35] "timeSeries" "" "" "" ...
  .. ..$ : chr [1:4] "" "S" "P" "SP"
 $ stats:List of 8
  ..$ startWallime      : chr "2012-10-31 10:39:47"
```

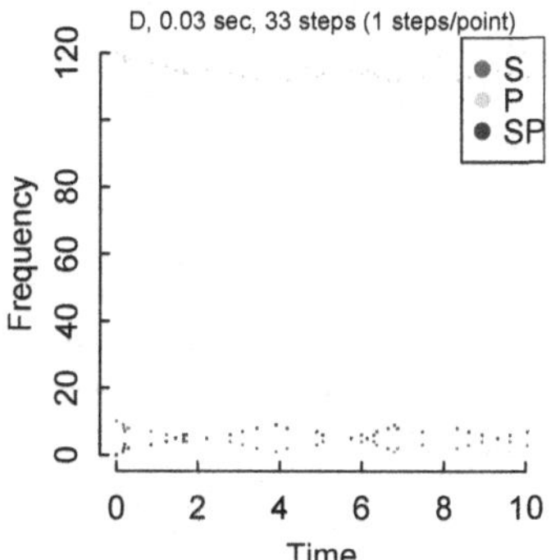

Figure 8.18: *Time dependence of the binding reaction $S + P = SP$ treated as a stochastic process.*

```
  ..$ endWallTime        : chr "2012-10-31 10:39:47"
  ..$ elapsedWallTime    : Named num 0.015
  .. ..- attr(*, "names")= chr "elapsed"
  ..$ terminationStatus : chr "finalTime"
  ..$ nSteps             : int 33
  ..$ meanStepSize       : num 0.304
  ..$ sdStepSize         : num 0.219
  ..$ nSuspendedTauLeaps: num 0
 $ args :List of 18
  ..$ x0                  : Named num [1:3] 10 120 0
  .. ..- attr(*, "names")= chr [1:3] "S" "P" "SP"
  ..$ a                   : chr [1:2] "kf*S*P" "kr*SP"
  ..$ nu                  : num [1:3, 1:2] -1 -1 1 1 1 -1
  ..$ parms               : Named num [1:2] 0.005 0.15
  .. ..- attr(*, "names")= chr [1:2] "kf" "kr"
  ..$ tf                  : num 10
  ..$ method              : chr "D"
  ..$ tau                 : num 0.3
  ..$ f                   : num 10
  ..$ epsilon             : num 0.03
  ..$ nc                  : num 10
  ..$ hor                 : num NaN
  ..$ dtf                 : num 10
  ..$ nd                  : num 100
  ..$ ignoreNegativeState: logi TRUE
  ..$ consoleInterval    : num 1
  ..$ censusInterval     : num 0
  ..$ verbose             : logi TRUE
  ..$ simName             : chr "Reversible Binding Reaction"
```

The vector of time is out$data[,1] and the vectors of concentrations are out$data[,2], out$data[,3], and out$data[,4] for S, P, and SP, respectively. We can use these vectors to calculate the time course of the fractional population of the sites that are bound:

$$f_{bound} = \frac{SP}{S+SP}. \tag{8.21}$$

We do so while comparing the direct (exact) Gillespie method with the three approximate tau-leap methods included as optional methods in ssa. These are intended to speed up the calculation by skipping some time steps according to their underlying algorithms. We plot the results (Figure 8.19) and include in the title of each plot the elapsed time as obtained from out$stats$elapsedWallTime.

```
> par(mfrow = c(2,2))  # Prepare for four plots
```

Direct method:

```
> set.seed(1)
> out = ssa(yini,a,nu,pars,tf,method="D",simName,
+ verbose=FALSE,consoleInterval=1)
> et = as.character(round(out$stats$elapsedWallTime,4)) #elapsed time
> time = out$data[,1]
> fractOcc = out$data[,4]/(out$data[,2] + out$data[,4])
> plot(time, fractOcc, pch = 16, cex = 0.5, main = paste("D ",et, " s"))
```

Explicit tau-leap method:

```
> set.seed(1)
> out = ssa(yini,a,nu,pars,tf,method="ETL",simName,
+   tau=0.003,verbose=FALSE,consoleInterval=1)
> et = as.character(round(out$stats$elapsedWallTime,4)) #elapsed time
> time = out$data[,1]
> fractOcc = out$data[,4]/(out$data[,2] + out$data[,4])
> plot(time, fractOcc, pch = 16, cex = 0.5,
+   main = paste("ETL ",et, "s"))
```

Binomial tau-leap method:

```
> set.seed(1)
> out = ssa(yini,a,nu,pars,tf,method="BTL",simName,
+   verbose=FALSE,consoleInterval=1)
> et = as.character(round(out$stats$elapsedWallTime,4)) #elapsed time
> time = out$data[,1]
> fractOcc = out$data[,4]/(out$data[,2] + out$data[,4])
> plot(time, fractOcc, pch = 16, cex = 0.5,
+   main = paste("BTL ",et, "s"))
```

Optimized tau-leap method:

```
> set.seed(1)
> out = ssa(yini,a,nu,pars,tf,method="OTL",simName,
+   verbose=FALSE,consoleInterval=1)
Warning messages:
1: In FUN(newX[, i], ...) : coercing argument of type 'double' to logical
2: In FUN(newX[, i], ...) : coercing argument of type 'double' to logical
3: In FUN(newX[, i], ...) : coercing argument of type 'double' to logical
```

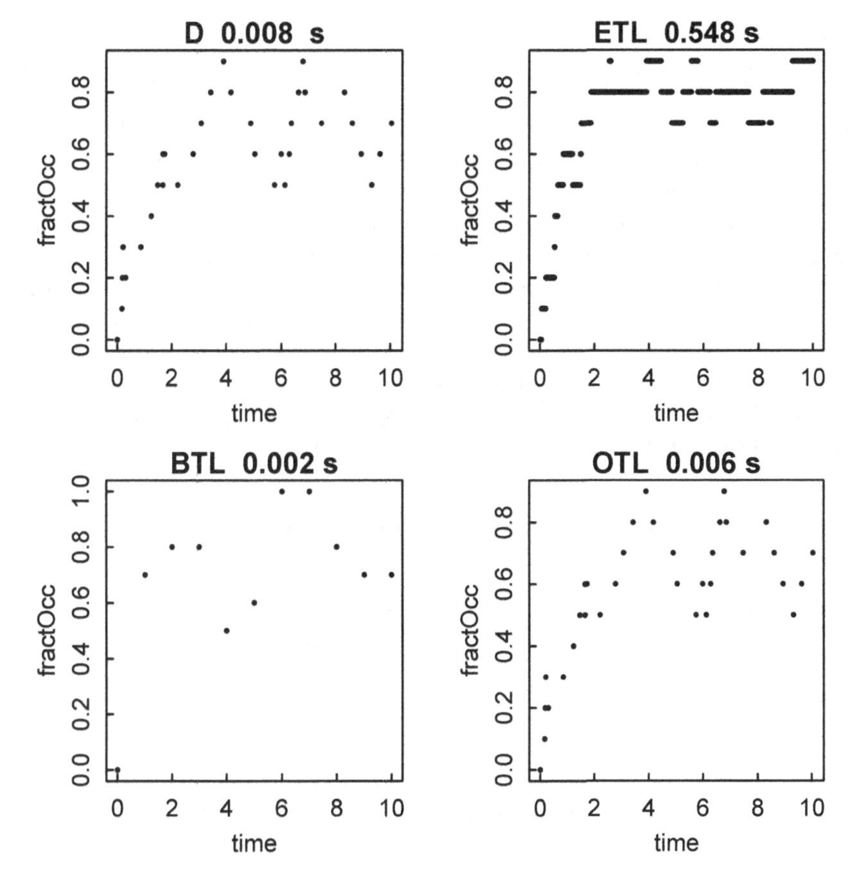

Figure 8.19: *Fractional occupancy of binding sites calculated according to the direct and three tau-leap methods of the Gillespie algorithm.*

```
> et = as.character(round(out$stats$elapsedWallTime,4)) #elapsed time
> time = out$data[,1]
> fractOcc = out$data[,4]/(out$data[,2] + out$data[,4])
> plot(time, fractOcc, pch = 16, cex = 0.5,
+   main = paste("OTL ",et, "s"))
```

There are considerable differences among the results from methods and, at least for this example, the direct method is not the slowest. In most cases, it will be the slowest (as seen in the examples referenced in the next paragraph), though it is somewhat unpredictable which method will be the fastest and most efficient.

The `GillespieSSA` package contains some instructive models, useful as modifiable templates, that can be called with `demo(GillespieSSA)`. They include a decaying-dimerization reaction set, a linear chain polymerization system, a logistic growth model, two predator–prey models, and two models of infectious processes. The help page for `ssa` provides some related models as examples.

8.12 Case studies

8.12.1 Launch of the space shuttle

Projectile motion is a good exercise ground for the numerical solution of ordinary differential equations. Here we consider the first two minutes of launch of the space shuttle, basing our treatment on that of VanWyk, 2008, p. 166.

The factors that would have to be taken into account if the vehicle lifted straight up are the mass of the shuttle, booster rockets and fuel that must be lifted against the nearly—but not quite—constant pull of gravity; the burn rate of the fuel that reduces the mass after ignition and lift-off; the thrust of the engines; and the air resistance that decreases with altitude. In addition, the shuttle does not rise straight up, since the thrust angle with respect to launch direction is changed at a constant rate to direct the vehicle over the ocean, from which its ejected booster rockets can be recovered. These considerations lead to the equations of motion

$$m\frac{d^2y}{dt^2} = (thrust - drag)\cos(\varepsilon t) - mg\left(\frac{R}{R+y}\right)^2 \tag{8.22}$$

and

$$m\frac{d^2x}{dt^2} = (thrust - drag)sin(\varepsilon t) \tag{8.23}$$

with the initial conditions

$$y(0) = x(0) = 0,\ y'(0) = x'(0) = 0. \tag{8.24}$$

The mass decreases with time according to

$$m = m_0 - burn.rate \times t \tag{8.25}$$

and the drag is calculated as

$$drag = \frac{1}{2}\rho A C_{drag} v^2 \tag{8.26}$$

where the air density decreases with altitude according to

$$\rho = \rho_0 e^{-y/8000}. \tag{8.27}$$

x and y are measured in meters while the other lengths are measured in km.

Values for the various factors are quantified in the parameters listed in the code below.

```
> # Parameters
> m0 = 2.04e6  # Initial mass, kg
> burn.rate = 9800  # kg/s
> R = 6371 # Radius of earth, km
> thrust = 28.6e6 # Newtons
> dens0 = 1.2 # kg/m^3 Density of air at earth surface
```

```
> A = 100  # m^2, cross-section of launch vehicle
> Cdrag = 0.3  # Drag coefficient
> eps = 0.007  # radians/s, rate of angular change
> g = 9.8 #
```

The equations of motion are expressed in terms of a vector y whose components are the x- and y- positions and velocities.

```
> # Equations of motion
> launch = function(t, y,parms) {
+  xpos = y[1]
+  xvel = y[2]
+  ypos = y[3]
+  yvel = y[4]
+  airdens = dens0*exp(-ypos/8000)
+  drag = 0.5*airdens*A*Cdrag*(xvel^2 + yvel^2)
+  m = m0-burn.rate*t
+  angle = eps*t
+  grav = g*(R/(R+ypos/1000))^2
+  xaccel = (thrust - drag)/m*sin(angle)
+  yaccel = (thrust - drag)/m*cos(angle) - grav
+  list(c(xvel, xaccel, yvel, yaccel))
+ }
```

We next specify the initial values of the positions and velocities, and the times over which the solution is to be calculated (every second for two minutes).

```
> # Initial values
> init = c(0,0,0,0)
>
> # Times
> times = 0:120
```

We load the `deSolve` package and, since the differential equations are not stiff, use the `"adams"` method of solution.

```
> # Solve with Adams method
> require(deSolve)
> out = ode(init, times, launch, parms=NULL, method="adams")
```

Finally, we plot the x-y coordinates, expressed in km, at each second of the launch (Figure 8.20).

```
> # Plot results
> time = out[,1]; x = out[,2]; y = out[,4]
> plot(x/1000,y/1000, cex=0.5, xlab="x/km", ylab="y/km")
```

8.12.2 Electrostatic potential of DNA solutions

DNA is perhaps the most highly charged molecule found in nature. As Watson and Crick showed, B-form DNA has two negative phosphate charges every 0.34 nm along

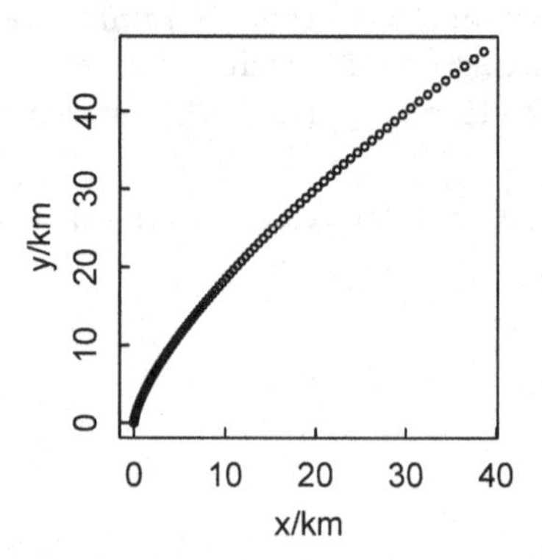

Figure 8.20: *Height vs. horizontal distance for the first 120 seconds of the space shuttle launch.*

a double helical backbone of radius 1 nm. Therefore, it interacts very strongly with other charged molecules, including other DNA molecules. To understand how DNA is tightly coiled and packaged in small volumes such as virus capsids, one needs to calculate the electrostatic repulsions between nearby DNA segments. The strength of electrostatic interactions is modulated by the concentration of small ions, such as salt, in the surrounding solution. The influence of ions on the electrostatic potential ϕ is given by the Debye–Hückel equation

$$\nabla^2\phi = -\frac{\kappa^2}{2I}\sum_i Z_i c_i e^{-Z_i\phi} \tag{8.28}$$

where κ is the inverse Debye length (nm), I is the ionic strength (molar), and Z_i and c_i are the charge and molar concentration of the ith ionic species:

$$I = \frac{1}{2}\sum_i c_i Z_i^2 \tag{8.29}$$

$$\kappa^{-1} = \frac{0.304}{\sqrt{I}} \tag{8.30}$$

We model DNA as a cylindrical rod with charge distributed uniformly on its surface. In cylindrical coordinates where there is no dependence on height or angle, the Laplacian operator can be written in terms of ρ, the distance from the rod axis to a point in solution, as

$$\nabla^2\phi = \frac{\partial^2\phi}{\partial\rho^2} + \frac{1}{\rho}\frac{\partial\phi}{\partial\rho} \tag{8.31}$$

Defining the dimensionless variable $x = \kappa\rho$ and $z = \ln x$, and confining our calculation to a uni-univalent salt such as NaCl at molar concentration c, Equation 8.28 can be written

$$\frac{\partial^2\phi}{\partial z^2} = -\frac{ce^{2z}}{2I}\left(e^{-\phi} - e^{\phi}\right) = -\frac{e^{2z}}{2}\left(e^{-\phi} - e^{\phi}\right). \tag{8.32}$$

Since this is a second-order differential equation, it needs two boundary conditions for a complete solution. One is the gradient of the potential at the helical rod surface, which can be written

$$\left(\frac{\partial \phi}{\partial z}\right)_{z=\ln \kappa a} = -4\pi\sigma/\varepsilon \tag{8.33}$$

where σ is the surface charge density and ε is the dielectric constant. For double-stranded DNA in the units we are using, $4\pi\sigma/\varepsilon$ = -0.84.

The second boundary condition depends on the environment in which the DNA finds itself. If it is effectively alone in dilute solution, then $\phi \to 0$ as $z \to \infty$. But if the DNA is in relatively concentrated solution, a different consideration holds. As stated by Bloomfield et al. (1980) "In an extensive array of parallel, equally spaced rods, a different boundary condition applies. Halfway between any two rods the potential will be a minimum, corresponding to equally balanced electrical forces perpendicular to the normal plane between the two rods. We then assume that we can approximate the polygonally shaped minimum potential surface surrounding any rod by a circular one with radius R/2, where R is the center-to-center distance between nearest neighbor rods." At that distance,

$$\left(\frac{\partial \phi}{\partial z}\right)_{z=\ln \kappa R/2} = 0 \tag{8.34}$$

We are now in a position to solve the boundary value problem for the potential as a function of distance from the surface of the DNA helix modeled as a cylindrical rod. We can first try the shooting method, but find that it fails. However, the functions `bvptwp()` and `bvpcol()` succeed, as shown in Figure 8.21. Note that to change from `bvptwp()` to `bvpcol()`, all that need be done is change the function name in the code.

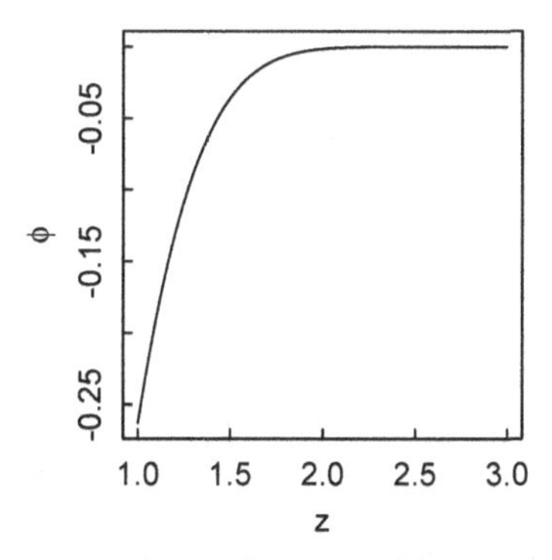

Figure 8.21: *Electrostatic potential as a function of distance from the surface of double-stranded DNA, surrounded by an array of parallel DNA molecules at an average distance of 3 nm center-to-center.*

```
> require(bvpSolve)
> fun = function(z,phi,parms) {
+  dphi1 = phi[2]
+  dphi2 = -1/2*exp(2*z)*(exp(-phi[1])-exp(phi[1]))
+  return(list(c(dphi1,dphi2)))
+ }
> init = c(phi=NA, dphi = 0.84)
> end = c(phi=NA, dphi = 0)
>
> sol = bvptwp(yini = init, x = seq(1,3,len=100), func=fun, yend = end)
>
> z = sol[,1]
> phi = sol[,2]
> plot(z,phi,type="l", ylab=expression(phi))
```

8.12.3 *Bifurcation analysis of Lotka–Volterra model*

Lotka–Volterra type models are instructive in elucidating predator–prey relations in ecology, and are also good models for analyzing the behavior of systems of differential equations. Here we follow very closely part of a 2003 article by Thomas Petzoldt, "R as a Simulation Platform in Ecological Modelling"[1], which constructs and analyzes a three-component system and takes the additional useful step of showing how to display the bifurcation behavior of the model. His treatment is based on a three-component food web model developed by Blasius et al. (1999) and Blasius and Stone (2000).

The model consists of three populations: plant resource u, herbivore v, and carnivore w. The set of differential equations describing the system is

$$\frac{du}{dt} = au - \alpha_1 f_1(u,v) \tag{8.35}$$

$$\frac{dv}{dt} = -bv + \alpha_1 f_1(u,v) - \alpha_2 f_2(v,w) \tag{8.36}$$

$$\frac{dw}{dt} = -c(w - w*) + \alpha_2 f_2(v,w) \tag{8.37}$$

with a logistic interaction term due to Holling

$$f_i(x,y) = \frac{xy}{1+k_i x}. \tag{8.38}$$

This interaction term, since it includes saturation, is probably more realistic than the simpler Lotka–Volterra $f_i(x,y) = xy$. Another refinement that enhances the realism of the model is $w*$, a minimum predator level that stabilizes the population when the prey population is low by recognizing that predators can consume alternative, albeit less desirable, prey.

We begin by loading the `deSolve` package. Petzoldt used the older `odesolve` package, which has since been removed from the R library.

[1] online at `http://www.r-project.org/doc/Rnews/Rnews_2003-3.pdf, pp. 8--16`

```
> library(deSolve)
```

We then proceed in the by now familiar way to define the functions for the interactions and for the time derivatives of the populations.

```
> f = function(x,y,k){x*y/(1+k*x)}
> model = function(t, xx, parms) {
+ u = xx[1]  # plant resource
+ v = xx[2]  # herbivore
+ w = xx[3]  # carnivore
+ with(as.list(parms),{
+ du = a*u - alpha1*f(u, v, k1)
+ dv = -b*v + alpha1*f(u, v, k1) - alpha2*f(v, w, k2)
+ dw = -c*(w - wstar) + alpha2*f(v, w, k2)
+ list(c(du, dv, dw))
+ })}
```

Next we define the times over which the simulation is to be carried out, the parameters in the calculation, and the starting values for the three populations.

```
> times = seq(0, 200, 0.1)
> parms = c(a=1, b=1, c=10, alpha1=0.2, alpha2=1,
+ k1=0.05, k2=0, wstar=0.006)
> xstart = c(u=10, v=5, w=0.1)
```

We then solve the model using the `lsoda` method as a function, and extract the time and population vectors for plotting.

```
> out = lsoda(xstart, times, model, parms)
> t = out[,1]
> u = out[,2]
> v = out[,3]
> w = out[,4]
```

We plot the three populations, which appear to oscillate in a rather unpredictable fashion but more or less in phase with one another (Figure 8.22). Blasius and coworkers call this UPCA (uniform phase, chaotic amplitude) behavior. Note how close w, the population of carnivores, comes to extinction at times, but is saved by $w*$.

```
> par(mfrow=c(1,3))
> plot(t, u, type="l", lty=1)
> plot(t, v, type="l", lty=1)
> plot(t, w, type="l", lty=1)
> par(mfrow = c(1,1))
```

We conclude by making a bifurcation diagram, which demonstrates how the dynamics of a process splits in two at certain values of a control parameter. In this case the predator–independent herbivore loss rate b is used as the control parameter. Bifurcations occur at the maxima or minima of the predator variable w. Thus, we first define a function to pick peaks and troughs, at which the amplitudes are greater than, or less than, their immediate neighbors to left and right.

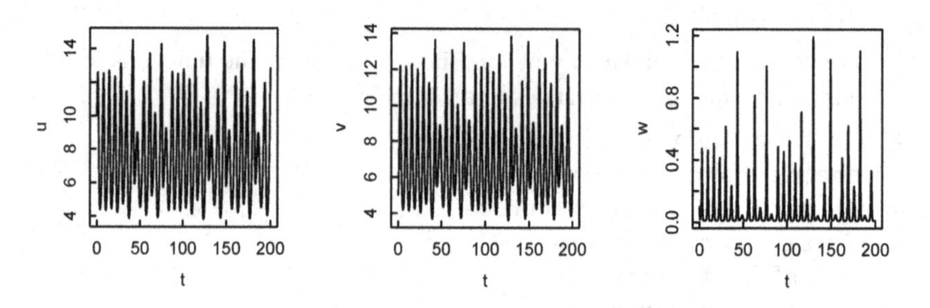

Figure 8.22: *Time course of the three-population model of resource u, consumer v, and predator w, illustrating uniform phase but chaotic amplitude behavior.*

```
> peaks = function(x) {
+ l = length(x)
+ xm1 = c(x[-1], x[l])
+ xp1 = c(x[1], x[-l])
+ x[x > xm1 & x > xp1 | x < xm1 & x < xp1] # Max or min
+ }
```

We next set up the axes, coordinates, and labeling of a plot, to be filled as the bifurcation modeling process proceeds.

```
> plot(0,0, xlim=c(0,2), ylim=c(0,1.5), type="n", xlab="b", ylab="w")
```

We embed the integration of the system of differential equations in a loop that varies b.

```
> for (b in seq(0.02,1.8,0.01)) {
+ parms["b"] = b
+ out = as.data.frame(lsoda(xstart, times,
+ model, parms, hmax=0.1))
```

Only the last third of the peaks are identified and plotted, to show the behavior at the end of each simulated time series (Figure 8.23)

```
+ l = length(out$w) %/% 3
+ out = out[(2*l):(3*l),]
+ p = peaks(out$w)
+ l = length(out$w)
+ xstart = c(u=out$u[l], v=out$v[l], w=out$w[l])
+ points(rep(b, length(p)), p, pch=".")
+ }
```

We can see that there is "a period-doubling route to chaos followed by a period-doubling reversal as the control parameter b is increased" (Blasius and Stone, 2000).

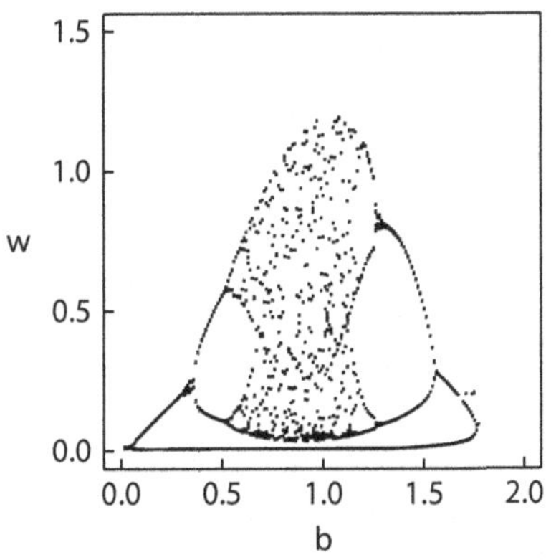

Figure 8.23: *Bifurcation diagram for the three-population model, with the predator–independent herbivore loss rate b as the control parameter. Bifurcations occur at the extrema of the predator variable w.*

To complete this series of solutions to the system of differential equations and plot the points on the bifurcation plot took 69 seconds on a ca. 2012 MacBook Air laptop.

Chapter 9

Partial differential equations

Partial differential equations (PDEs) arise in all fields of science and engineering. In contrast to ordinary differential equations, they involve more than one independent variable, often time and one or more position variables, or several spatial variables.

The most common approach to solving PDEs numerically is the method of lines: one discretizes the spatial derivatives and leaves the time variable continuous. This leads to a system of ordinary differential equations to which one of the methods discussed in the previous chapter for initial value ODEs can be applied.

R has three packages, `ReacTran`, `deSolve`, and `rootSolve`, that together contain most of the tools needed to solve most commonly encountered PDEs. The task view DifferentialEquations lists resources for PDEs as well as for the various types of ODEs discussed in the previous chapter.

PDEs are commonly classified into three types: parabolic (time-dependent and diffusive), hyperbolic (time-dependent and wavelike), and elliptic (time-independent). We shall give examples of how each of these may be solved with explicit R code, before showing how the functions in `ReacTran`, `deSolve`, and `rootSolve` can be used to solve such problems concisely and efficiently.

In preparing the first part of this chapter I have drawn heavily on Garcia, *Numerical Methods for Physics*, Chs. 6-9. The latter part of the chapter, focusing on the `ReacTran` package, is based on the work of Soetaert and coworkers, *Solving Differential Equations in R* and *A Practical Guide to Ecological Modelling: Using R as a Simulation Platform*, which—along with the help pages and vignettes for the package—should be consulted for more details and interesting examples.

9.1 Diffusion equation

The diffusion equation (Fick's 2nd law) in one spatial dimension,

$$\frac{\partial C}{\partial t} = D\frac{\partial^2 C}{\partial x^2}, \tag{9.1}$$

is, like the heat conduction equation, a parabolic differential equation. (In the heat conduction equation, the concentration C is replaced by the temperature T, and the diffusion coefficient D is replaced by the thermal diffusion coefficient κ.)

To solve the diffusion equation numerically, a common procedure is to discretize the time derivative using the Euler approximation

$$\frac{\partial C}{\partial t} \Rightarrow \frac{C(t_i + \triangle t, x_j) - C(t_i, x_j)}{\triangle t} \tag{9.2}$$

and the spatial second derivative using the centered approximation.

$$\frac{\partial^2 C}{\partial x^2} \Rightarrow \frac{C(t_i, x_j + \triangle x) + C(t_i, x_j - \triangle x) - 2C(t_i, x_j)}{\triangle x^2} \tag{9.3}$$

Rearranging, we find that the concentration at time point $i+1$ can be computed as follows.

$$C(i+1, j) = C(i, j) + A[C(i, j+1) + C(i, j-1) - 2C(i, j)] \tag{9.4}$$

where

$$A = \frac{D\triangle t}{\triangle x^2} \tag{9.5}$$

This is the equation, along with suitable boundary conditions, that we shall use to compute the time-evolution of the concentration profile.

The analytic solution to the one-dimensional diffusion equation, in which the concentration is initially a spike of magnitude C_0 at the origin x_0 and zero everywhere else, is well-known to be

$$C(t, x) = \frac{C_0}{\sqrt{2\pi\sigma^2}} \exp\left[-\frac{(x - x_0)^2}{2\sigma^2}\right] \tag{9.6}$$

where the standard deviation σ is

$$\sigma = \left\langle (x - x_0)^2 \right\rangle^{1/2} = \sqrt{2Dt}. \tag{9.7}$$

In other words, the initially very sharp peak broadens with the square root of the elapsed time. It is this behavior that we shall demonstrate in R. In the code below, note that the initialization and updating of C maintains the boundary conditions of $C = 0$ at the boundaries.

Set the parameters of the diffusion process. An important consideration in choosing the time and distance increments is that the coefficient $A = D\triangle t/\triangle x^2$ must be $\leq 1/2$ for the computation to be stable.

```
> dt=3 #Timestep,s
> dx = .1 # Distance step, cm
> D = 1e-4 # Diffusion coeff, cm^2/s
> (A = D*dt/dx^2) # Coefficient should be < 0.5 for stability
[1] 0.03
```

Discretize the spatial grid and set the number of time iterations.

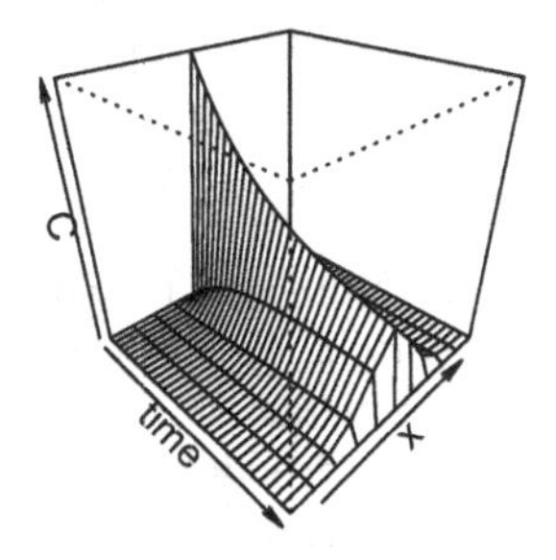

Figure 9.1: *Perspective plot of the evolution of a sharp concentration spike due to diffusion.*

```
> L=1 #Length from -L/2 to L/2
> n = L/dx + 1 # Number of grid points
> x = seq(-L/2,L/2,dx) # Location of grid points
> steps = 30 # Number of iterations
> time = 0:steps
```

Initialize concentrations to 0 except for the spike at the center of the grid.

```
> C = matrix(rep(0, (steps+1)*n), nrow = steps+1, ncol = n)
> C[1, round(n/2)] = 1/dx # Initial spike at central point
```

Loop over time and space variables, building a matrix for the subsequent perspective plot.

```
> # Loop over desired number of time steps
> for(i in 1:(steps-1)) {
+ # Compute new concentration profile at each time #
+   for(j in 2:(n-1)) {
+   C[i+1,j] = C[i,j] + A*(C[i,j+1] + C[i,j-1] - 2*C[i,j])
+   }
+ }
```

Finally, plot a perspective view of the concentration evolution in space and time (Figure 9.1).

```
> persp(time, x, C, theta = 45, phi = 30)
```

9.2 Wave equation

The one-dimensional wave equation

$$\frac{\partial^2 W}{\partial t^2} = c^2 \frac{\partial^2 W}{\partial x^2}, \tag{9.8}$$

where W is the displacement and c the wave speed, is a typical example of a hyperbolic PDE. A simplified version (see Garcia, p. 216) is the advection equation

$$\frac{\partial y}{\partial t} = -c\frac{\partial y}{\partial x}, \tag{9.9}$$

which describes the evolution of the scalar field $y(t,x)$ carried along by a flow of constant speed c moving to the right if $c > 0$.The advection equation is the simplest example of a flux conservation equation.

The analytical solution of the advection equation, with initial condition $y(0,x) = y_0(x)$ is simply $y(t,x) = y_0(x-ct)$. However, the numerical solution is by no means trivial, and in fact the forward- in-t, centered-in-x approach that worked for parabolic equations does not work for the advection equation.

As in the previous section, we replace the time derivative by its forward Euler approximation

$$\frac{\partial y}{\partial t} \Rightarrow \frac{y(t_i+\Delta t, x_j) - y(t_i, x_j)}{\Delta t} \tag{9.10}$$

and the space derivative by the centered discretized approximation

$$\frac{\partial y}{\partial x} \Rightarrow \frac{y(t_i, x_j+\Delta x) - y(t_i, x_j-\Delta x)}{2\Delta x} \tag{9.11}$$

Combining and rearranging leads to the equation for y at timepoint $i+1$,

$$y(i+1,j) = y(i,j) - \frac{c\Delta t}{2\Delta x}[y(i,j+1) - y(i,j-1)] \tag{9.12}$$

once we provide the initial condition and boundary conditions. We use as initial condition a Gaussian pulse, and impose cyclic boundary conditions, so that grid points x_n and x_1 are adjacent.

9.2.1 FTCS method

We first try the forward-in-time, centered-in-space (FTCS) method. Set the parameters to be used in the calculation.

```
> dt=.002 #Timestep,s
> n = 50 # number of grid points
> L=1 # Length from -L/2 to L/2, cm
> (dx = L/n) # Distance step, cm
[1] 0.02
> v=1 #Wavespeed, cm/s
> (A = v*dt/(2*dx)) # Coefficient
[1] 0.05
> (steps = L/(v*dt)) # Number of iterations
[1] 500
> time = 0:steps
> (tw = dx/v) # Characteristic time to move one step
[1] 0.02
```

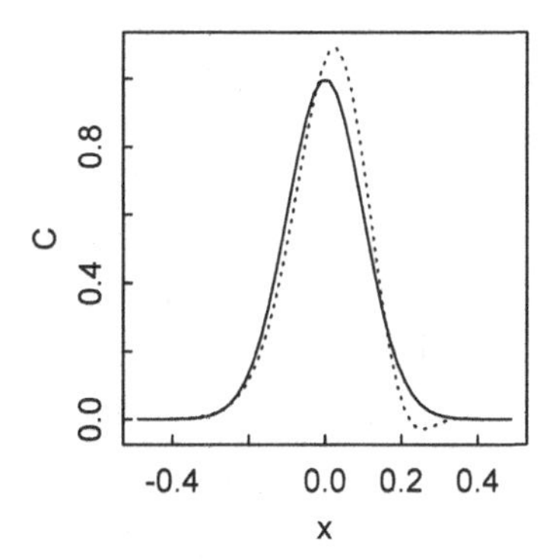

Figure 9.2: *Advection of a Gaussian pulse calculated according to the FTCS method.*

Set the locations of the grid points and initialize the space-time matrix of concentration values.

```
> x = (1:n - 0.5)*dx - L/2 # Location of grid points
> sig = 0.1 # Standard deviation of initial Gaussian wave
> amp0 = exp(-x^2/(2*sig^2)) # Initial Gaussian amplitude
> C = matrix(rep(0, (steps+1)*n), nrow = steps+1, ncol = n)
> C[1,] = amp0 # Initial concentration distribution
```

Establish periodic boundary conditions.

```
> jplus1 = c(2:n,1)
> jminus1 = c(n,1:(n-1))
```

For the body of the calculation, loop over the desired number of time steps and compute the new concentration profile at each time.

```
> for(i in 1:steps) {  # Loop over desired number of steps
+   for(j in 1:n) { # Compute new C profile at each time
+     C[i+1,j] = C[i,j] + A*( C[i,jplus1[j]] - C[i,jminus1[j]] )
+     }
+    }
```

Finally, plot the initial and final concentration profiles (Figure 9.2).

```
> plot(x, C[1,], type = "l", ylab = "C", ylim = c(min(C), max(C)))
> lines(x, C[steps, ], lty = 3)
```

If the advection equation were properly solved by this method, the two waveforms should be superimposable. Instead, distortion occurs as the wave propagates. It can be shown that in fact there is no stable solution for any value of the characteristic time dx/v.

9.2.2 Lax method

A more successful method, due to Lax, is to replace $C[i,j]$ with the average of its left and right neighbors. Also, the best result is obtained if the time step is neither

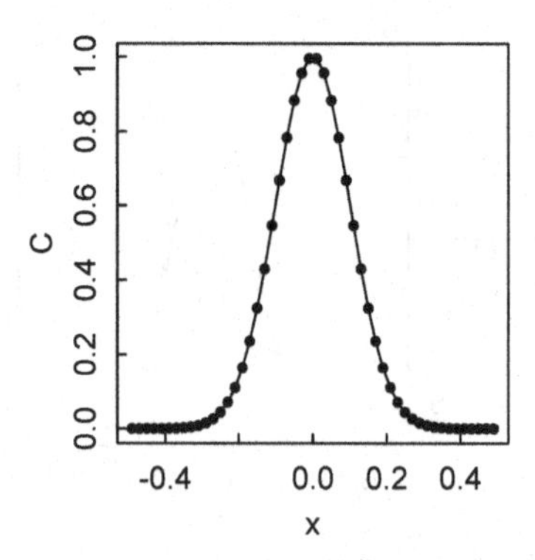

Figure 9.3: *Advection of a Gaussian pulse calculated according to the Lax method.*

too large (the calculation becomes unstable) nor too small (the pulse decays as it progresses). It can be shown that the optimum time step is $dt = dx/v$. The code is exactly the same as for the FTCS method, except for the body of the calculation, where the looping over the desired number of time steps and computation of the new concentration profile at each time takes place. The result is shown in Figure 9.3.

```
> # Loop over desired number of steps #
> for(i in 1:steps) {
+   # Compute new concentration profile at each time #
+   for(j in 1:n) {
+     C[i+1,j] = 0.5*(C[i,jplus1[j]] + C[i, jminus1[j]]) +
+     A*(C[i,jplus1[j]] - C[i,jminus1[j]])
+     }
+   }
```

A still better approach, as explained by Garcia (pp. 222–4), is the Lax–Wendorff method, which uses a second-order finite difference scheme to treat the time derivative. This yields Equation 9.13 for the updating of the advection equation:

$$C_j^{i+1} = C_j^i - A\left(C_{j+1}^i - C_{j-1}^i\right) + 2A^2\left(C_{j+1}^i + C_{j-1}^i - 2C_j^i\right) \tag{9.13}$$

9.3 Laplace's equation

The Laplace equation in two dimensions

$$\frac{\partial^2 V}{\partial x^2} + \frac{\partial^2 V}{\partial y^2} = 0 \tag{9.14}$$

is an example of the third type of PDE, an elliptic equation. It arises frequently in electrostatics, gravitation, and other fields in which the potential V is to be calculated as a function of position. If there are charges or masses in the space, and if we generalize to three dimensions, the equation becomes the Poisson equation

$$\frac{\partial^2 V}{\partial x^2} + \frac{\partial^2 V}{\partial y^2} + \frac{\partial^2 V}{\partial z^2} = f(x,y,z) \tag{9.15}$$

Depending on the geometry of the problem, the equation may also be written in spherical, cylindrical, or other coordinates.

To solve an elliptic equation of this type, one must be given the boundary conditions. Typically, these specify that certain points, lines, or surfaces are held at constant values of the potential. Then the potentials at other points are adjusted until the equation is satisfied to some desired approximation. (In rare cases, the equation with boundary conditions can be solved exactly analytically; but usually an approximate solution must suffice.)

There are many approaches to numerical solution of the Laplace equation. Perhaps the simplest is that due to Jacobi, in which the interior points are successively approximated by the mean of their surrounding points, while the boundary points are held at their fixed, specified values. We consider as an example a square plane, bounded by (0,1) in the x and y directions, in which the edge at $y = 1$ is held at $V = 1$ and the other three edges are held at $V = 0$. We make a rather arbitrary initial guess for the potentials at the interior points, but these will be evened out as the solution converges.

In the following code we solve the Laplace equation on a square lattice using the Jacobi method. We begin by setting the parameters

```
> n = 30 # Number of grid points per side
> L=1  # Length of a side
> dx = L/(n-1) # Grid spacing
> x = y = 0:(n-1)*dx # x and y coordinates
```

and making a rather arbitrary initial guess for the voltage profile.

```
> V0 = 1
> V = matrix(V0/2*sin(2*pi*x/L)*sin(2*pi*y/L),
+   nrow = n, ncol = n, byrow = TRUE)
```

We set the boundary conditions ($V = 0$ on three edges of the plate, $V = 1$ on the fourth edge:

```
> V[1,] = 0
> V[n,] = 0
> V[,1] = 0
> V[,n] = V0*rep(1,n)
```

We make a perspective plot of the initial guess,

```
> par(mfrow = c(1,2))
> persp(x,y,V, theta = -45, phi = 15)
```

then proceed with the Jacobi-method calculation.

```
> ## Loop until desired tolerance is obtained
> newV = V
> itmax = n^2 # Hope that solution converges within n^2 iterations
```

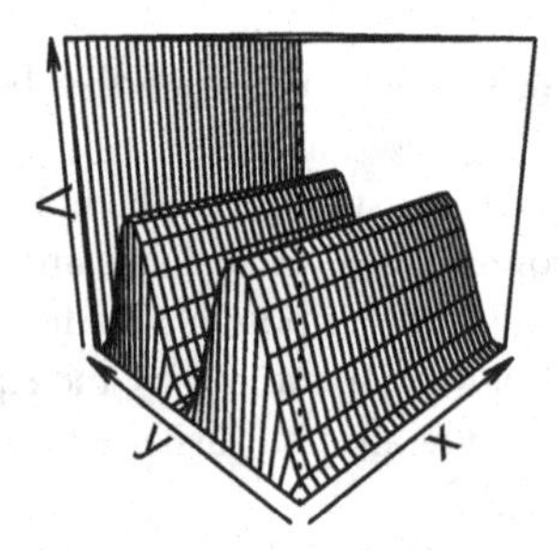

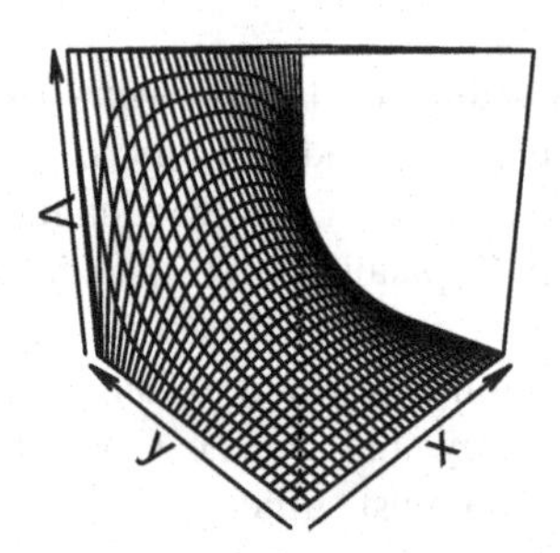

Figure 9.4: *Solution to the Laplace equation with the Jacobi method.*

```
> tol = 1e-4
> for (it in 1:itmax) {
+   dVsum = 0
+   for (i in 2:(n-1)) {
+     for (j in 2:(n-1)) {
+       newV[i,j] = 0.25*(V[i-1,j] + V[i+1,j] + V[i,j-1] + V[i,j+1])
+       dVsum = dVsum + abs(1-V[i,j]/newV[i,j])
+       }
+     }
+   V=newV
+   dV = dVsum/(n-2)^2 # Average deviation from previous value
+   if (dV < tol) break # Desired tolerance achieved
+   }
>
> it # Iterations to achieve convergence to tol
[1] 419
> dV
[1] 9.908314e-05
```

Finally, we plot the converged solution alongside the initial guess (Figure 9.4).

```
> persp(x,y,V, theta = -45, phi = 15)
> par(mfrow = c(1,1))
```

9.4 Solving PDEs with the `ReacTran` package

Solving of partial differential equations in R can also be done with the `ReacTran` package and ancillary packages that it calls. Package `ReacTran` facilitates modeling of reactive transport in 1, 2, and 3 dimensions. It "contains routines that enable the development of reactive transport models in aquatic systems (rivers, lakes, oceans), porous media (floc aggregates, sediments, ...) and even idealized organisms (spherical cells, cylindrical worms, ...)." Although ReacTran was developed largely to support the authors' research interests in ecological hydrology, its methods are useful for numerically solving all the standard types of PDEs.

The package contains:

- Functions to set up a finite-difference grid (1D or 2D)
- Functions to attach parameters and properties to this grid (1D or 2D)
- Functions to calculate the advective-diffusive transport term over the grid (1D, 2D, 3D)
- Various utility functions

When `ReacTran` is loaded, it also loads two support packages that we have previously encountered: `rootSolve` and `deSolve`. To quote from their help pages, the `rootSolve` package "solves the steady-state conditions for uni-and multicomponent 1-D, 2-D and 3-D partial differential equations, that have been converted to ODEs by numerical differencing (using the method-of-lines approach)." The `deSolve` package provides "functions that solve initial value problems of a system of first-order ordinary differential equations (ODE), of partial differential equations (PDE), of differential algebraic equations (DAE) and delay differential equations."

ReacTran also loads the `shape` package, which provides "functions for plotting graphical shapes such as ellipses, circles, cylinders, arrows, ..." However, we shall not use `shape` in what follows.

9.4.1 setup.grid.1D

Use of `ReacTran` generally proceeds in three or four steps. First, the function `setup.grid.1D` is used to establish the grid. In the simplest case, this function subdivides the one-dimensional space of length `L`, between `x.up` and `x.down`, into `N` grid cells of size `dx.1`. The calling usage is

```
setup.grid.1D(x.up=0, x.down=NULL, L=NULL, N=NULL, dx.1=NULL, p.dx.1=
rep(1,length(L)), max.dx.1=L, dx.N=NULL, p.dx.N=rep(1,length(L)),
max.dx.N=L)
```

where

- `x.up` is the position of the upstream interface
- `x.down` is the position of the downstream interface
- `L = x.down - x.up`
- `N` is the number of grid cells = `L/dx.1`

In more complex situations, the size of the cells can vary, or there may be more than one zone. These situations are described in the help page for `setup.grid.1D`.

The values returned by `setup.grid.1D` include `x.mid`, a vector of length `N`, which specifies the positions of the midpoints of the grid cells at which the concentrations are measured, and `x.int`, a vector of length `(N+1)`, which specifies the positions of the interfaces between grid cells, at which the fluxes are measured.

The plot function for `grid.1D` plots both the positions of the cells and the box thicknesses, showing both `x.mid` and `x.int`. The examples on the help page demonstrate this behavior.

`setup.grid.1D` serves as the starting point for `setup.grid.2D`, which creates a grid over a rectangular domain defined by two orthogonal 1D grids.

9.4.2 *setup.prop.1D*

Many transport models will involve grids with constant properties. But if some property that affects diffusion or advection varies with position in the grid, the variation can be incorporated with the function `setup.prop.1D` (or `setup.prop.2D` in two dimensions).

Given either a mathematical function or a data matrix, the `setup.prop.1D` function calculates the value of the property of interest at the middle of the grid cells and at the interfaces between cells. The function is called with

```
setup.prop.1D(func=NULL, value=NULL, xy=NULL, interpolate="spline",
grid, ...)
```

where

- `func` is a function that governs the spatial dependency of the property
- `value` is the constant value given to the property if there is no spatial dependency
- `xy` is a data matrix in which the first column gives the position, and the second column gives the values which are interpolated over the grid
- `interpolate` is the interpolation method (spline or linear)
- `grid` is the object defined with `setup.grid.1D`
- ... are additional arguments to be passed to `func`

9.4.3 *tran.1D*

This function calculates the transport terms—the rate of change of concentration due to diffusion and advection—in a 1D model of a liquid (volume fraction = 1) or a porous solid (volume fraction may be variable and < 1).

`tran.1D` is also used for problems in spherical or cylindrical geometries, though in these cases the grid cell interfaces will have variable areas.

The calling usage for tran.1D is

```
tran.1D(C, C.up = C[1], C.down = C[length(C)], flux.up = NULL, flux.down
= NULL, a.bl.up = NULL, a.bl.down = NULL, D = 0, v = 0, AFDW = 1, VF = 1,
A = 1, dx, full.check = FALSE, full.output = FALSE)
```

where

- `C` is a vector of concentrations at the midpoints of the grid cells.
- `C.up` and `C.down` are the concentrations at the upstream and downstream boundaries.
- `flux.up` and `flux.down` are the fluxes into and out of the system at the upstream and downstream boundaries.

- If there is convective transfer across the upstream and downstream boundary layers, `a.bl.up` and `a.bl.down` are the coefficients.
- `D` is the diffusion coefficient, and `v` is the advective velocity.
- `ADFW` is the weight used in the finite difference scheme for advection.
- `VF` and `A` are the volume fraction and area at the grid cell interfaces.
- `dx` is the thickness of the grid cells, either a constant value or a vector.
- `full.check` and `full.output` are logical flags to check consistency and regulate output of the calculation. Both are `FALSE` by default.

See the help page for details on these inputs.

When `full.output = FALSE`, the values returned by `trans.1D` are `dC`, the rate of change of `C` at the center of each grid cell due to transport, and `flux.up` and `flux.down`, the fluxes into and out of the model at the upstream and downstream boundaries.

`ReacTran` also has functions for estimating the diffusion and advection terms in two- and three-dimensional models, and in cylindrical and polar coordinates. The number of inputs grows with dimension, but the inputs are essentially the same as in the 1D case. See the help pages for `tran.2D`, `tran.3D`, `tran.cylindrical`, and `tran.polar`.

Yet another refinement is the function `tran.volume.1D`, which estimates the volumetric transport term in a 1D model. In contrast to `tran.1D`, which uses fluxes (mass per unit area per unit time), `tran.volume.1D` uses flows (mass per unit time). It is useful for modeling channels for which the cross-sectional area changes, when the change in area need not be explicitly modeled. It also allows lateral input from side channels.

9.4.4 *Calling `ode.1D` or `steady.1D`*

Once the grid has been set up and properties assigned to it, and the transport model has been formulated with `tran.1D` (or its 2D or 3D analogs), then `ReacTran` calls upon `ode.1D` from the `deSolve` package if a time-dependent solution is needed, or `steady.1D` from the `rootSolve` package if a steady-state solution is desired. The system of ODEs resulting from the method of lines approach is typically both sparse and stiff. The integrators in `deSolve`, such as "`lsoda`" (the 1D default method) are particularly well suited to deal with such systems of equations. If the system of ODEs is not stiff, then "`adams`" is generally a good choice of method.

9.5 Examples with the `ReacTran` package

9.5.1 *1-D diffusion-advection equation*

Here is a modification of the 1-dimensional diffusion equation solved earlier, done using the functions in the `ReacTran` package, and including an advection term. This might represent, for example, a narrow layer of a small molecule at the top of a solution column, subject both to diffusion and to an electrophoretic field driving it with velocity v.

Load ReacTran, which also causes loading of its ancillary packages.

```
> require(ReacTran)
Loading required package: ReacTran
Loading required package: rootSolve
Loading required package: deSolve
Loading required package: shape
```

Establish the grid, using the `setup.grid.1D()` function, and supply values for the parameters.

```
> N = 100 # Number of grid cells
> xgrid = setup.grid.1D(x.up = 0, x.down = 1, N = N) # Between 0 and 1
> x = xgrid$x.mid # Midpoints of grid cells
> D = 1e-4 # Diffusion coefficient
> v = 0.1 # Advection velocity
```

Construct the function that defines the diffusion-advection equation.

```
> Diffusion = function(t, Y, parms) {
+   tran=tran.1D(C=Y,C.up=0,C.down=0,D=D,v=v,dx= xgrid)
+   list(dY = tran$dC, flux.up = tran$flux.up,
+   flux.down = tran $flux.down)
+   }
```

Initialize the concentration on the grid.

```
> Yini = rep(0,N) # Initial concentration = 0
> Yini[2] = 100 # Except in the second cell
```

Now run the calculation for five time units, with a time step of 0.01.

```
> # Calculate for 5 time units
> times = seq(from = 0, to = 5, by = 0.01)
> out = ode.1D(y = Yini, times = times, func = Diffusion,
+ parms = NULL,dimens = N)
```

Finally, plot the initial concentration spike and the subsequent concentration distributions at intervals of 50 time steps (Figure 9.5).

```
> plot(x,out[1,2:(N+1)], type = "l", lwd = 1,  xlab = "x", ylab = "Y")
> # Plot subsequent conc distributions, every 50 time intervals
> for(i in seq(2, length(times), by = 50))  lines(x, out[i, 2:(N+1)])
```

9.5.2 1-D wave equation

The wave equation 9.8 can be solved in the same way as the diffusion equation by setting $c^2 = D$, letting $W = u$ and $\partial u/\partial t = v$, and solving in the now familiar way for the pair of variables (u, v). Here we consider the 1-D wave equation for a plucked string, held initially at 0 amplitude for $x < -25$ and $x > 25$, and stretched linearly to a maximum at $x = 0$. `ode.1D` is used to solve the set of simultaneous ODEs with $c = 1$.

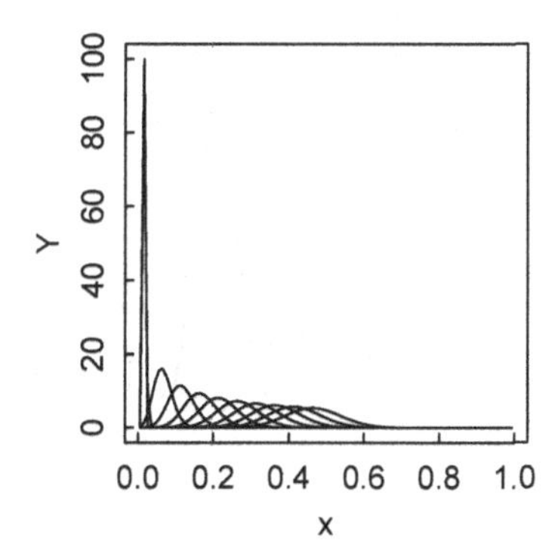

Figure 9.5: *Advection and diffusion of an initially sharp concentration layer.*

Load `ReacTran` and set up the grid.

```
> require(ReacTran)
> dx = 0.2  # Spacing of grid cells
> # String extends from -100 to +100
> xgrid = setup.grid.1D(x.up = -100, x.down = 100, dx.1 = dx)
> x = xgrid$x.mid  # midpoints of grid cells
> N = xgrid$N  # number of grid cells
```

Set initial conditions on string height profile and velocity.

```
> uini = rep(0,N)  # String height vector before stretching
> vini = rep(0,N)  # Initial string velocity vector
> displ = 10  # Initial displacement at center of string
> # Impose initial triangular height profile on string between +/- 25
> for(i in 1:N) {
+     if (x[i] > -25 & x[i] <= 0) uini[i] = displ/25*(25 + x[i]) else
+     if (x[i] > 0 & x[i] < 25) uini[i] = displ/25*(25 - x[i])
+   }
> yini = c(uini, vini)
```

Set the time sequence over which to compute the solution

```
> times = seq(from = 0, to = 50, by = 1)
```

Define the function that establishes the displacement and velocity vectors

```
> wave = function(t,y,parms) {
+     u = y[1:N] # Separate displacement and velocity vectors
+     v = y[(N+1):(2*N)]
+   du=v
+   dv=tran.1D(C=u,C.up=0,C.down=0,D=1,dx=xgrid)$dC
+   return(list(c(du, dv))) }
```

Solve the equations using `ode.1D` with the "adams" method. Note the use of the `subset()` function to extract the displacement vector `u` from the result vector.

```
> out = ode.1D(func = wave, y = yini, times = times,
```

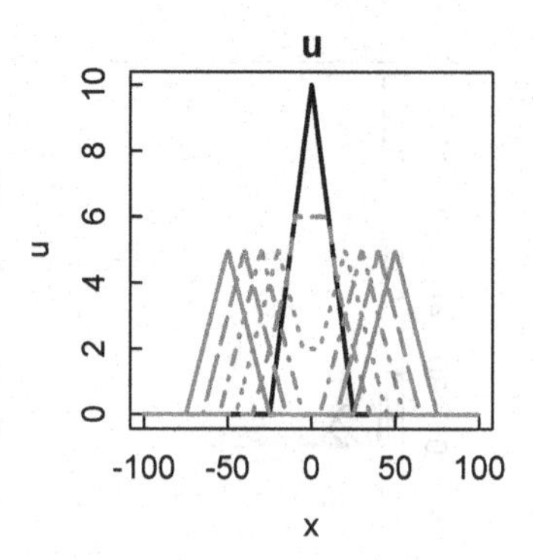

Figure 9.6: *Behavior of a plucked string.*

```
+       parms = NULL, method = "adams",
+       dimens = N, names = c("u", "v"))
> u = subset(out, which = "u") # Extract displacement vector
```

Finally, plot the displacement every 10th time interval (Figure 9.6).

```
> outtime = seq(from = 0, to = 50, by = 10)
> matplot.1D(out, which = "u", subset = time %in% outtime,
+ grid=x,xlab="x",ylab="u",type="l",
+ lwd = 2, xlim = c(-100,100), col = c("black", rep("darkgrey",5)))
```

We see that the initial displacement splits in two and propagates symmetrically to left and right.

9.5.3 *Laplace equation*

Here we use `ReacTran` to solve the 2D Laplace equation, treated earlier in this chapter by a different method. In this example the gradient in the y-direction is -1. (The gradient is just the flux, $D(\partial C/\partial x)$, with D set equal to 1. The solver is `steady.2D`, because there is no time dependence in the equation. As arbitrary initial conditions, we use $N_x \times N_y$ uniformly distributed random numbers. We must also specify `nspec`, the number of species in the model (just one, the potential, in this case), `dimens`, a 2-valued vector with the number of cells in the x and y directions, and `lrw`, the length of the real work array. See the help page for `steady.2D` for more details.

Load `ReacTran` and set up the grid.

```
> require(ReacTran)
> Nx = 100
> Ny = 100
> xgrid = setup.grid.1D(x.up = 0, x.down = 1, N = Nx)
> ygrid = setup.grid.1D(x.up = 0, x.down = 1, N = Ny)
> x = xgrid$x.mid
> y = ygrid$x.mid
```

Specify the function that calculates the evolution of the variables.

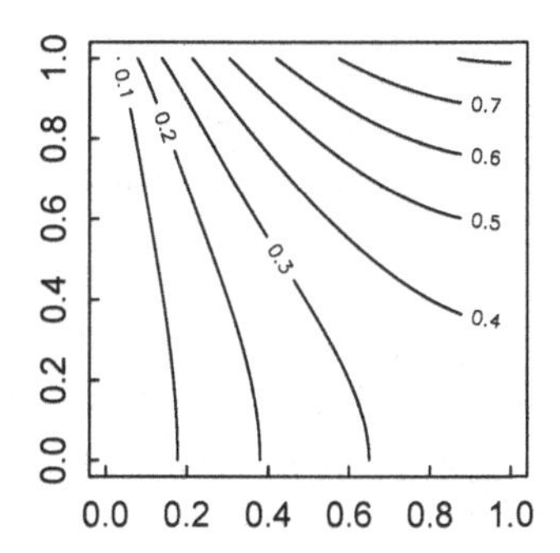

Figure 9.7: *Contour plot of solution to Laplace equation with gradient $\partial w/\partial y = -1$.*

```
> laplace = function(t, U, parms) {
+   w = matrix(nrow = Nx, ncol = Ny, data = U)
+   dw = tran.2D(C = w, C.x.up = 0, C.y.down = 0,
+     flux.y.up = 0,
+     flux.y.down = -1,
+     D.x = 1, D.y = 1,
+     dx = xgrid, dy = ygrid)$dC
+   list(dw) }
```

Start with uniformly distributed random numbers as initial conditions, then solve for the steady-state values and make a contour plot of the result (Figure 9.7).

```
> out = steady.2D(y = runif(Nx*Ny), func = laplace, parms = NULL,
+   nspec = 1, dimens = c(Nx, Ny), lrw = 1e7)
>
> z <- matrix(nr = Nx, nc = Ny, data = out$y)
> contour(z)
```

9.5.4 Poisson equation for a dipole

Finally, we solve the 2D Poisson equation

$$\frac{\partial^2 w}{\partial x^2} + \frac{\partial^2 w}{\partial y^2} = -\frac{\rho}{\varepsilon_0} \tag{9.16}$$

for a dipole located in the middle of a square sheet otherwise at 0 potential. For simplicity, we set all scale factors equal to one. In the definition of the `poisson` function, the values in the $N_x \times N_y$ matrix $\mathbf{w}$ are input through the data vector `U`. As in the Laplace equation above, we set the initial values of $\mathbf{w}$ at the grid cells equal to uniformly distributed random numbers.

Load `ReacTran` and establish the grid.

```
> require(ReacTran)
```

```
> Nx = 100
> Ny = 100
> xgrid = setup.grid.1D(x.up = 0, x.down = 1, N = Nx)
> ygrid = setup.grid.1D(x.up = 0, x.down = 1, N = Ny)
> x = xgrid$x.mid
> y = ygrid$x.mid
```

Find the x and y grid points closest to (0.4, 0.5) for the positive charges, and the (x,y) grid points closest to (0.6, 0.5) for the negative charges.

```
> # x and y coordinates of positive and negative charges
> ipos = which.min(abs(x - 0.4))
> jpos = which.min(abs(y - 0.50))
>
> ineg = which.min(abs(x - 0.6))
> jneg = which.min(abs(y - 0.50))
```

Define the `poisson` function for the potential and its derivatives.

```
> poisson = function(t, U, parms) {
+   w = matrix(nrow = Nx, ncol = Ny, data = U)
+   dw = tran.2D(C = w, C.x.up = 0, C.y.down = 0,
+   flux.y.up = 0,
+   flux.y.down = 0,
+   D.x = 1, D.y = 1,
+   dx = xgrid, dy = ygrid)$dC
+   dw[ipos,jpos] = dw[ipos,jpos] + 1
+   dw[ineg,jneg] = dw[ineg,jneg] - 1
+   list(dw) }
```

Solve for the steady-state potential distribution, and make a contour plot of the result (Figure 9.8).

```
> out = steady.2D(y = runif(Nx*Ny), func = poisson, parms = NULL,
+      nspec = 1, dimens = c(Nx, Ny), lrw = 1e7)
>
> z <- matrix(nr = Nx, nc = Ny, data = out$y)
> contour(z, nlevels = 30)
```

9.6 Case studies

9.6.1 Diffusion in a viscosity gradient

Biochemists and molecular biologists often use sucrose gradients to separate nucleic acid molecules of different composition. The gradient of sucrose produces gradients of both density and viscosity. Both of these are important in separation by sedimentation, but here we consider only the effect of viscosity on diffusional flux. Our aim is to show how to introduce nonuniformity into the properties of the grid. The diffusion coefficient D of a molecule, modeled as a sphere of radius R, is given by the

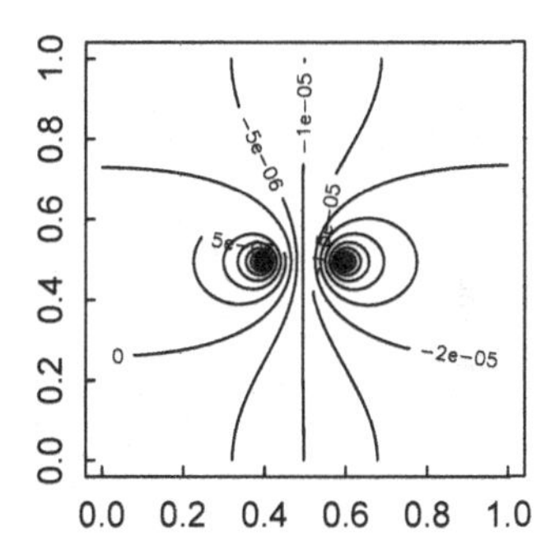

Figure 9.8: *Contour plot of solution to Poisson equation for a dipole.*

Stokes–Einstein equation

$$D = \frac{k_B T}{6\pi\eta R} \tag{9.17}$$

where k_B is the Boltzmann constant, T the Kelvin temperature, and η the viscosity.

We use the functions in the `ReacTran` package to show how the viscosity gradient leads to an asymmetry in the concentration profile of a diffusing molecule in one dimension.

```
> require(ReacTran)
```

We set up a grid in the x-direction with $N = 100$ cells and 101 interfaces including the left and right (or up and down) boundaries.

```
> N=100
> xgrid = setup.grid.1D(x.up=0,x.down=1,N=N)
> x = xgrid$x.mid  # Coordinates of cell midpoints
> xint = xgrid$x.int  # Coordiates of interfaces
```

We set the average value of the diffusion coefficient equal to an arbitrary value of 1, and specify a linear viscosity gradient so that the diffusion coefficients at the left and right sides are 1/4 and 4 times the average value:

```
> Davg = 1
> D.coeff = Davg*(0.25 +3.75*xint)
```

A similar linear dependence could be imposed with the `ReacTran` function `p.lin()`, and exponentially or sigmoidally decreasing dependence with `p.exp()` or `p.sig`. See the help pages for details.

We set the initial concentration to a band of width 10 with concentration 0.1 in the middle of the solution, and concentration 0 elsewhere.

```
> Yini = rep(0,N); Yini[45:55] = 0.1
```

We set the time scale using the result, established by Einstein in his theory of Brownian motion, that the mean-square distance diffused by a Brownian particle in time t is

$$< x^2 >= 2Dt. \tag{9.18}$$

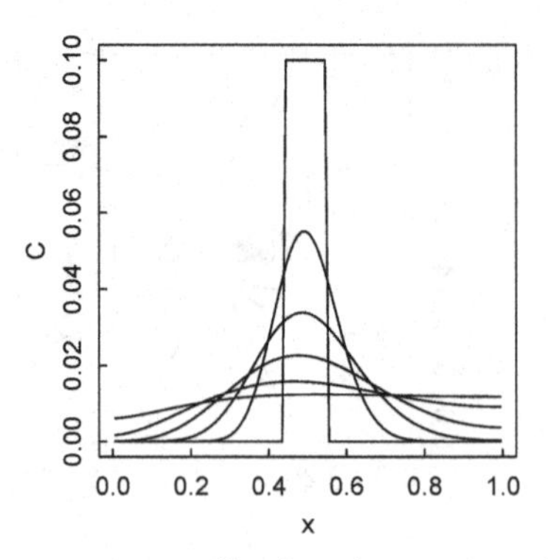

Figure 9.9: *Concentration profile of a substance in a viscosity gradient.*

In our case, the mean-square distance from the middle to either end of the solution is 1/4, so we set the maximum time for the simulation as $t_{max} = 1/8$. We then divide the simulation into 100 time steps.

```
> tmin = 0; tmax = 1/(8*Davg)
> times = seq(tmin, tmax,len=100)
```

We now define the function, `Diffusion()`, that gives the time-derivatives of the concentration (the fluxes):

```
> Diffusion = function(t,Y,parms){
+  tran = tran.1D(C=Y,D=D.coeff, dx=xgrid)
+  list(dY = tran$dC, flux.up = tran$flux.up,
    flux.down=tran$flux.down)
+ }
```

Having made all the necessary preparations, we invoke the differential equation solver `ode.1D()`, which most likely calls its default method, `lsoda`.

```
> out = ode.1D(y=Yini, times=times, func=Diffusion, parms=NULL,
  dimens=N)
```

The result, `out`, is a matrix in which column 1 gives the time and columns 2 to $N+1$ the concentrations at the midpoints of the N cells. We first plot the initial concentration profile in row 1 of `out`. We then use `lines()` plot the concentration profiles at subsequent times spaced to give roughly equal diffusion distances, considering the square-root dependence of average diffusion distance on time (Figure 9.9).

```
> plot(x, out[1,2:(N+1)],type="l",xlab="x",ylab="C",
  ylim=c(0,0.1))
> for (i in c(2,4,8,16,32)) lines(x,out[i,2:(N+1)])
```

Note the asymmetry in the concentration profile, with more material accumulating to the right, where the viscosity is lower and the diffusion coefficient higher.

9.6.2 *Evolution of a Gaussian wave packet*

The familiar time-dependent Schrödinger equation in one dimension,

$$i\hbar \frac{\partial \psi(x,t)}{\partial t} = \mathbf{H}\psi = -\frac{\hbar^2}{2m}\frac{\partial^2 \psi}{\partial x^2} + V(x)\psi \tag{9.19}$$

is an example of a diffusion-advection equation. **H** is the Hamiltonian. With the potential $V(x) = 0$, Equation 9.19 has the form of Fick's second law of diffusion, with the diffusion coefficient $i\hbar/2m$.

We show how this equation can be solved numerically using the `ReacTran` package to calculate the evolution of probability density of a Gaussian wave packet in free space. Part of the interest in this calculation is in showing how complex numbers are handled in R. Our treatment is adapted from Garcia (2000), pp. 287–293.

We begin by loading `ReacTran` and defining the constants and the lattice on which the calculation will be carried out.

```
> hbar = 1; m = 1
> D = 1i*hbar/(2*m)
> require(ReacTran)
> N = 131
> L = N-1
> xgrid = setup.grid.1D(-30,100,N=N)
> x = xgrid$x.mid
```

Next we define the function, `Schrodinger`, by which the derivative will be calculated and updated.

```
> Schrodinger = function(t,u,parms) {
+   du = tran.1D(C = u, D = D, dx = xgrid)$dC
+   list(du)
+ }
```

For the simplest calculation, we choose a Gaussian wave packet

$$\psi(x,t=0) = (\sigma_0\sqrt{\pi})^{-1/2} e^{ik_0 x} e^{-(x-x_0)^2/2\sigma_0^2} \tag{9.20}$$

initially centered at x_0, moving in the positive direction with wave number $k_0 = mv/\hbar$, and standard deviation of the packet width σ_0. The wave function is appropriately normalized. We give values for these parameters in arbitrary units:

```
> # Initialize wave function
> x0 = 0  # Center of wave packet
> vel = 0.5  # Mean velocity
> k0 = m*vel/hbar  # Mean wave number
> sig0 = L/10  # Std of wave function
```

We then calculate the normalization and the initial magnitude of the wave function as a function of x, and plot the result, showing both real and imaginary parts (Figure 9.10).

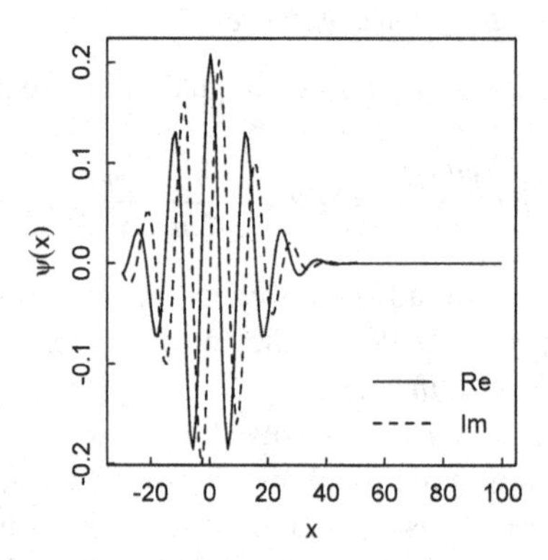

Figure 9.10: *Real and imaginary parts of a Gaussian wave packet.*

```
> A = 1/sqrt(sig0*sqrt(pi)) # Normalization coeff
> psi = A*exp(1i*k0*x)*exp(-(x-x0)^2/(2*sig0^2))
> # Plot initial wavefunction
> Re_psi = Re(psi); Im_psi = Im(psi)
> plot(x,Re_psi,type="l", lty=1,ylab=expression(psi(x))
> lines(x,Im_psi,lty=2)
> legend("bottomright", bty="n", legend=c("Re","Im"), lty=1:2)
```

All of this is preliminary to our ultimate goal, calculating the time-dependent probability density of the Gaussian wave packet. This we do by solving the diffusion equation with `ode.1D`, using the "adams" method because it is more efficient for this non-stiff equation.

```
> times = 0:120
> print(system.time(
+   out <- ode.1D(y=psi0, parms=NULL, func=Schrodinger,
+   times=times, dimens=130, method = "adams")
+ ))
   user  system elapsed
  0.189   0.001   0.192
```

We then plot the probability density of the wave packet, $P(x,t) = \psi^*(x,t)\psi(x,t)$, using the `Conj()` function in R to get the complex conjugate of the wave function vector.

```
> pdens0 = Re(out[1,2:(N+1)]*Conj(out[1,2:(N+1)]))
> plot(x, pdens0, type = "l",
+ ylim = c(0, 1.05*max(pdens0)), xlab="x",
+ ylab = "P(x,t)", xaxs="i", yaxs="i")
```

and then plot every 20th curve thereafter.

```
> for (j in seq(20,120,20)) {
+   pdens = Re(out[j,2:(N+1)]*Conj(out[j,2:(N+1)]))
```

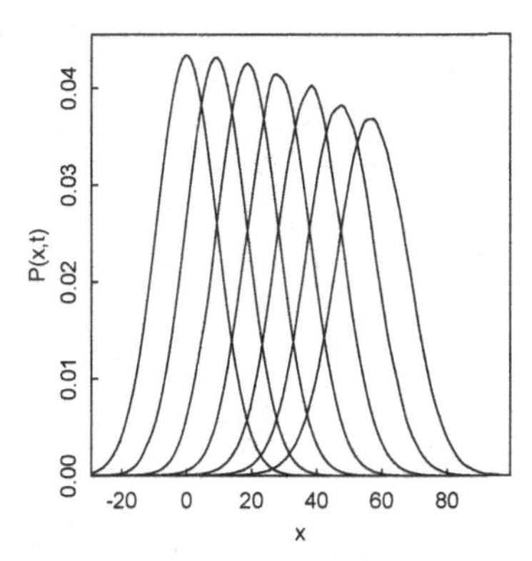

Figure 9.11: *Time evolution of the probability density of a Gaussian wave packet.*

```
+   lines(x, pdens)
+ }
```

Note that the xaxs="i" and yaxs="i" options set the limits of the plot (Figure 9.11) equal to the numerical limits, rather than leaving a little space at each margin. However, we set the upper y-axis limit as slightly larger than the amplitude of the zero-time probability density.

9.6.3 Burgers equation

The Burgers equation for the time and space dependence of the fluid velocity u,

$$\frac{\partial u}{\partial t} = D\frac{\partial^2 u}{\partial x^2} - vu\frac{\partial u}{\partial x}, \tag{9.21}$$

arises in fluid mechanics modeling of nonlinear phenomena such as gas dynamics and traffic flow. Formally, it resembles a diffusion-advection equation, but with the advection term multiplied by the velocity. We show how to solve it numerically with ReacTran, for simplicity setting the dispersion coefficient D and viscosity v equal to one.

```
> require(ReacTran)
> D = 1; v = 1
```

We set up the grid in the now familiar way, to be used in both the diffusion and advection parts of the calculation.

```
> N = 100
> xgrid = setup.grid.1D(x.up = -5, x.down = 5, N = N)
> x = xgrid$x.mid
```

We set the initial velocity equal to +1 for $x < 0$, and to -1 for $x > 0$, and consider only the early portion of the process with a small time increment.

```
> uini = c(rep(1,N/2), rep(-1,N/2))
> times = seq(0,1,by = .01)
```

We now define the function, `Burgers()`, that calculates the derivative of u for passage to the ode solver. The boundary conditions `C.up` and `C.down` are consistent with the initial conditions. Note how we have calculated the diffusion and advection contributions separately, and combined them at the end.

```
> Burgers = function(t,u,parms) {
+    tran = tran.1D(C = u, C.up = 1, C.down = -1, D = D,
     dx = xgrid)
+    advec =  advection.1D(C = u, C.up = 1, C.down = -1, v = v,
     dx = xgrid)
+    list(du = tran$dC + u*advec$dC)
+    }
```

We feed the results from `Burgers()`, along with the initial conditions, into the `ode.1D` solver, accepting the default `lsoda` method, to generate the matrix `out`.

```
> print(system.time(
+ out <- ode.1D(y = uini, parms = NULL, func = Burgers,
  times = times, dimens = N)
+ ))
   user  system elapsed
  0.226   0.013   0.245
```

Each row of `out` corresponds to a time increment, with the first column containing the time and the next N rows the velocity at the positions specified by `xgrid`. We set up a 1×2 plot layout, so we can compare the results of the `ReacTran` calculation with those of an analytical result to follow (Figure 9.12). We use `plot()` in the left panel to display the initial velocity distribution, and then `lines()` at four subsequent times to display the evolving distribution.

```
> par(mfrow=c(1,2))
> plot(x, out[1,2:(N+1)], type="l",
+     xlab = "x", ylab = "u")
> for (i in c(10,20,50,80))
+     lines(x, out[i,2:(N+1)])
```

Our numerical result can be compared with the exact solution in the limit $L \to \infty$ (Garcia, 2000, p. 294):

$$u(x,t) = v\frac{F(x,t) - F(-x,t)}{F(x,t) + F(-x,t)} \tag{9.22}$$

where

$$F(x,t) = \frac{1}{2}e^{t-x}\left[1 - \operatorname{erf}\left(\frac{x-2t}{2\sqrt{t}}\right)\right]. \tag{9.23}$$

and `erf(x)` is the error function

$$\operatorname{erf}(x) = \frac{2}{\sqrt{\pi}}\int_0^x e^{-t^2}\,dt \tag{9.24}$$

which is calculated in the `pracma` package as

```
erf(x) = 2*pnorm(sqrt(2)*x) - 1
```

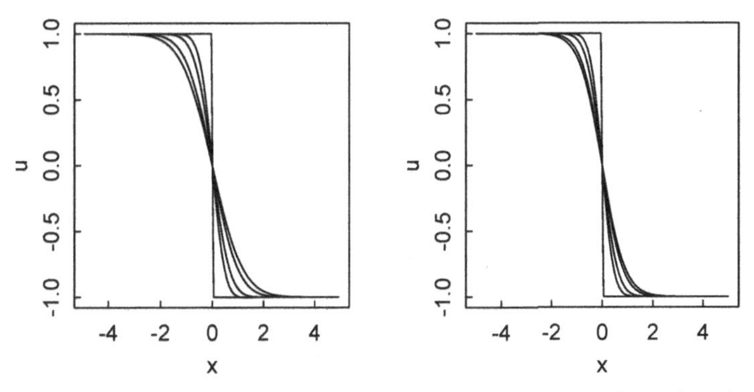

Figure 9.12: *Solution of the Burgers Equation 9.21 with* `ReacTran` *(left) and exact solution for* $L \to \infty$ *(right).*

where `pnorm` in **R** is the distribution function for the normal distribution. We load `pracma`, define the functions in equations 9.22 and 9.23,

```
> require(pracma)
> Fn = function(t,x) 1/2*exp(t-x)*(1-erf((x-2*t)/(2*sqrt(t))))
> u = function(t,x) (Fn(t,x)-Fn(t,-x))/(Fn(t,x)+Fn(t,-x))
```

set up the time and space array as above,

```
> t = seq(0,1,.01)
> L = 10
> x = seq(-L/2,L/2,len=100)
```

initialize the matrix **M** to hold the results,

```
> M = matrix(rep(0,length(t)*length(x)),nrow=length(t))
```

perform the calculations,

```
> for (i in 1:length(t)) {
+  for (j in (1:length(x))) {
+  M[i,j] = u(t[i],x[j])
+  }
+ }
```

and plot the results in the right panel of Figure 9.12.

```
> plot(x, M[1,], type = "l", ylab="u")
> for (i in c(10,20,50,80)) lines(x, M[i,])
```

Agreement between the two modes of calculation is excellent at first, but the results diverge slightly as time proceeds. This may be both because of accumulating numerical imprecision in the `ReacTran` calculation, and because Equation 9.22 is no longer exact as the initial discontinuity spreads toward the limits.

Chapter 10

Analyzing data

In the final two chapters we focus on data analysis, a topic for which R is particularly well-suited—indeed, for which it was initially developed and about which most of the literature on R is concerned. However, rather than refer the reader to other resources, it seems reasonable to present here at least a brief survey of some of the major topics, recognizing that scientists and engineers generally spend much of their time dealing with real data, not just developing numerical simulations.

We begin in this chapter by showing how to get data into R from external files, and how to structure data in data frames. We then turn to standard statistical topics of characterizing a univariate dataset, comparing two datasets, determining goodness of fit to a theoretical model, and determining the correlation of two variables. Finally, we introduce two methods of exploratory data analysis—principal component analysis and cluster analysis—which are crucial in making sense of large datasets.

10.1 Getting data into R

The first task is to get the data into R. Small datasets can simply be entered by hand as vectors representing the independent and dependent variables. But some datasets are quite large, and if they already exist in digitized form, in spreadsheets or on the Web, effort and errors will be minimized if they can be read into R directly. Since most such data are probably available in tabular form, the key R function is `read.table()`.

To use this function requires consideration of where the data file is stored and in what format. By default, R puts files in the user's home directory, which—unless instructed otherwise—considers the working directory. To find out the address of the working directory, type `getwd()` at the R prompt. The working directory can be changed with `setwd()`. For example, the sequence of commands

```
> getwd()
[1] "/Users/victor"
> setwd("~/Desktop")
> getwd()
[1] "/Users/victor/Desktop"
> setwd("~/")
```

shows that the working directory on my Macintosh is the same as my home directory, sets the new working directory to my desktop, verifies the change, and changes back to the home directory.

To maintain the current working directory, but to access a file in another directory, give the path to the file from the working directory, e.g., `~/Desktop/NIST/lanczos3.txt` if the desired file `lanczos3.txt` is located in the `NIST` folder on my desktop.

If the entries in the file are in tabular form separated by spaces, and the columns have headers, then the file can be read into R as a data frame (see later in this chapter) by the command

```
lan = read.table("~/Desktop/NIST/lanczos3.txt", header=TRUE)
```

The default is `header = FALSE`, with entries separated by spaces. If the entries were separated by tabs or commas, include the option `sep = "\t"` or `sep = ","` in `read.table()`. Alternatively, since comma-separated (csv) files are a common format of files exported from spreadsheets, one may use `read.csv()` for those files. Consult the help file `?read.table` for a complete description of the usage of these commands.

Conversely, if we have calculated a vector, matrix, or other array of data called `my.data`, and wish to save it in the file `my_file` on the desktop, we do so with the function

```
> write.table(my.data, file="~/Desktop/my_file")
```

Such a file can be imported by a spreadsheet.

10.2 Data frames

Experimental studies commonly arrange data in tables, with each row corresponding to a single experimental instance (subject, time point, etc.) and each column specifying a given type of measurement or condition. In R, such a construct is called a "data frame." Each column is a vector containing entries of the same class (numeric, logical, or character), and all columns must be of the same length (i.e., the same measurements were performed on all subjects). (If an entry is missing, it is generally replaced by `NA`.) A column may contain either data or factors: categorial variables that indicate subdivisions of the dataset.

For example, `chickwts`, in the package `datasets` installed with base R, is a data frame with 71 observations on 2 variables: `weight`, a numeric variable giving the chick weight, and `feed`: a factor giving the feed type.

```
> head(chickwts)
  weight      feed
1    179 horsebean
2    160 horsebean
3    136 horsebean
4    227 horsebean
5    217 horsebean
6    168 horsebean
```

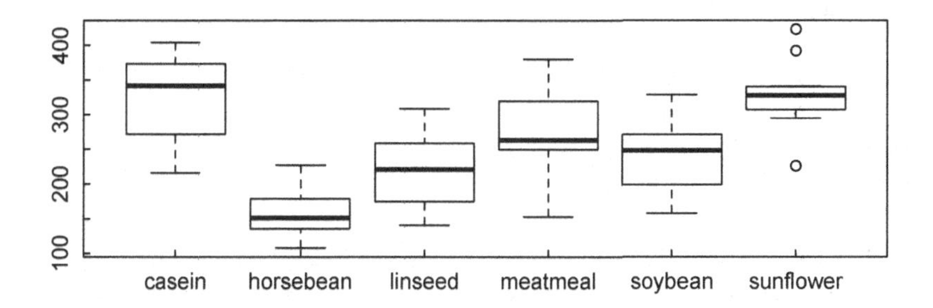

Figure 10.1: *Box plot of chick weights according to feed type.*

In this example, the `head()` function displays just the first six rows of the data frame. In general, `head(x,n)` displays the first *n* (default = 6) rows of the object `x`, which may be a vector, matrix, or data frame. Likewise, the `tail()` function displays the last rows of the object.

The columns of a data frame may be specified with the $ operator:

```
> class(chickwts$feed)
[1] "factor"
> class(chickwts$weight)
[1] "numeric"
```

A handy function to summarize measurements grouped by factor is `tapply`, in which the first argument is the measurement to be summarized, the second is the factor on which grouping is to be done, and the third is the function to be applied (`mean`, `summary`, `sum`, etc.).

```
> options(digits=1)
> tapply(chickwts$weight, chickwts$feed, mean)
   casein horsebean   linseed  meatmeal   soybean sunflower
      324       160       219       277       246       329
```

The `boxplot` function provides a handy graphical overview of the distribution of measurements grouped by factor (Figure 10.1).

```
> boxplot(chickwts$weight ~ chickwts$feed)
```

10.3 Summary statistics for a single dataset

Investigators often make repeated measurements of a quantity, to determine some sort of average and distribution of values. R provides powerful tools to characterize such a dataset. As an example, consider the classical data of Michelson and Morley on the measurement of the speed of light. These data are found in the data frame `morley` in the base R installation. The data consists of five experiments, each consisting of 20 consecutive runs. The data frame reports the experiment number (a factor), the run number (a factor), and a quantity proportional to the speed of light (numeric).

```
> head(morley)
    Expt Run Speed
001    1   1   850
002    1   2   740
003    1   3   900
004    1   4  1070
005    1   5   930
006    1   6   850
```

We will later compare individual experiments, but for now consider all measurements of Speed as constituting a single vector speed, which we want to characterize statistically.

```
> speed = morley$Speed
```

The summary function gives the range (minimum, maximum), the first and third quartiles, the median and the mean. Unfortunately, it does not give the standard deviation sd, which must be calculated separately.

```
> summary(speed)
   Min. 1st Qu.  Median    Mean 3rd Qu.    Max.
    620     808     850     852     892    1070
> sd(speed)
[1] 79.01
```

To get a visual impression of the distribution of speed measurements, we plot the histogram (Figure 10.2). To see how closely the distribution approximates a normal distribution, we use a qqnorm plot, which plots the quantiles from the observed distribution against the quantiles of a theoretical distribution (a normal distribution in this case). If the approximation is good, the points should lie on a line (qqline) running at 45 degrees from lower left to upper right.

```
> par(mfrow=c(1,2))
> hist(speed)
```

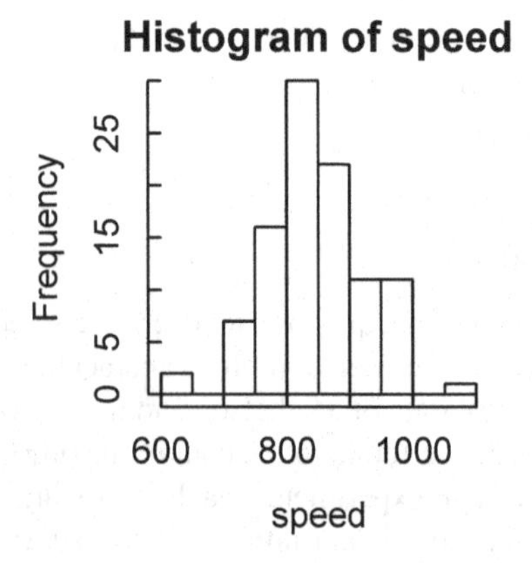

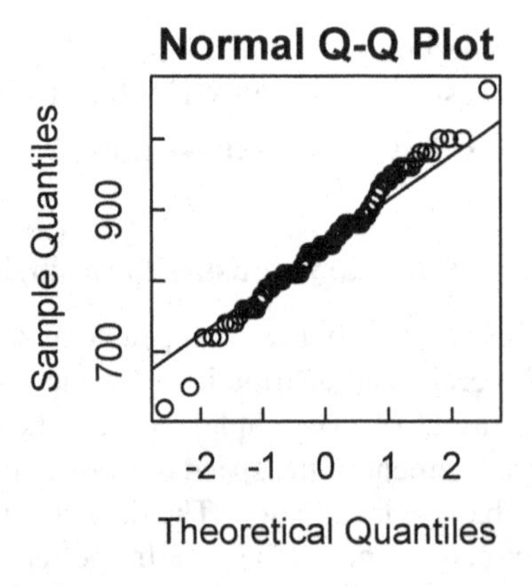

Figure 10.2: *Histogram and qqplot of Michelson–Morley data.*

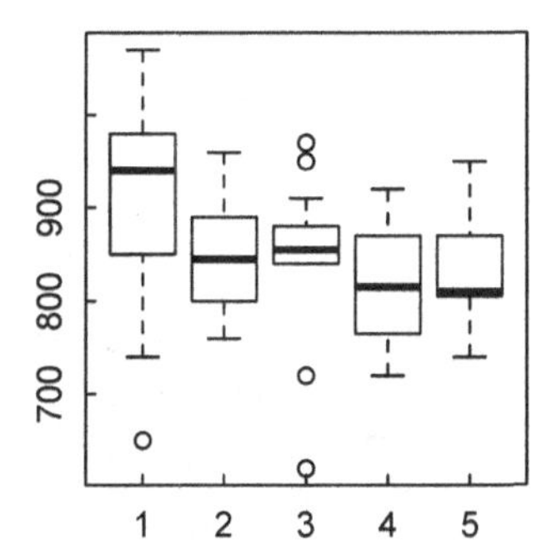

Figure 10.3: *Comparison of speed measurements in five sets of Michelson–Morley experiments.*

```
> qqnorm(speed)
> qqline(speed)
```

10.4 Statistical comparison of two samples

A common statistical task is to judge whether two samples are significantly different from one another (e.g., the weight gains of two sets of animals raised on different feeds, corrosion resistance of samples of a treated metal relative to untreated controls, etc.) We can use different experiment sets in the `morley` data to illustrate. We use the `boxplot` function to visualize the distribution of `Speed` in each of the five `Expt` sets (Figure 10.3):

```
> boxplot(morley$Speed ~ morley$Expt)
```

Sets 1 and 5 look the most different, so we separate them out from the complete data frame using the `subset()` function,

```
> morley1 = subset(morley, Expt == 1, Speed)
> morley5 = subset(morley, Expt == 5, Speed)
```

and apply Student's t-test—which tests the null hypothesis that the difference in means of the two datasets is equal to 0—to the `speed` vectors of each subsetted data frame.

```
> t.test(morley1$Speed, morley5$Speed)

Welch Two Sample t-test

data:  morley1$Speed and morley5$Speed
t = 2.935, df = 28.47, p-value = 0.006538
alternative hypothesis: true difference in means is not equal to 0
95 percent confidence interval:
  23.44 131.56
sample estimates:
```

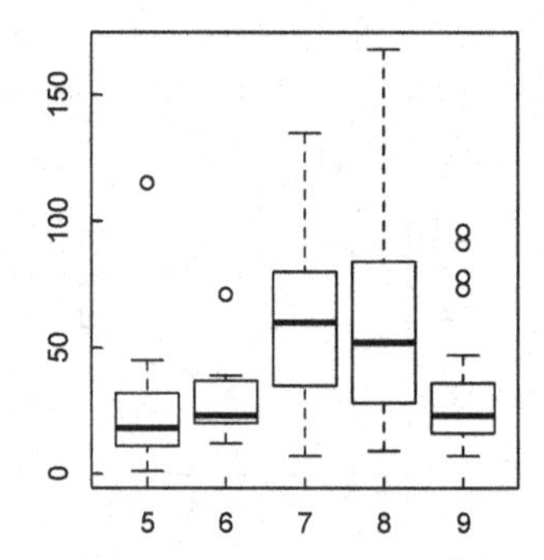

Figure 10.4: *Box plots of ozone level by months 5–9.*

```
mean of x mean of y
    909.0     831.5
```

The test indicates that the means of the two experimental sets are significantly different at the p = 0.0065 level; that is, the null hypothesis has only a probability of 0.0065 of being correct by chance.

Several variants of the t test should be noted. The example above is a two-sample, unpaired, two-sided test. A one-sample t test compares a single sample against a hypothetical mean `mu`, e.g. `t.test(morley$Speed, mu = 850)`. In a paired t test, the individuals in each sample are related in some way (e.g., IQ of identical twins, Young's modulus of several steel bars before and after heat treatment, etc.). In such a case, the argument `paired = TRUE` should be specified. A two-sided test is one in which the mean of one sample can be either greater or less than that of the other. If it is desired to test whether the mean of sample 1 is greater than that of sample 2, use `alternative = "greater"`, and similarly for `"less"`. See `?t.test` for details.

The t test applies rigorously only if the variation in the vectors is normally distributed. We saw that was essentially the case with the `morley` data, but not all data behave so nicely. Consider, for example, the `airquality` dataset in the base R installation (Figure 10.4).

```
> boxplot(Ozone ~ Month, data = airquality)
```

Suppose we want to test the hypothesis that the mean ozone levels in months 5 and 8 are equal. A histogram and qqnorm plot of the month 5 data show a distinctly non-normal distribution of ozone level occurrences (Figure 10.5); the same is true for month 8.

```
> airq5 = subset(airquality, Month == 5)
> par(mfrow=c(1,2))
> hist(airq5$Ozone)
> qqnorm(airq5$Ozone)
```

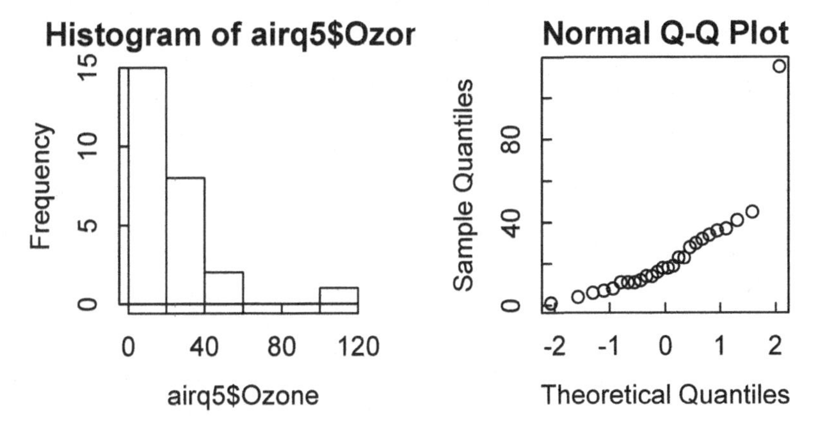

Figure 10.5: *Histogram and qqplot of ozone levels in month 5.*

In this case, the Wilcoxon (also known as Mann–Whitney) rank-sum test is more appropriate than the t test. Executing the example in the help page for `wilcox.test`, we obtain

```
> wilcox.test(Ozone ~ Month, data = airquality,
+     subset = Month %in% c(5, 8))
Wilcoxon rank sum test with continuity correction
data:  Ozone by Month
W = 127.5, p-value = 0.0001208
alternative hypothesis: true location shift is not equal to 0
Warning message:
In wilcox.test.default(x = c(41L, 36L, 12L, 18L, 28L, 23L, 19L,  :
  cannot compute exact p-value with ties
```

so there is only a probability of one part in 10^4 that the means are equal.

10.5 Chi-squared test for goodness of fit

Pearson's chi-square test examines the null hypothesis that the frequency distribution of certain events observed in a sample is consistent with a particular theoretical distribution. For example, suppose that a biochemist measures the number of DNA base pairs (A,T,G,C) in a 100-base pair sample and comes up with the values in x:

```
> x = c(20,30,28,22)
```

In the DNA solution overall, the probability of each of the four bases is 1/4.

```
> p = rep(1/4,4)
```

Is the sample representative of the overall solution?

```
> chisq.test(x, p = p)
  Chi-squared test for given probabilities
data:  x
```

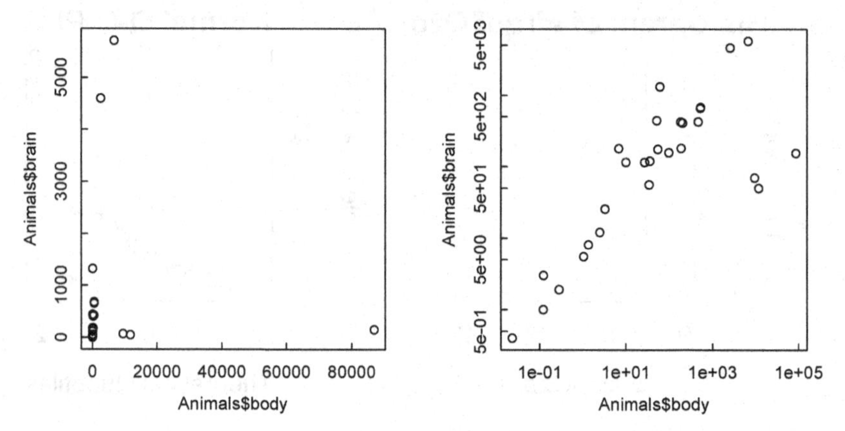

Figure 10.6: *Linear and log-log plots of brain weight vs. body weight, from MASS dataset* `Animals`.

```
X-squared = 2.72, df = 3, p-value = 0.4368
```

The sample appears to be adequately representative.

10.6 Correlation

We are often interested in whether, and to what extent, two sets of data are correlated with one another. Correlation may, but need not, imply a causal relation between the variables. There are three standard measures of correlation: Pearson's product-moment coefficient, and rank correlation coefficients due to Spearman and Kendall. R gives access to all of these via the `cor.test` function, with Pearson's as the default.

We demonstrate the use of the `cor.test` function via the `Animals` dataset in the `MASS` package. It is almost always useful to first graph the data (Figure 10.6).

```
> require(MASS)
> par(mfrow=c(1,2))
> plot(Animals$body, Animals$brain)
> plot(Animals$body, Animals$brain, log="xy")
```

We see that because of a few outliers (elephants, humans), the linear plot is not very informative, but the log-log plot shows a strong correlation between body weight and brain weight. However, when we use the linear data with the default (Pearson) cor.test, we find virtually no correlation because of the strong influence of the outliers.

```
> cor.test(Animals$body, Animals$brain)

Pearson's product-moment correlation

data:  Animals$body and Animals$brain
```

```
t = -0.0272, df = 26, p-value = 0.9785
alternative hypothesis: true correlation is not equal to 0
95 percent confidence interval:
 -0.3777  0.3685
sample estimates:
      cor
-0.005341
```

On the other hand, the rank correlation methods give more sensible results.

```
> cor.test(Animals$body, Animals$brain, method="spearman")

Spearman's rank correlation rho

data:  Animals$body and Animals$brain
S = 1037, p-value = 1.813e-05
alternative hypothesis: true rho is not equal to 0
sample estimates:
   rho
0.7163

Warning message:
In cor.test.default(Animals$body, Animals$brain, method = "spearman") :
  Cannot compute exact p-values with ties
> cor.test(Animals$body, Animals$brain, method="kendall")

Kendall's rank correlation tau

data:  Animals$body and Animals$brain
z = 4.604, p-value = 4.141e-06
alternative hypothesis: true tau is not equal to 0
sample estimates:
   tau
0.6172

Warning message:
In cor.test.default(Animals$body, Animals$brain, method = "kendall") :
  Cannot compute exact p-value with ties
```

10.7 Principal component analysis

Principal component analysis uses an orthogonal transformation (generally singular value or eigenvalue decomposition) to convert a set of observations of possibly correlated variables into a set of uncorrelated (orthogonal) variables called principal components. The transformation is defined such that the first principal component has as high a variance as possible (i.e., accounts for as much of the variability in the data as possible), and each succeeding component in turn has the highest variance possible under the constraint that it be orthogonal to the preceding components.

In R, principal component analysis is generally carried out with the `prcomp()` function. We illustrate its use with the `iris` dataset in the base R installation. (Type `?iris` for a description of the dataset.) The output below shows how the four numerical variables are transformed into four principal components. Scaling the data is probably not necessary in this case, since all four measurements have the same units and are of similar magnitudes. However, it is generally a good practice.

```
> iris1 = iris[, -5] # Remove the non-numeric species column.
> iris1_pca = prcomp(iris1, scale = T)
> iris1_pca
Standard deviations:
[1] 1.7084 0.9560 0.3831 0.1439

Rotation:
                 PC1      PC2     PC3     PC4
Sepal.Length  0.5211 -0.37742  0.7196  0.2613
Sepal.Width  -0.2693 -0.92330 -0.2444 -0.1235
Petal.Length  0.5804 -0.02449 -0.1421 -0.8014
Petal.Width   0.5649 -0.06694 -0.6343  0.5236
```

The `summary` function gives the proportion of the total variance attributable to each of the principal components, and the cumulative proportion as each component is added in. We see that the first two components account for more than 95% of the total variance.

```
> summary(iris1_pca)
Importance of components:
                        PC1   PC2    PC3     PC4
Standard deviation     1.71 0.956 0.3831 0.14393
Proportion of Variance 0.73 0.229 0.0367 0.00518
Cumulative Proportion  0.73 0.958 0.9948 1.00000
```

The histogram (the result of `plot` in a `prcomp` analysis) graphically recapitulates the proportions of the variance contributed by each principal component, while the `biplot` shows how the initial variables are projected on the first two principal components (Figure 10.7). It also shows (albeit illegibly at the printed scale) the coordinates of each sample in the (PC1, PC2) space. One species of iris (which turns out to be *setosa* from the cluster analysis below) is distinctly separated from the other two species in this coordinate space.

```
> par(mfrow=c(1,2))
> plot(iris1_pca)
> biplot(iris1_pca, col = c("gray", "black"))
> par(mfrow=c(1,1))
```

See the Multivariate Statistics task view in CRAN for more information and options.

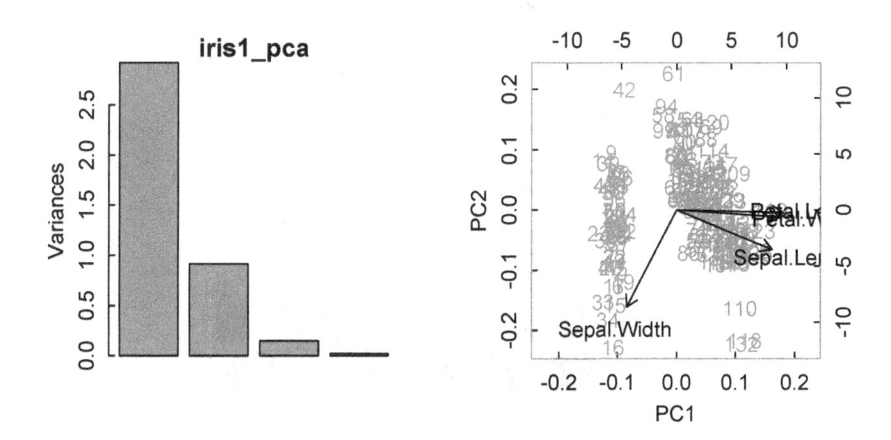

Figure 10.7: *Principal component (`prcomp`) analysis of `iris` data.*

10.8 Cluster analysis

Cluster analysis attempts to sort a set of objects into groups (clusters) such that objects in the same cluster are more similar to each other than to those in other clusters. It is used for exploratory analysis via data mining in many fields, such as bioinformatics, evolutionary biology, image analysis, and machine learning.

According to Wikipedia: "Cluster analysis itself is not one specific algorithm, but the general task to be solved. It can be achieved by various algorithms that differ significantly in their notion of what constitutes a cluster and how to efficiently find them. Popular notions of clusters include groups with low distances among the cluster members, dense areas of the data space, intervals or particular statistical distributions. The appropriate clustering algorithm and parameter settings (including values such as the distance function to use, a density threshold or the number of expected clusters) depend on the individual dataset and intended use of the results. Cluster analysis as such is not an automatic task, but an iterative process of knowledge discovery that involves trial and failure. It will often be necessary to modify preprocessing and parameters until the result achieves the desired properties."

The Cluster (Cluster Analysis & Finite Mixture Models) task view in CRAN divides clustering methods into three main approaches: hierarchical, partitioning, and model-based. We give examples of the first two approaches.

10.8.1 Using `hclust` for agglomerative hierarchical clustering

Hierarchical clustering builds a hierarchy of clusters, where the metric of hierarchy is some measure of dissimilarity between clusters. According to the help page for `hclust`, an agglomerative hierarchical clustering method, "This function performs

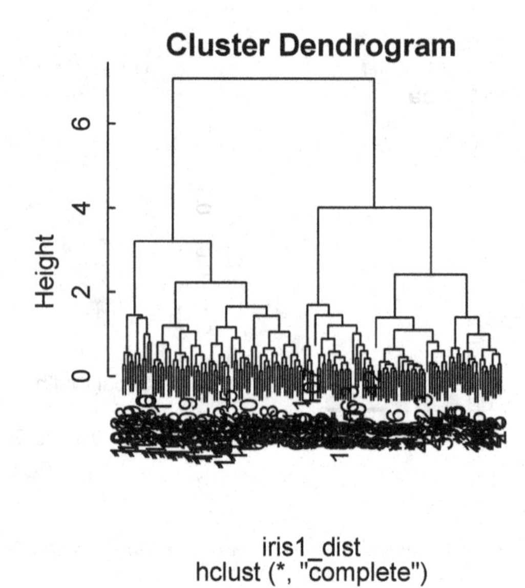

Figure 10.8: *Hierarchical cluster analysis of `iris` data using `hclust`.*

a hierarchical cluster analysis using a set of dissimilarities for the n objects being clustered. Initially, each object is assigned to its own cluster and then the algorithm proceeds iteratively, at each stage joining the two most similar clusters, continuing until there is just a single cluster. At each stage distances between clusters are recomputed by the Lance–Williams dissimilarity update formula according to the particular clustering method being used." There are seven agglomeration methods available, with `complete`—which searches for compact, spherical clusters—as the default. See `help(hclust)` for details.

```
> iris1_dist = dist(iris1) # Uses default method
> plot(hclust(iris1_dist))
```

10.8.2 Using `diana` for divisive hierarchical clustering

According to the `diana` (DIvisive ANAlysis Clustering) help page in the `cluster` package, "The diana-algorithm constructs a hierarchy of clusterings, starting with one large cluster containing all n observations. Clusters are divided until each cluster contains only a single observation. At each stage, the cluster with the largest diameter is selected. (The diameter of a cluster is the largest dissimilarity between any two of its observations.)" (See Figure 10.9)

```
> library(cluster)
> hierclust = diana(iris1)
> plot(hierclust,which.plots=2, main="DIANA for iris")
```

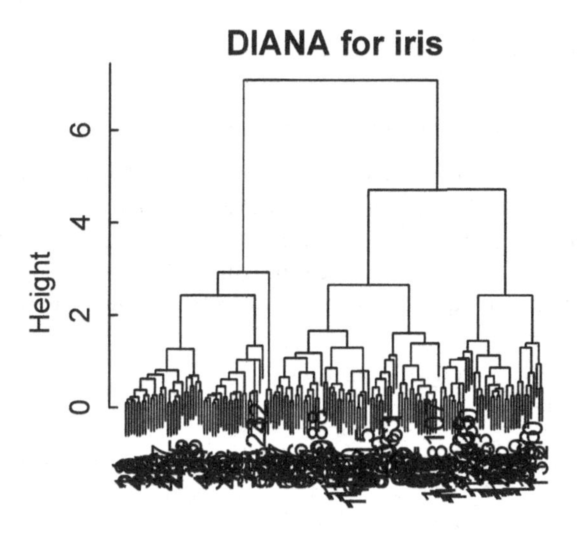

Figure 10.9: *Divisive hierarchical cluster analysis of* `iris` *data using* `diana`.

10.8.3 Using `kmeans` for partitioning clustering

k-means clustering partitions n observations into k clusters in which each observation belongs to the cluster with the nearest mean. The user must specify the number of centers (clusters) desired as output.

```
> iris1_kmeans3 = kmeans(iris1, centers = 3)
> table(iris1_kmeans3$cluster)

 1  2  3
96 21 33
> ccent = function(cl) {
+   f = function(i) colMeans(iris1[cl==i,])
+   x = sapply(sort(unique(cl)), f)
+   colnames(x) = sort(unique(cl))
+   return(x)
+   }
> ccent(iris1_kmeans3$cluster)
                 1      2      3
Sepal.Length 6.315 4.7381 5.1758
```

```
Sepal.Width  2.896 2.9048 3.6242
Petal.Length 4.974 1.7905 1.4727
Petal.Width  1.703 0.3524 0.2727
```

10.8.4 Using `pam` *for partitioning around medoids*

`pam` partitions the data into k clusters around medoids. The medoid of a finite set of data is the data point whose average dissimilarity to all the data points is a minimum. That is, it is the most centrally located point in the set. According to the `pam` help page, the k-medoids approach is more robust than the k-means approach "because it minimizes a sum of dissimilarities instead of a sum of squared euclidean distances"

```
> require(cluster)
Loading required package: cluster
> pam(iris1, k=3)
Medoids:
      ID Sepal.Length Sepal.Width Petal.Length Petal.Width
[1,]   8          5.0         3.4          1.5         0.2
[2,]  79          6.0         2.9          4.5         1.5
[3,] 113          6.8         3.0          5.5         2.1
Clustering vector:
  [1] 1 1 1 1 1 1 1 1 1 1 1 1 1 1 1 1 1 1 1 1 1 1 1 1 1 1 1 1 1 1 1 1 1
 [34] 1 1 1 1 1 1 1 1 1 1 1 1 1 1 1 1 1 2 2 3 2 2 2 2 2 2 2 2 2 2 2 2 2
 [67] 2 2 2 2 2 2 2 2 2 2 2 3 2 2 2 2 2 2 2 2 2 2 2 2 2 2 2 2 2 2 2 2 2
[100] 2 3 2 3 3 3 3 2 3 3 3 3 3 3 2 2 3 3 3 3 2 3 2 3 2 3 3 2 2 3 3 3 3
[133] 3 2 3 3 3 3 2 3 3 3 2 3 3 3 2 3 3 2
Objective function:
 build    swap
0.6709 0.6542
Available components:
 [1] "medoids"    "id.med"     "clustering" "objective"  "isolation"
"clusinfo"
 [7] "silinfo"    "diss"       "call"       "data"
> plot(pam(iris1, k=3),which.plots=1,labels=3,main="PAM for iris")
```

Components 1 and 2 together explain 95.81% of the point variability.

10.9 Case studies

10.9.1 Chi square analysis of radioactive decay

In a 2013 blog post[1] "The Chemical Statistician" Eric Chan showed how one can use a chi-squared test in R to examine the hypothesis that the distribution of alpha particle decay counts from ^{241}Americium obeys a Poisson distribution. The data were initially analyzed by Berkson (1966) and were later used by Rice (1995) as an example in his text. They are available online in tab-separated text format at

[1]http://chemicalstatistician.wordpress.com/2013/04/14/checking-the-goodness-of-fit-of-the-poisson-distribution-for-alpha-decay-by-americium-241/#more-612, accessed 2013-08-30

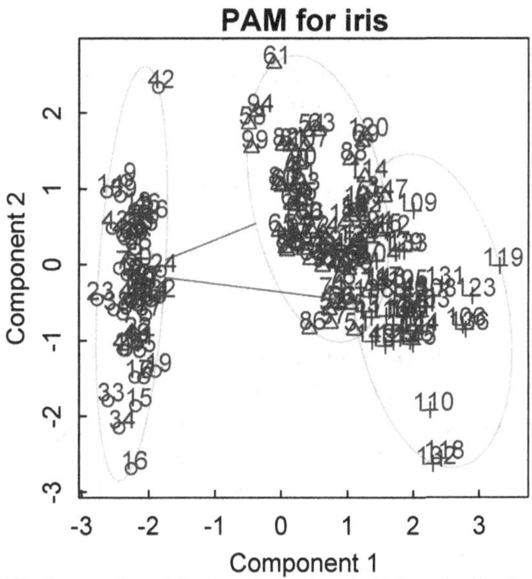

Figure 10.10: *pam (partitioning around medoids) analysis of* `iris` *data.*

`http://www.math.uah.edu/stat/data/Alpha.txt`. Chan used this dataset for his exposition, and our treatment is adapted from his.

We downloaded the dataset and saved it to the Desktop as `alpha.txt`. We then read it in as `alpha`, a data frame.

```
> alpha = read.table("~/Desktop/alpha.txt", header=TRUE)
```

The first column is the number of emissions observed in a 10-second interval, from 0 to 19. The second column is the number of intervals in which that number of emissions was observed.

```
> (emissions = alpha[,1])
 [1]  0  1  2  3  4  5  6  7  8  9 10 11 12 13 14 15 16 17 18 19
> (obsCounts = alpha[,2])
 [1]   1   4  13  28  56 105 126 146 164 161 123 101  74  53  23  15   9
[18]   3   1   1
```

The total number of alpha particle decays is the sum of the element-by-element products (i.e., the dot product) of the `emissions` vector with the `obsCounts` vector, and the total number of 10-second intervals is the sum of `obsCounts`. The average number of decays per 10-second interval, λ, is the quotient of these two values.

```
> (totEmissions = emissions%*%obsCounts)
      [,1]
[1,] 10099
> (totIntervals = sum(obsCounts))
[1] 1207
```

```
> (lambda = totEmissions/totIntervals)
          [,1]
[1,] 8.367026
```

If the distribution of decays is to be described by a Poisson distribution, the probability of observing k emissions in a 10-second interval is

$$f(k) = \frac{\lambda^k e^{-\lambda}}{k!} \tag{10.1}$$

and the expected number of occurences of k emissions is this probability multiplied by the total number of intervals.

```
> k = emissions
> expCounts = totIntervals*lambda^k*exp(-lambda)/factorial(k)
> expCounts = round(expCounts,2)
```

The chi-squared test for goodness of fit demands an expected count of at least five in each interval. Therefore, the first three intervals are combined into one, as are the last three. We can then display the observed (O) and expected (E) counts as a table.

```
> O = c(sum(obsCounts[1:3]),obsCounts[4:17],sum(obsCounts[18:20]))
> E = c(sum(expCounts[1:3]),expCounts[4:17],sum(expCounts[18:20]))
> cbind(O,E)
        O      E
 [1,]  18  12.45
 [2,]  28  27.39
 [3,]  56  57.28
 [4,] 105  95.86
 [5,] 126 133.67
 [6,] 146 159.78
 [7,] 164 167.11
 [8,] 161 155.36
 [9,] 123 129.99
[10,] 101  98.87
[11,]  74  68.94
[12,]  53  44.37
[13,]  23  26.52
[14,]  15  14.79
[15,]   9   7.74
[16,]   5   6.36
```

The Pearson chi-square test statistic is

$$\chi^2 = \sum_{k=1}^{n} (O_k - E_k)^2 / E_k. \tag{10.2}$$

```
> chisq = sum((O-E)^2/E)
> round(chisq,3)
[1] 8.717
```

The number of degrees of freedom, df, is the number of bins minus the number of independent parameters fitted (λ) minus 1.

```
> df = length(O)-2
```

Then the p.value of the test statistic may be calculated with the pchisq() function in R, the distribution function for the chi-squared distribution with df degrees of freedom. The option lower.tail = F specifies that probabilities $P/X > x$.

```
> p.value = pchisq(chisq, df, lower.tail = F)
> p.value
[1] 0.8487564
```

Thus, there is strong evidence that the Poisson distribution is a good fit.

10.9.2 Principal component analysis of quasars

The ninth data release of the Sloan Digital Sky Survey Quasar Catalog (http://www.sdss3.org/dr9/algorithms/qso_catalog.php) contains a file with 87,822 quasars that have been identified up to 2012. An earlier and smaller set, with only(!) 46,420 quasars was used in the Summer School in Statistics for Astronomers V, June 1–6, 2009, at the Penn State Center for Astrostatistics. We shall use that file, named SDSS_quasar.dat and located at http://astrostatistics.psu.edu/su09/lecturenotes/SDSS_quasar.dat, in our example.

We downloaded the file, saved it on the desktop as a text file, and read it into R with read.table as a data frame:

```
> quasar = read.table("~/Desktop/SDSS_quasar.dat.txt",head=T)
```

We check the size of the quasar data frame, get the names of its 23 columns, and check that there are no missing data.

```
> dim(quasar)
[1] 46420    23
> names(quasar)
 [1] "SDSS_J" "R.A."   "Dec."   "z"      "u_mag"  "sig_u"  "g_mag"
 [8] "sig_g"  "r_mag"  "sig_r"  "i_mag"  "sig_i"  "z_mag"  "sig_z"
[15] "Radio"  "X.ray"  "J_mag"  "sig_J"  "H_mag"  "sig_H"  "K_mag"
[22] "sig_K"  "M_i"
> quasar = na.omit(quasar)
> dim(quasar)
[1] 46420    23
```

The first column, "SDSS_J", simply names the object, and the second and third columns, "R.A." and "Dec.", give its angular position in the sky. The remaining 20 columns code for physical properties, from which we will derive the principal components. Because these properties are of quite different magnitudes, we use the scale = TRUE option to normalize each to unit variance. The results of the scaled calculation are quite different from those of the default, unscaled calculation. After performing the calculation of pc (prcomp uses singular value decomposition to get the eigenvalues), summary(pc) gives the importance of the components.

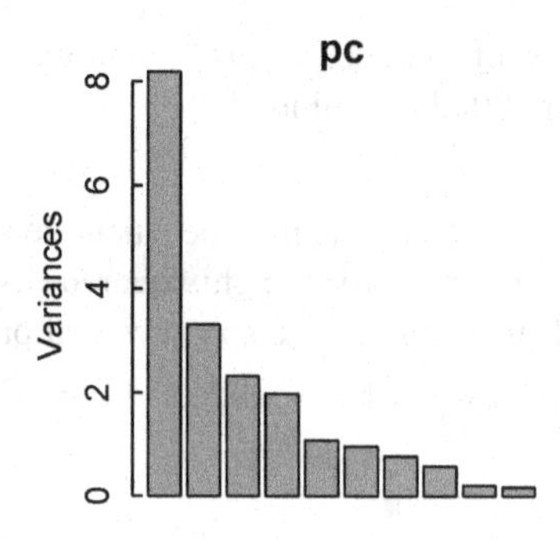

Figure 10.11: *Screeplot of quasar data.*

```
> pc = prcomp(quasar[,-(1:3)], scale=T)
> summary(pc)
Importance of components:
                         PC1   PC2   PC3   PC4    PC5    PC6    PC7
Standard deviation     2.861 1.821 1.523 1.407 1.0331 0.9768 0.8743
Proportion of Variance 0.409 0.166 0.116 0.099 0.0534 0.0477 0.0382
Cumulative Proportion  0.409 0.575 0.691 0.790 0.8434 0.8911 0.9293
                          PC8   PC9    PC10    PC11    PC12    PC13
Standard deviation     0.7592 0.447 0.41251 0.36537 0.30197 0.25569
Proportion of Variance 0.0288 0.010 0.00851 0.00667 0.00456 0.00327
Cumulative Proportion  0.9581 0.968 0.97663 0.98331 0.98787 0.99114
                         PC14    PC15    PC16    PC17    PC18
Standard deviation     0.2408 0.21940 0.19617 0.14188 0.11003
Proportion of Variance 0.0029 0.00241 0.00192 0.00101 0.00061
Cumulative Proportion  0.9940 0.99644 0.99837 0.99938 0.99998
                          PC19    PC20
Standard deviation     0.01534 0.01212
Proportion of Variance 0.00001 0.00001
Cumulative Proportion  0.99999 1.00000

> screeplot(pc)
```

The first eight principal components contribute most of the variance. This is made visually apparent with `screeplot`, which plots the variances against the number of the principal component (Figure 10.11).

We learn which properties contribute most to the major principal components by calling the `rotation` element of the `prcomp` list. (`princomp` calls this the `loadings` element.)

```
> round(pc$rotation[,1:8],2)
       PC1  PC2   PC3   PC4   PC5   PC6  PC7   PC8
z     0.16 0.17 -0.38  0.37 -0.05 -0.05 0.17  0.28
u_mag 0.23 0.30 -0.25 -0.04  0.08  0.03 0.17 -0.25
```

```
sig_u   0.12 0.31 -0.23  0.17  0.22  0.18  0.19 -0.63
g_mag   0.27 0.26 -0.12 -0.20  0.02  0.01 -0.07  0.05
sig_g   0.08 0.23 -0.14  0.06  0.33  0.35 -0.77  0.22
r_mag   0.29 0.20 -0.02 -0.28 -0.09 -0.07  0.02  0.18
sig_r   0.05 0.25  0.44  0.32 -0.01  0.00 -0.02 -0.03
i_mag   0.29 0.17  0.01 -0.29 -0.13 -0.09  0.08  0.16
sig_i   0.06 0.26  0.45  0.29 -0.01 -0.01 -0.02  0.01
z_mag   0.29 0.14  0.02 -0.26 -0.15 -0.10  0.09  0.19
sig_z   0.14 0.29  0.40  0.16 -0.07 -0.06  0.08  0.08
Radio  -0.03 0.04 -0.01 -0.01  0.58 -0.80 -0.10 -0.02
X.ray  -0.11 0.01  0.12 -0.12  0.62  0.39  0.51  0.40
J_mag  -0.31 0.23 -0.05 -0.06 -0.03 -0.02  0.01  0.02
sig_J  -0.29 0.26 -0.05 -0.11 -0.11 -0.04  0.01  0.05
H_mag  -0.31 0.23 -0.05 -0.06 -0.04 -0.02  0.01  0.02
sig_H  -0.29 0.25 -0.06 -0.09 -0.11 -0.04  0.01  0.06
K_mag  -0.31 0.23 -0.05 -0.06 -0.04 -0.02  0.01  0.02
sig_K  -0.29 0.25 -0.07 -0.07 -0.13 -0.06  0.01  0.08
M_i    -0.03 0.01  0.34 -0.55  0.12  0.12 -0.11 -0.37
```

Chapter 11

Fitting models to data

A large part of scientific computation involves using data to determine the parameters in theoretical or empirical model equations. Not surprisingly, given its statistical roots, R has powerful tools for fitting functions to data. In this chapter we discuss the most important of these tools: linear and nonlinear least-squares fitting, and polynomial and spline interpolation. We also show how these methods can be used to accelerate the convergence of slowly convergent series with Padé and Shanks approximations. We then consider the related topics of time series, Fourier analysis of periodic data, spectrum analysis, and signal processing, with a focus on extracting signal from noise.

11.1 Fitting data with linear models

Perhaps the most common data-analysis task in science and engineering is to make a series of measurements of property y, assumed to be a linear function of x, and to determine the slope and intercept of y vs. x using least squares. In R, the function that performs this analysis is `lm()`, for linear model. Consider, for example, the following simulated data and analysis, in which the y measurements are afflicted with a small amount of normally distributed random error.

```
> x = 0:10
> set.seed(333)
> y = 3*x + 4 + rnorm(n = length(x), mean = 0, sd = 0.3)
```

We then fit the data to a linear model and call the result.

```
> yfit = lm(y~x)
> yfit

Call:
lm(formula = y ~ x)

Coefficients:
(Intercept)            x
      4.029        2.988
```

The intercept and slope are recovered within a few percent of the original.

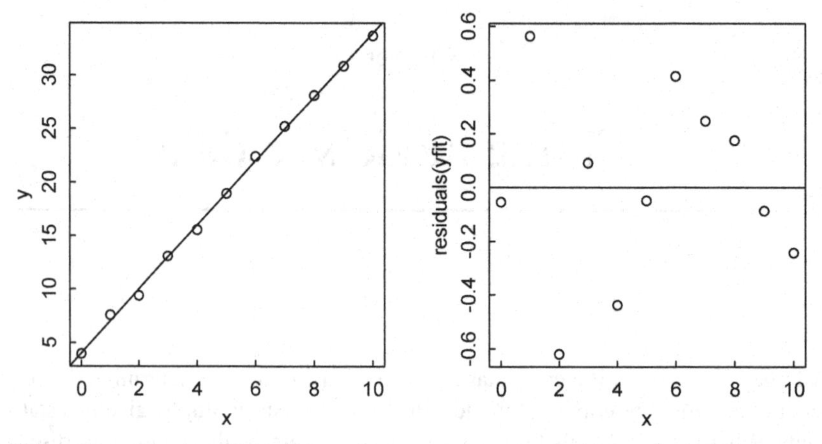

Figure 11.1: *Linear fit (left) and residuals (right) for simulated data with random error.*

Note that `lm()` enables one to draw the fitted line with the `abline(h,v)` function, in which `h` and `v` are taken from the fitted coefficients. The `lm()` function also calculates the residuals, convenient for visual inspection of the quality of the fit (Figure 11.1).

```
> par(mfrow=c(1,2))
> plot(x,y)
> abline(yfit)
> plot(x,residuals(yfit))
> abline(0,0)
```

If appropriate, the measurements may be accompanied by a vector of weights, in which case weighted least squares is used. See `?lm` for further details.

11.1.1 Polynomial fitting with lm

Linear models may also be used for polynomial fitting, since y depends linearly on the polynomial coefficients. Consider, for example, the synthetic data produced by

```
> set.seed(66)
> x=0:20
> y=1+x/10+x^2/100+rnorm(length(x),0,.5)
```

where we have added some normally distributed random noise onto a quadratic function of x. We call for a linear model fit with

```
> y2fit = lm(y ~ 1 + x + I(x^2))
```

or equivalently `y2fit = lm(y ~ poly(x,2,raw=TRUE))`, where the `I()` in the formula enforces identity, so that the function remains unchanged. (Note that `y2fit = lm(y ~ poly(x,2) is not equivalent, since \texttt{poly()} uses orthonormal polynomials).`) Then `summary()` gives the results.

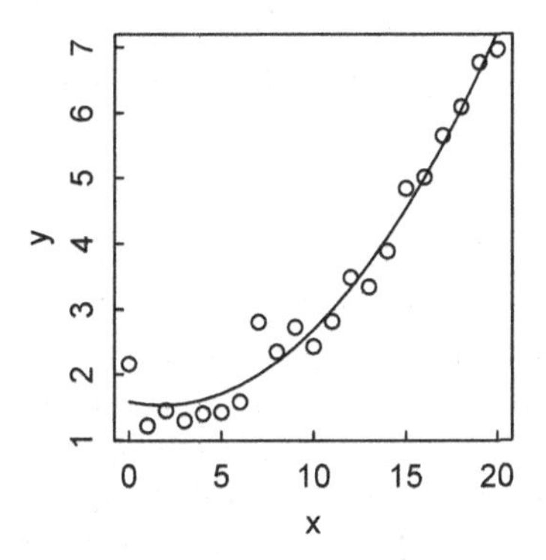

Figure 11.2: *lm() fit to a quadratic polynomial with random error.*

```
> summary(y2fit)

Call:
lm(formula = y ~ 1 + x + I(x^2))

Residuals:
     Min       1Q   Median       3Q      Max
-0.34951 -0.25683 -0.08032  0.15884  0.80823

Coefficients:
             Estimate Std. Error t value Pr(>|t|)
(Intercept)  1.583913   0.197273   8.029 2.33e-07 ***
x           -0.061677   0.045711  -1.349    0.194
I(x^2)       0.017214   0.002207   7.801 3.50e-07 ***
---
Signif. codes:  0 *** 0.001 ** 0.01 * 0.05 . 0.1   1

Residual standard error: 0.3305 on 18 degrees of freedom
Multiple R-squared: 0.972,Adjusted R-squared: 0.9688
F-statistic: 311.9 on 2 and 18 DF,  p-value: 1.073e-14
```

The coefficients are of the right order of magnitude, but deviate significantly from the input (1,.1.,01) because of the large random term. The data and fit are plotted (Figure 11.2) with

```
> plot(x,y)
> points(x,predict(y2fit),type="l")
```

where `predict()` gives a vector of predicted y values corresponding to the x vector values.

The same calculation can be done by specifying the degree of the polynomial with

```
ypoly2 = lm(y ~ poly(x,degree=2, raw=TRUE))
```

where `raw = TRUE` is required since we don't want orthogonal polynomials (the default is `raw = FALSE`).

11.2 Fitting data with nonlinear models

Fitting to nonlinear models is done in base R with the `nls()` function, which uses a Gauss–Newton algorithm. The Gauss–Newton method assumes that the least squares function is locally quadratic, and finds the minimum of the quadratic. However, this approach can fail if the starting guess is too far from the true minimum. Therefore, the more commonly used method in the scientific literature for nonlinear least-squares minimization is the Levenberg–Marquardt (LM) method. The LM method combines two minimization methods: gradient descent (steepest descent) and Gauss–Newton. The gradient descent method reduces the sum of squared deviations by updating the unknown parameters in the direction of the steepest gradient of the least squares objective function. The LM method favors the gradient descent method when the sum of squared deviations is large, and favors the Gauss–Newton approach as the optimal value is approached.

The Levenberg–Marquardt method is not available in base R (although it probably should be), but the package `minpack.lm` provides it. As the description in the `minpack.lm` documentation states, the package "provides R interface to `lmder` and `lmdif` from the MINPACK library, for solving nonlinear least-squares problems by a modification of the Levenberg–Marquardt algorithm, with support for lower and upper parameter bounds." The function that is called to do this work in `minpack.lm` is `nls.lm`.

The LM method can be implemented directly with `nls.lm`, but perhaps more conveniently with a `nls`-like call to the `nlsLM` function that uses `nls.lm` for fitting. As the help page states, "Since an object of class 'nls' is returned, all generic functions such as anova, coef, confint, deviance, df.residual, fitted, formula, logLik, predict, print, profile, residuals, summary, update, vcov and weights are applicable."

We test these nonlinear fitting functions with several datasets from the NIST StRD Nonlinear Regression Data Sets at `http://www.itl.nist.gov/div898/strd/nls/nls_main.shtml`. We begin with an exponential model in the lower level of difficulty category.

As noted at the beginning of this chapter, our first task is to get the data into R. The data are copied from the web page `http://www.itl.nist.gov/div898/strd/nls/data/LINKS/DATA/Misra1a.dat`, with `"Data:"` cut, pasted into the file `Misra1a.txt` in the `NIST` folder on my desktop, then brought into R with

```
> misra1a = read.table(file="~/Desktop/NIST/Misra1a.txt",header=T)
> misra1a
      y     x
1  10.07  77.6
2  14.73 114.9
3  17.94 141.1
4  23.93 190.8
```

```
5  29.61 239.9
6  35.18 289.0
7  40.02 332.8
8  44.82 378.4
9  50.76 434.8
10 55.05 477.3
11 61.01 536.8
12 66.40 593.1
13 75.47 689.1
14 81.78 760.0
```

The result of `read.table` is a data frame, whose components can be dissected as follows:

```
> x=misra1a$x
> y=misra1a$y
```

A plot of the data looks almost linear, so for fun we first try a linear model:

```
> lmfit = lm(y~x)
> summary(lmfit)
Call:
lm(formula = y ~ x)
Residuals:
    Min      1Q  Median      3Q     Max
-2.1063 -0.8814  0.3314  0.9620  1.1703
Coefficients:
            Estimate Std. Error t value Pr(>|t|)
(Intercept) 3.764972   0.661522   5.691    1e-04 ***
x           0.105423   0.001541  68.410   <2e-16 ***
---
Signif. codes:  0 '***' 0.001 '**' 0.01 '*' 0.05 '.' 0.1 ' ' 1
Residual standard error: 1.2 on 12 degrees of freedom
Multiple R-squared: 0.9974,
Adjusted R-squared: 0.9972 F-statistic: 4680 on 1 and 12 DF,
p-value: < 2.2e-16
```

The best-fit line shows clearly that y is a slightly nonlinear function of x, as is evident from the residuals plot.

```
> par(mfrow=c(1,2))
> plot(x,y)
> abline(lmfit)
> plot(x,residuals(lmfit))
> par(mfrow=c(1,1))
```

The NIST site tells us that y is in fact an exponential function of x, so we go to a nonlinear model and begin with the `nls` function. The usage for `nls` is

```
nls(formula, data, start, control, algorithm,
    trace, subset, weights, na.action, model,
```

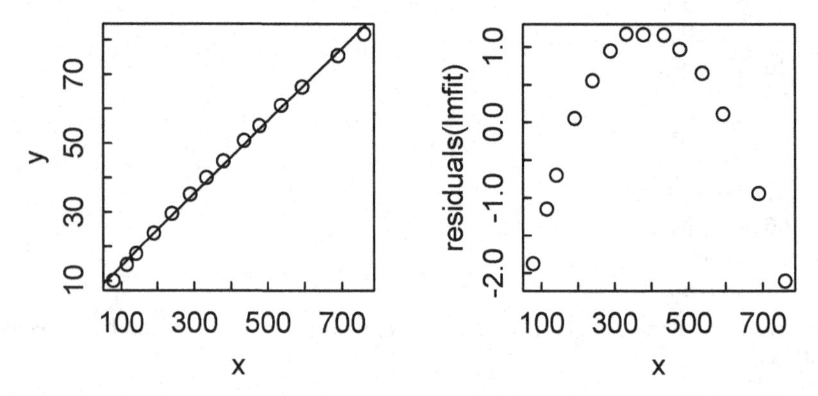

Figure 11.3: *(left) Plot of* `misra1a` *data with abline of linear fit; (right) Residuals of linear fit to* `misra1a` *data.*

```
    lower, upper, ...)
```

`formula` is a nonlinear model formula including variables and parameters. `data` is typically a data frame with which to evaluate the variables, but may be omitted if the variables have already been established. `start` is a named list or named numeric vector of starting values for the parameters in the model. The other arguments will be discussed as needed, or consult the help page for details.

Applying `nls` to the x,y data from `misra1a`, we obtain

```
> nlsfit = nls(y ~ b1*(1-exp(-b2*x)), start=list(b1=500,b2=1e-4))
> summary(nlsfit)
Formula: y ~ b1 * (1 - exp(-b2 * x))
Parameters:
    Estimate Std. Error t value Pr(>|t|)
b1 2.389e+02  2.707e+00   88.27   <2e-16 ***
b2 5.502e-04  7.267e-06   75.71   <2e-16 ***
---
Signif. codes:  0 '***' 0.001 '**' 0.01 '*' 0.05 '.' 0.1 ' ' 1
Residual standard error: 0.1019 on 12 degrees of freedom
Number of iterations to convergence: 11
Achieved convergence tolerance: 4.14e-06
```

We plot the (x,y) points again, then draw the `predict()` line through them (Figure 11.3):

```
> par(mfrow=c(1,2)) # Anticipate adding residuals plot
> plot(x,y)
> lines(x,predict(nlsfit))
```

Starting with values much closer to the Certified Values given on the website, we arrive at the same values for the parameters, but with fewer iterations. In both cases, the results agree to within the displayed number of significant figures with the Certified Values on the Misra1a page on the NIST website.

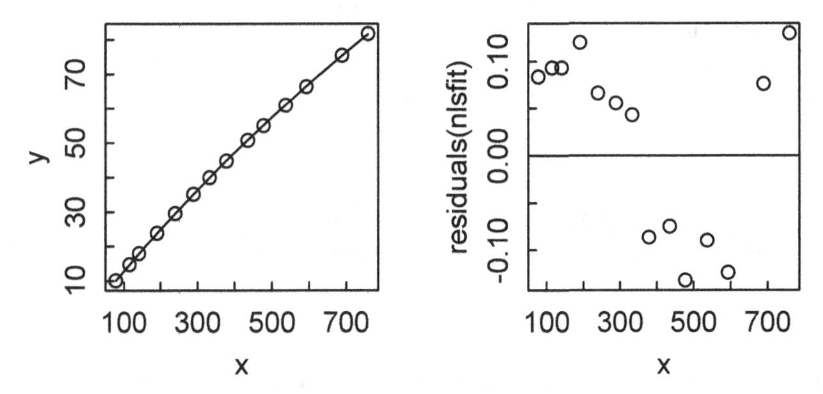

Figure 11.4: *(left) nls() exponential fit to* `misra1a` *data; (right) Residuals of nls() exponential fit to* `misra1a` *data.*

```
> nlsfit2 = nls(y~b1*(1-exp(-b2*x)), start=list(b1=250,b2=5e-4))
> summary(nlsfit2)

Formula: y ~ b1 * (1 - exp(-b2 * x))

Parameters:
    Estimate Std. Error t value Pr(>|t|)
b1 2.389e+02  2.707e+00   88.27   <2e-16 ***
b2 5.502e-04  7.267e-06   75.71   <2e-16 ***
---
Signif. codes:  0 '***' 0.001 '**' 0.01 '*' 0.05 '.' 0.1 ' ' 1

Residual standard error: 0.1019 on 12 degrees of freedom

Number of iterations to convergence: 3
Achieved convergence tolerance: 5.842e-07
```

A plot of the residuals shows that the errors, even in the fit to this function, do not appear quite random, although the probability of flipping a fair coin and finding runs of 7 heads, 5 tails, and 2 heads is not infinitesimal. Note, however, that the residuals are an order of magnitude smaller than in the linear fit (Figure 11.4).

```
> plot(x,residuals(nlsfit))
> abline(0,0)
```

Applying the `nlsLM` function to the data (having first loaded `minpack.lm`), we use the same function call as for `nls` and get the same result. The only difference is that, behind the scenes, the Levenberg–Marquardt method has been used instead of the Gauss–Newton method.

```
> require(minpack.lm)
```

```
> nlsLMfit = nls(y~b1*(1-exp(-b2*x)), start=list(b1=500,b2=1e-4))
> summary(nlsLMfit)

Formula: y ~ b1 * (1 - exp(-b2 * x))

Parameters:
    Estimate Std. Error t value Pr(>|t|)
b1 2.389e+02  2.707e+00   88.27   <2e-16 ***
b2 5.502e-04  7.267e-06   75.71   <2e-16 ***
---
Signif. codes:  0 '***' 0.001 '**' 0.01 '*' 0.05 '.' 0.1 ' ' 1

Residual standard error: 0.1019 on 12 degrees of freedom

Number of iterations to convergence: 11
Achieved convergence tolerance: 4.14e-06
```

Our third nonlinear minimization method is `nls.lm`, which is called with the usage

```
nls.lm(par, lower=NULL, upper=NULL, fn, jac = NULL, control =
  nls.lm.control(), ...)
```

`par` is a list or numeric vector of starting estimates for the parameters in the formula. `lower` and `upper` are numeric vectors of lower and upper bounds on each parameter (set to $\pm$ Inf if not given). `fn` is a function that returns a numeric vector of residuals, the sum of squares of which is to be minimized. The first argument of `fn` must be `par`. `jac`, if given, is a function to return the Jacobian for `fn`. `control` is an optional list of control settings (such as tolerances, maximum number of iterations, whether iterates are to be printed, etc.), whose names and effects are given in `nls.lm.control`. As usual, ... stands for other arguments to be passed to `fn` and `jac`.

Note that `nls.lm` calls a vector function whose value it seeks to minimize in a sum of squares sense, while `nls` and `nlsLM` call a formula of the form `y ~ f(x)` which they attempt to satisfy as closely as possible.

We apply `nls.lm` to the `misra1a` dataset, with results identical to those achieved previously. (Remember that since `nlsLM` calls `nls.lm` to do the heavy lifting, this should not be surprising.)

```
> install.packages(minpack.lm)
> require(minpack.lm)

> ## model based on a list of parameters
> modFun = function(param, x) param$b1 * (1 - exp(-x*param$b2))
>
> ## residual function is the function to be minimized
> residFun = function(p, observed, x) observed - modFun(p,x)
>
> ## starting values for parameters
```

```
> initParams = list(b1 = 500, b2 = 1e-4)
>
> ## perform fit
> nls.lm.out = nls.lm(par=initParams, fn = residFun, observed =
+ y, x = x, control = nls.lm.control(nprint=0))
>
> summary(nls.lm.out)

Parameters:
   Estimate Std. Error t value Pr(>|t|)
b1 2.389e+02  2.707e+00   88.27   <2e-16 ***
b2 5.502e-04  7.267e-06   75.71   <2e-16 ***
---
Signif. codes:  0 '***' 0.001 '**' 0.01 '*' 0.05 '.' 0.1 ' ' 1

Residual standard error: 0.1019 on 12 degrees of freedom
Number of iterations to termination: 15
Reason for termination: Relative error in the sum of squares is
  at most 'ftol'.
```

It is sometimes of interest to follow the progress of the iterative calculations with these methods. In `nls`, this may be done by setting the `trace` argument to `TRUE`; the default is `FALSE`. In `nlsLM` and `nls.lm`, it is accomplished by setting `control = nls.lm.control(nprint=1)` where 1 may be any positive number.

A somewhat more challenging problem is a fit to the sum of three rather tightly spaced decaying exponentials. The data were generated with the function (Lanczos, 1956)

$$f(x) = 0.0951e^{-x} + 0.8607e^{-3x} + 1.5576e^{-5x} \tag{11.1}$$

with results to 5 significant figures tabulated on the NIST website. As before, we copy them to a file, read the file into R, and attempt a nonlinear least squares fit.

```
> lanczos3 = read.table(file="~/Desktop/NIST/Lanczos3.txt", header=T)
> x = lanczos3$x; y = lanczos3$y
> nls_lan3 = nls(y~b1*exp(-b2*x)+b3*exp(-b4*x)+b5*exp(-
b6*x),start=list(b1=1.2,b2=0.3,b3=5.6,b4=5.5,b5=6.5,b6=7.6))
Error in nls(y ~ b1 * exp(-b2 * x) + b3 * exp(-b4 * x) + b5 * exp(-b6 *:
  step factor 0.000488281 reduced below 'minFactor' of 0.000976562
```

This time we get an error message, but readily correct the error by adjusting two of the control options, `tol` and `minFactor` (see `?nls` for details). Such adjustments require a trial and error approach.

```
> nls_lan3 = nls(y~b1*exp(-b2*x)+b3*exp(-b4*x)+b5*exp(-
+ b6*x),start=list(b1=1.2,b2=0.3,b3=5.6,b4=5.5,b5=6.5,b6=7.6),
+ control=list(tol=1e-4, minFactor=1e-6))
> summary(nls_lan3)

Formula: y ~ b1 * exp(-b2 * x) + b3 * exp(-b4 * x) + b5 * exp(-b6 * x)
```

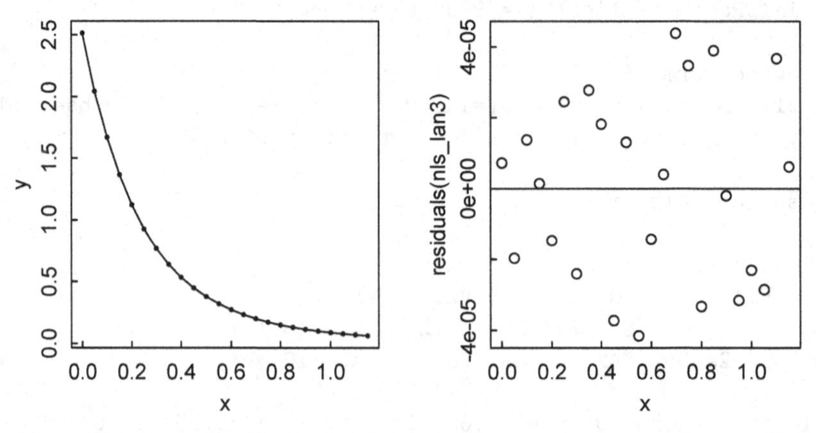

Figure 11.5: *Fit and residuals of nls() fit to the 3-exponential Lanczos function 11.1.*

```
Parameters:
   Estimate Std. Error t value Pr(>|t|)
b1  0.08682    0.01720   5.048 8.37e-05 ***
b2  0.95498    0.09704   9.841 1.14e-08 ***
b3  0.84401    0.04149  20.343 7.18e-14 ***
b4  2.95160    0.10766  27.416 3.93e-16 ***
b5  1.58257    0.05837  27.112 4.77e-16 ***
b6  4.98636    0.03444 144.801  < 2e-16 ***
---
Signif. codes:  0 '***' 0.001 '**' 0.01 '*' 0.05 '.' 0.1 ' ' 1

Residual standard error: 2.992e-05 on 18 degrees of freedom

Number of iterations to convergence: 12
Achieved convergence tolerance: 1.95e-05
```

Plots of data, fit. and residuals now look good, and agreement of fitted with input parameters is satisfactory (Figure 11.5).

```
> par(mfrow=c(1,2))
> plot(x,y,pch=16,cex=0.5)
> lines(x,predict(nls_lan3))
> plot(x,residuals(nls_lan3))
> abline(0,0)
```

Using the Levenberg–Marquardt proceeds as before, most simply with `nlsLM`, giving results not dissimilar from those with `nls`.

```
> nlsLM_lan3 = nlsLM(y~b1*exp(-b2*x)+b3*exp(-b4*x)+b5*exp(-
+ b6*x),start=list(b1=1.2,b2=0.3,b3=5.6,b4=5.5,b5=6.5,b6=7.6))
Warning message:
In nls.lm(par = start, fn = FCT, jac = jac, control = control,
  lower = lower,  :
```

```
  lmdif: info = -1. Number of iterations has reached
   'maxiter' == 50.

> summary(nlsLM_lan3)

Formula: y ~ b1 * exp(-b2*x) + b3 * exp(-b4*x) + b5 * exp(-b6*x)

Parameters:
   Estimate Std. Error t value Pr(>|t|)
b1  0.10963    0.01939   5.656 2.30e-05 ***
b2  1.06938    0.08704  12.286 3.45e-10 ***
b3  0.90322    0.05645  16.001 4.35e-12 ***
b4  3.09411    0.12182  25.399 1.50e-15 ***
b5  1.50055    0.07550  19.874 1.07e-13 ***
b6  5.03437    0.04458 112.930  < 2e-16 ***
---
Signif. codes:  0 '***' 0.001 '**' 0.01 '*' 0.05 '.' 0.1 ' ' 1

Residual standard error: 3.137e-05 on 18 degrees of freedom

Number of iterations till stop: 50
Achieved convergence tolerance: 1.49e-08
Reason stopped: Number of iterations has reached 'maxiter' == 50.
```

If we increase maxiter to 100, we get convergence after 76 iterations, with slightly better agreement with the starting variables:

```
> nlsLM_lan3 = nlsLM(y~b1*exp(-b2*x)+b3*exp(-b4*x)+b5*exp(-
+ b6*x),start=list(b1=1.2,b2=0.3,b3=5.6,b4=5.5,b5=6.5,b6=7.6),
+ control = nls.lm.control(maxiter=100))
>
> summary(nlsLM_lan3)

Formula: y ~ b1 * exp(-b2*x) + b3 * exp(-b4*x) + b5 * exp(-b6*x)

Parameters:
   Estimate Std. Error t value Pr(>|t|)
b1  0.08682    0.01720   5.048 8.36e-05 ***
b2  0.95499    0.09703   9.842 1.14e-08 ***
b3  0.84401    0.04149  20.344 7.17e-14 ***
b4  2.95161    0.10766  27.417 3.92e-16 ***
b5  1.58256    0.05837  27.113 4.77e-16 ***
b6  4.98636    0.03443 144.807  < 2e-16 ***
---
Signif. codes:  0 '***' 0.001 '**' 0.01 '*' 0.05 '.' 0.1 ' ' 1

Residual standard error: 2.992e-05 on 18 degrees of freedom
```

```
Number of iterations to convergence: 76
Achieved convergence tolerance: 1.49e-08
```

`nls.lm` would, of course, give the same results, which are not significantly different from those of `nls` in this case. These estimated values and standard errors of the parameters agree to within the displayed number of significant figures with the Certified Values on the Lanczos3.dat page of the NIST website.

The reader is urged to attempt fitting several other samples from the NIST datasets. The conclusion seems likely to be that, with suitable adjustment of controls, the `nls()` or `nls.lm` functions in R are adequate to handle a wide range of rather difficult nonlinear data fitting problems.

11.3 Inverse modeling of ODEs with the FME package

In Chapter 8 we were concerned with showing how to use R to solve ordinary differential equations, given initial or boundary conditions and certain parameters (numerical coefficients of rate terms, such as rate constants in chemical kinetics). However, sometimes we don't know the parameters, and want to determine them by fitting to data. This is known as inverse modeling, and can be done in R with the `FME` package (`http://CRAN.R-project.org/package=FME`).

As described in the `FME` vignette (Soetaert and Petzoldt (2010) *J. Stat. Software*, `http://www.jstatsoft.org/v33/i03`), estimation of parameters for a complex dynamical system is a nonlinear optimization problem. That is, "the objective is to find parameter values that minimize a measure of badness of fit, usually a least squares function or a weighted sum of squared residuals." The `FME` package takes advantage of R's powerful facilities for nonlinear optimization, and adds some functions of its own.

We illustrate the basics by simulating the kinetic behavior of a simple reversible chemical reaction

$$A + B \underset{k_r}{\overset{k_f}{\rightleftharpoons}} C. \tag{11.2}$$

The initial concentrations of the three species are A_0, B_0, C_0, and the amount of A and B converted to C after the reaction begins is x. The differential equation that describes the time evolution of x is

$$\frac{dx}{dt} = k_f(A_0 - x)(B_0 - x) - k_r(C_0 + x). \tag{11.3}$$

We suppose that C has a characteristic spectral signature which enables its concentration to be followed as a function of time. In the laboratory, the measurement of C would have some uncertainty, or "noise," associated with it. Our task would be to determine k_f and k_r from the time dependence of this noisy signal. We simulate that process by generating a reaction curve and then adding some random noise to it.

We begin by loading the deSolve package and defining the function rxn(pars) which numerically solves the differential equation given the parameters specified in pars, k_f and k_r, for which we will eventually try to find the best values.

```
> require(deSolve)
>
> rxn = function(pars) {
+ derivs = function(times, init, pars) {
+ with(as.list(c(pars, init)), {
+ dx = kf*(A0-x)*(B0-x) - kr*(C0+x)
+ list(dx)
+ })
+ }
+ # Initial condition and time sequence
+ init = c(x = 0)
+ times = seq(0, 10, .1)
+
+ # Solve using ode()
+ out = ode(y=init, parms=pars, times=times, func=derivs)
+
+ # Output the result as a data frame with time in column 1,
  x in column 2
+ as.data.frame(out)
+ }
```

We next use the rxn() function with the known rate constants and starting concentrations to solve for the value of x as a function of time. We add x to C_0 to get C, which is the quantity measured, and plot the result.

```
> pars = c(kf = 0.2, kr = 0.3) # Rate constant parameters
> A0 = 2; B0 = 3; C0 = 0.5 # Initial concentrations
>  # Solve the equation
> out = rxn(pars = pars)
> # Extract time and concentration variables
> time = out$time
> x = out$x
> # Plot C vs time
> plot(time, x+C0, xlab = "time", ylab = "C", type = "l",
+ ylim = c(0,1.5))
```

Suppose that the measurement of the concentration of C has an uncertainty of 10% of C_0. Therefore, we generate a set of "experimental" points by adding normally distributed random noise with an amplitude of $0.1C_0$ to each point, and superimposing the points on the theoretical plot (Figure 11.6).

```
> dataC = cbind(time, x = x + 0.1*C0*rnorm(length(C)))
> points(time, dataC[,2] + C0)
```

Now we invoke FME (which must, of course, already be installed in R) to gain access to two of its functions: modCost() and modFit().

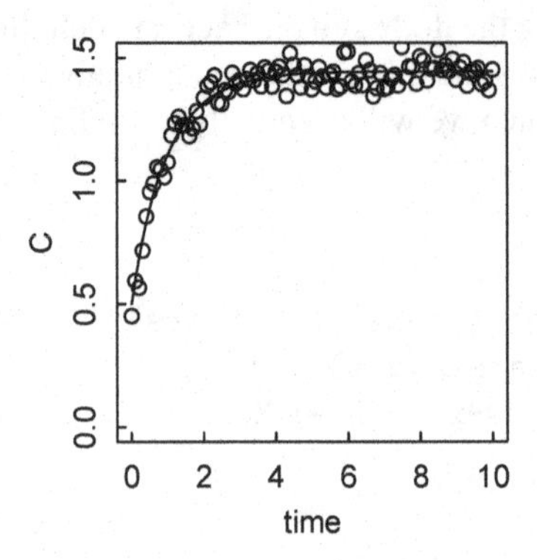

Figure 11.6: *Concentration of product C of reversible reaction with points reflecting measurement errors.*

Given a solution of a model and observed data, modCost estimates the residuals, and the variable and model costs (sum of squared residuals). The function is called with

```
modCost(model, obs, x = "time", y = NULL, err = NULL,
        weight = "none", scaleVar = FALSE, cost = NULL,   ...)
```

where the arguments are (see the help page for details):

model model output, as generated by the integration routine or the steady-state solver, a matrix or a data.frame, with one column per dependent and independent variable.

obs the observed data, either in long (database) format (name, x, y), a data.frame, or in wide (crosstable, or matrix) format.

x the name of the independent variable; it should be a name occurring both in the obs and model data structures.

y either NULL, the name of the column with the dependent variable values,or an index to the dependent variable values; if NULL then the observations are assumed to be in crosstable (matrix) format, and the names of the independent variables are given by the column names of this matrix.

cost if not NULL, the output of a previous call to modCost; in this case, the new output will combine both.

weight only if err=NULL: how to weigh the residuals, one of "none," "std," 'mean."

scaleVar if TRUE, then the residuals of one observed variable are scaled respectively to the number of observations.

... additional arguments passed to R-function approx.

In our case, model is the data frame out, and obs is the data frame dataC. x and y are picked up from the names in the data frames, and the other arguments are handled as defaults.

```
> require(FME)
> rxnCost = function(pars) {
+   out = rxn(pars)
+   cost = modCost(model = out, obs = dataC)
+ }
```

modFit performs constrained fitting of a model to data, in many ways like the other nonlinear optimization routines we have considered, and is called as follows:

```
modFit(f, p, ..., lower = -Inf, upper = Inf,
  method = c("Marq", "Port", "Newton",
         "Nelder-Mead", "BFGS", "CG", "L-BFGS-B", "SANN",
         "Pseudo"), jac = NULL,
  control = list(), hessian = TRUE)
```

Its arguments are

f a function to be minimized, with first argument the vector of parameters over which minimization is to take place. It should return either a vector of residuals (of model versus data) or an element of class modCost (as returned by a call to modCost).

p initial values for the parameters to be optimized over.

... additional arguments passed to function f (modFit) or passed to the methods.

lower, upper lower and upper bounds on the parameters; if unbounded set equal to $\pm$Inf.

method the method to be used, one of "Marq," "Port," "Newton," "Nelder-Mead," "BFGS," "CG," "L-BFGS-B," "SANN," "Pseudo"—see the help page for details. Note that the Levenberg–Marquardt method is the default method.

jac a function that calculates the Jacobian; it should be called as jac(x, ...) and return the matrix with derivatives of the model residuals as a function of the parameters. Supplying the Jacobian can substantially improve performance; see last example.

hessian TRUE if Hessian is to be estimated. Note that, if set to FALSE, then a summary cannot be estimated.

control additional control arguments passed to the optimization routine.

Applying modFit to our fitting problem with guesses for the parameters that are not too far from the real values, and using rxnCost as the function to be minimized, we obtain

```
> Fit = modFit(p = c(kf=.5, kr=.5), f = rxnCost)
> summary(Fit)

Parameters:
   Estimate Std. Error t value Pr(>|t|)
kf 0.196391   0.007781   25.24   <2e-16 ***
kr 0.293774   0.013954   21.05   <2e-16 ***
---
Signif. codes:  0 *** 0.001 ** 0.01 * 0.05 . 0.1   1
```

```
Residual standard error: 0.05233 on 99 degrees of freedom

Parameter correlation:
       kf     kr
kf 1.0000 0.9618
kr 0.9618 1.0000
```

If the guesses for the parameters are too far from the correct values, an error message may be returned stating effectively that the calculation did not converge after the maximum number of iterations, but that "Results are accurate, as far as they go." In that case, those results may be used to start a new calculation, or the maximum number of iterations may be increased.

Proper inverse modeling involves a number of subtleties and complexities beyond just nonlinear optimization. The `FME` vignette uses a relatively simple model of HIV infection to demonstrate these points. We urge the reader to work through the vignette, and summarize its contents as follows.

1. The model is formulated as a function containing a set of ODEs with given parameters, and solved using the `deSolve` function `ode`, with output going to a data frame. The function is initially coded in R (HIV_R) but then in Fortran for speed (HIV), since at later stages the calculation must be repeated thousands of times. The process for writing code in Fortran, C, or C++ is described in the vignette "deSolve: Writing Code in Compiled Languages" available at `http://cran.r-project.org/web/packages/deSolve/vignettes/compiledCode.pdf`.
2. The output is compared with the "data," actually artificial data to which random noise has been applied. The weighted and scaled residuals are converted into a cost (`HIVcost`) using the `modCost` function of `FME`.
3. Local sensitivity (sensitivity to the specific parameters in the model) is then calculated using the `senFun` function of `FME` which takes as input `HIVcost` and the parameters. This process identifies parameters that have little effect on the cost when they are varied, and parameters that have strongly similar effects, indicating that they may not be independent. Parameters that are not strongly pairwise correlated are termed "identifiable," and are the ones that are important to the model.
4. Further examination of identifiability comes from multivariate parameter analysis using the `collin` function of `FME`. This yields a set of collinearity indices, indicating the extent to which a change in one parameter can be undone by appropriate changes in the other parameters. `semFun` and `collin` together enable selection of the set of parameters with the smallest collinearities for subsequent fitting.
5. To find the "best" values of the remaining parameters, nonlinear data fitting is carried out with the `modFit` function of `FME`. This is a wrapper for the optimization functions in `optim`, `nls`, and `nlminb` from the R base packages, with the addition of the Levenberg–Marquardt algorithm from the `minpack.lm` package and a pseudo-random search algorithm implemented in `FME`.

6. The steps up to this point have provided values for the identifiable parameters that are optimal in the least squares sense. However, it is important to estimate the effect of uncertainties in the parameters on the fit between the model and the data. This is done with the `modMCMC` function in `FME`, using a Markov chain Monte Carlo method with probabilities drawn from the target distribution, as described in the vignette. It is at this stage that the large number of runs of the model are carried out, making desirable coding of the HIV function in a fast, low-level language.
7. The function `sensRange` of `FME` is then used to generate graphs and summary data on the effect of parameter uncertainty on the output of the model.
8. An extension of this approach to global parameter sensitivity is made by the function `modCRL` which tests the effect of parameter variation on a single output variable (such as mean viral load), rather than on a time series.

`FME` is applicable not just to simulations of dynamic processes, but also to steady states (vignette "`FMEsteady`") and nonlinear equilibrium models (vignette "`FMEother`"). The vignettes "`FMEdyna`" and "`FMEmcmc`" demonstrate additional aspects of the `FME` package. All of these vignettes are available at `http://CRAN.R-project.org/package=FME`.

11.4 Improving the convergence of series: Padé and Shanks

A task related to that of fitting a function to a set of points is finding an approximation to the function that is better than the first few terms of a Taylor's series. We demonstrate two approaches here, both using rational functions (ratios of polynomials).

We begin with Padé approximants, which use the "data" provided by the first few terms in a Taylor's series to construct the best approximation of the desired function by a rational function of given order. The Padé approximant often gives a better approximation of the desired function than the Taylor's series, and may work even if the power series does not converge. For an example, consider the function $\ln(1+x)$ whose Taylor's series expansion to 4th order is $x - x^2/2 + x^3/3 - x^4/4$.

Calculation of Padé approximants in R is carried out in the `pracma` package, which must be installed before it can be loaded and used.

```
> require(pracma)
Loading required package: pracma
PRACMA 1.1.6:
Renamed some functions to avoid shadowing R base functions.
```

The `pade` function takes as arguments the vector of coefficients of the series in descending order, and the desired degrees of the numerator and denominator polynomials. In this case we use second-order polynomials.

```
> P = pade(c(-1/4,1/3,-1/2,1,0),d1=2,d2=2)
> r1 = P$r1; r2 = P$r2
```

We now define functions for the original, series, and Padé results, and plot them for comparison (Figure 11.7). The improvement in accuracy of the Padé result over an extended range is striking.

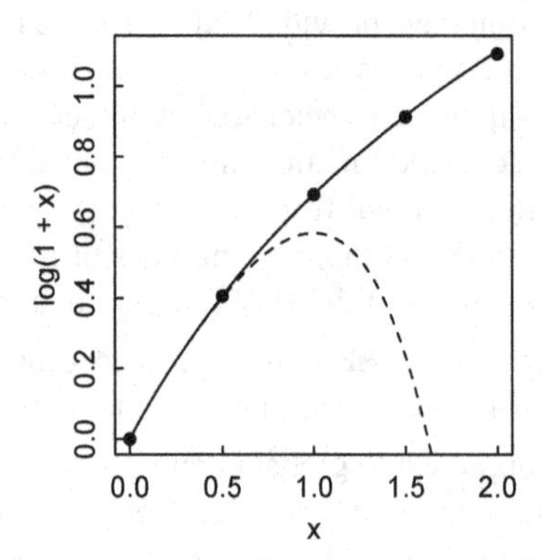

Figure 11.7: *Approximations of ln(1+x): Solid line, true function; dashed line, Taylor's series; points, Padé approximation.*

```
> origfn = function(x) log(1+x)
> taylorfn = function(x) x-x^2/2+x^3/3-x^4/4
> padefn = function(x) polyval(r1,x)/polyval(r2,x)
>
> curve(log(1+x),0,2)
> curve(taylorfn, add=T,lty=2)
> x=seq(0,2,.5)
> points(x,padefn(x),pch=16)
```

The Shanks transformation is used to accelerate the convergence of a slowly convergent series. In many common cases, the true value of the series, S, can be represented as the nth partial sum plus an "error" term decreasing geometrically with n: $S = S + C\lambda^n$. Manipulation of this equation shows that

$$S = S_{n+1} + \frac{\lambda}{1-\lambda}(S_{n+1} - S_n)$$

where

$$\lambda = \frac{S_{n+1} - S_n}{S_n - S_{n-1}}$$

Consider the application of this approach to the Riemann zeta function $\zeta(2)$:

$$\zeta(2) = \sum_{x=1}^{\infty} \frac{1}{x^2} = \frac{\pi^2}{6}$$

where the last equality was proved by Euler. It takes 60 terms of the series for the cumulative sum to come within 1% of the correct answer (Figure 11.8).

```
> x = 1:60
> y = 1/x^2
> csy=cumsum(y)
> plot(x,csy,type="l", ylim=c(1,1.8))
> abline(h=pi^2/6, lty=3)
```

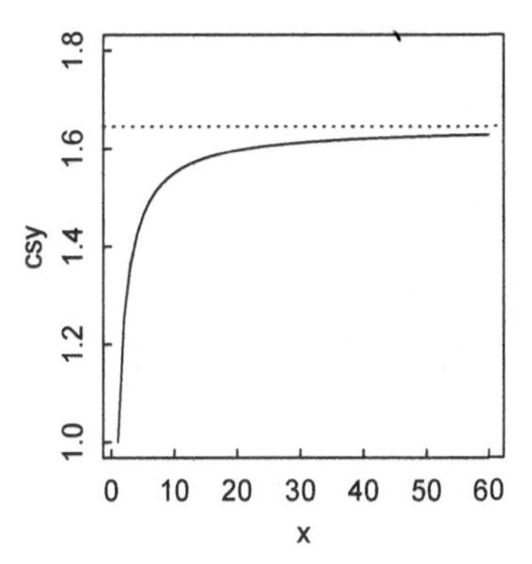

Figure 11.8: *Approximation to* $\zeta(2)$ *by direct summation of* $1/x^2$.

The R program below, which applies the Shanks transformation three times in succession, comes within 1.6% using only the first seven terms of the series, while direct summation takes 37 terms to get that close.

```
> S = function(w,n) {
+   lam = (w[n+1]-w[n])/(w[n]-w[n-1])
+   return(w[n+1]+lam/(1-lam)*(w[n+1]-w[n]))
+   }
> # Use terms (1,2,3) to get S(csy,2), ...
> # (5,6,7) to get S(csy,6)
> S1 = c(S(csy,2),S(csy,3),S(csy,4),S(csy,5),S(csy,6))
> S1
[1] 1.450000 1.503968 1.534722 1.554520 1.568312
> # Now use the previous five values to get three new values
> S2 = c(S(S1,2),S(S1,3),S(S1,4))
> S2
[1] 1.575465 1.590296 1.599981
> # Use those three values to get one new value
> S3 = S(S2,2);
> S3
[1] 1.618209
> pi^2/6
[1] 1.644934
```

11.5 Interpolation

Often one has tabulated values of a property as a function of some condition, but wants the value at some other conditions than those tabulated. If the desired condition lies within the tabulated range, the value can be estimated by interpolation. (Extrapolating beyond the tabulated range is a much riskier business.) R has several functions for doing such interpolation.

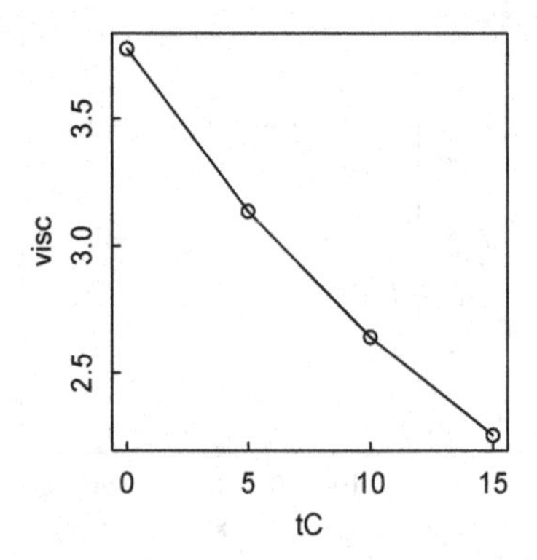

Figure 11.9: *Viscosity of 20% solutions of sucrose in water as a function of temperature.*

For example, biochemists often sediment proteins and nucleic acids through aqueous sucrose solutions. Tables of the viscosity of such solutions are available at 5 deg C temperature increments (0, 5, 10, 15, etc.). But suppose sedimentation measurements are to be done at other temperatures, e.g., 4, 7, and 12 deg C. See Figure 11.9.

```
> # Known values
> tC = c(0,5,10,15)
> visc = c(3.774, 3.135, 2.642, 2.255)
> plot(tC,visc, type="o")
```

11.5.1 Linear interpolation

The simplest interpolation, but generally not the most appropriate, is a linear extrapolation between neighboring tabulated points bounding the temperature of interest. This is handled in R by the `approx()` or `approxfun()` functions.

```
> # Desired temperatures
> tExp = c(4,7,12)
> # Linear approximation
> approx(tC,visc,tExp)
$x
[1]  4  7 12

$y
[1] 3.2628 2.9378 2.4872

> # Linear approximation using approxfun
> apf = approxfun(tC,visc)
> apf(tExp)
[1] 3.2628 2.9378 2.4872
```

11.5.2 Polynomial interpolation

Given the small but noticeable curvature in the tC-visc plot, a polynomial plot might be slightly more accurate. `poly.calc` computes the Lagrange interpolating polynomial, from which the values at the desired conditions can be obtained.

```
> require(PolynomF)
Loading required package: PolynomF
> polyf = poly.calc(tC, visc)
> polyf(tExp)
[1] 3.24984 2.92252 2.47672
```

A variant of the Lagrange interpolation procedure is Barycentric Lagrange interpolation, implemented in the `pracma` package, which states "Barycentric interpolation is preferred because of its numerical stability."

```
> require(pracma)
> barylag(tC,visc,tExp)
[1] 3.24984 2.92252 2.47672
```

11.5.3 Spline interpolation

For this only mildly curved dataset, identical results are obtained with cubic spline interpolation, using either `spline()` or the perhaps preferable `splinefun()`, which gives the function over the full range of inputs.

```
> spline(tC,visc,xout=tExp)
$x
[1] 4 712
$y
[1] 3.24984 2.92252 2.47672
> spf = splinefun(tC, visc)
> spf(tExp)
[1] 3.24984 2.92252 2.47672
```

Polynomial interpolation functions may often oscillate substantially and inappropriately. Spline functions are generally better behaved, but even they may exhibit inappropriate non-monotonic behavior. In such circumstances, `splinefun` has the method `"monoH.FC"`, which guarantees that the spline will be monotonic increasing or decreasing if the data points are. This behavior is demonstrated in the following example (Figure 11.10).

```
> options(digits=4)
> x=c(0,.5,1,2,3,4)
> y=c(0,.93,1,1.1,1.15,1.2)
> require(PolynomF)
> polyfit = poly.calc(x,y)
> polyfit
3.638*x - 4.794*x^2 + 2.828*x^3 - 0.7438*x^4 + 0.07105*x^5
> plot(x,y)  # Plot of points
```

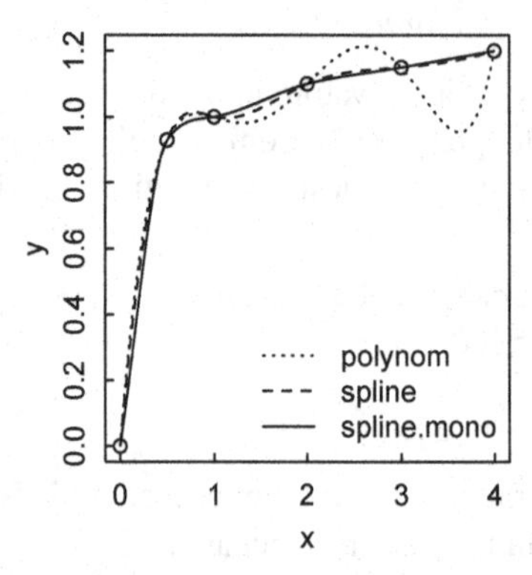

Figure 11.10: *Examples of non-monotonic and monotonic fitting to a set of points.*

```
> curve(polyfit,add=T,lty=3)  # Polynomial curve fit
> splinefit=splinefun(x,y)
> curve(splinefit,add=T,lty=2)  # Spline fit
> splinefit.mono = splinefun(x,y,method="mono")
> curve(splinefit.mono,add=T,lty=1)  # Monotonic spline fit
> legend("bottomright",legend=c("polynom","spline",
+ "spline.mono"), lty=c(3:1),bty="n")
```

11.5.3.1 Integration and differentiation with splines

`integrate()` (see Chapter 6) can be combined with spline fitting to find the area under a set of points, using `splinefun()`. (To get the coordinates of the spline fit points themselves, rather than the function that determines them, use `spline()`.) For example, suppose that one simulates the UV spectrum of a mixture of three compounds, each of which is characterized by a Gaussian band shape with maximum at `x0` and standard deviation `sig`, with the amplitude being measured every 5 nm between 180 nm and 400 nm.

```
> fn = function(x,x0,sig) exp(-(x-x0)^2/(2*sig^2))
> x = seq(180,400,4)
> y = 1*fn(x,220,15) + 1.3*fn(x,280,12) + .8*fn(x,320,15)
> fsp = splinefun(x,y)
> integrate(fsp,180,400)
106.6383 with absolute error < 0.011
> plot(x,y,pch=16, cex=0.5, ylim=c(-1,1.4))
> curve(fsp(x), add = T)
```

One can also use the spline function to numerically differentiate the data. This can be useful to emphasize maxima and minima in the data: they turn into zero crossings when differentiated once.

```
> curve(10*fsp(x,deriv=1), add=T, lty="dashed")
```

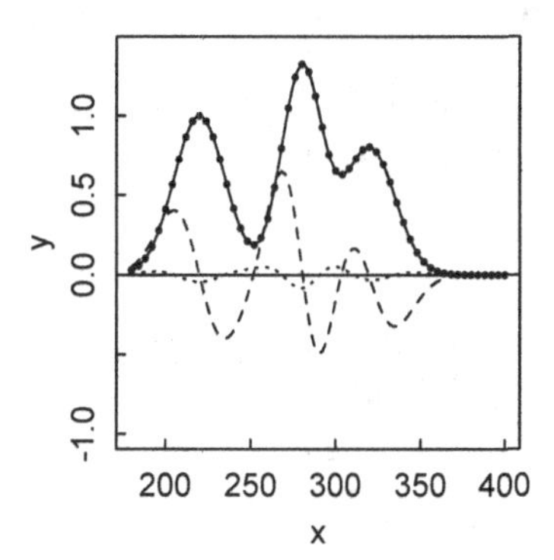

Figure 11.11: *Fit of a spline function to a simulated spectrum, along with first and second derivative curves.*

Second and higher order derivatives can also be calculated (Figure 11.11).

```
> curve(10*fsp(x, deriv=2),add=T, lty="dotted")
> abline(0,0)
```

11.5.4 Rational interpolation

Rational interpolation, implemented in `pracma`, is less commonly employed than polynomial or spline methods, but it may be the most reliable, especially for functions with poles (see Press et al. (2007), p. 124). The procedure, giving a function that is the ratio of two polynomials, is essentially the same as for calculating Padé approximants.

```
> require(pracma)
> ratinterp(tC,visc,tExp)
[1] 3.249560 2.922859 2.476251
```

Polynomial and spline interpolating functions will often diverge or oscillate markedly if applied outside the range for which they were calculated. Therefore, they are generally very unreliable for extrapolation. Rational approximation, on the other hand, can often be used for extrapolation of real-life data. Consider, for example, extrapolation of the aqueous sucrose data to 20 deg C.

```
> ratinterp(tC,visc,20)  # rational interpolation
[1] 1.946371
> polyf(20)  # polynomial fit
[1] 1.934
> spf(20)  # spline fit
[1] 1.934
```

1.946 is the tabulated experimental value.

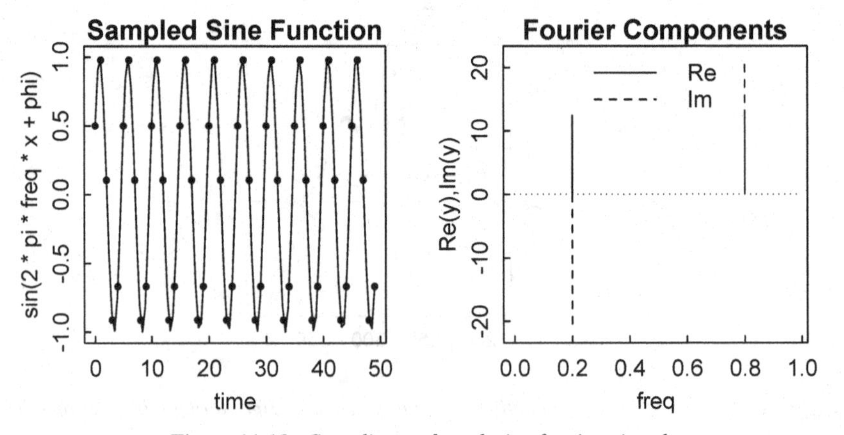

Figure 11.12: *Sampling and analysis of a sine signal.*

11.6 Time series, spectrum analysis, and signal processing

Scientists and engineers often need to make sense out of a series of data points measured at successive times, a topic collectively denoted as "time series." Often the signal oscillates in time, but the data are complicated by non-constant baselines and random noise. A common task is to determine the frequency or frequencies of the underlying signal. The basic tools in R to accomplish this task are Fourier analysis, carried out with the `fft()` (fast Fourier transform) function, and power spectrum analysis, carried out with the `spectrum()` function. We also consider the `signal` package, which gives access to a broader range of signal processing and filtering functions.

11.6.1 Fast Fourier transform: `fft()` function

We begin with a simple sine wave, with frequency `freq`, amplitude `A`, phase `phi`, sampled `N` times at interval `tau`. See Figure 11.12.

```
> # Parameters
> N = 50; freq = 1/5; A = 1; phi = pi/6; tau = 1
> par(mfrow=c(1,2)) # To display various features side-by-side
> # Draw the smooth underlying sine wave
> curve(sin(2*pi*freq*x + phi),0,N-1, xlab="time",
+ main="Sampled Sine Function")
> # Plot the points at which sampling will occur
> j=0:(N-1)
> y = sin(2*pi*freq*j*tau + phi)
> points(j,y,pch=16,cex=0.7)
> # Calculate the real and imaginary parts of the fft
> ry = Re(fft(y)); iy = Im(fft(y))
> # Set the infinitesimal components to zero
```

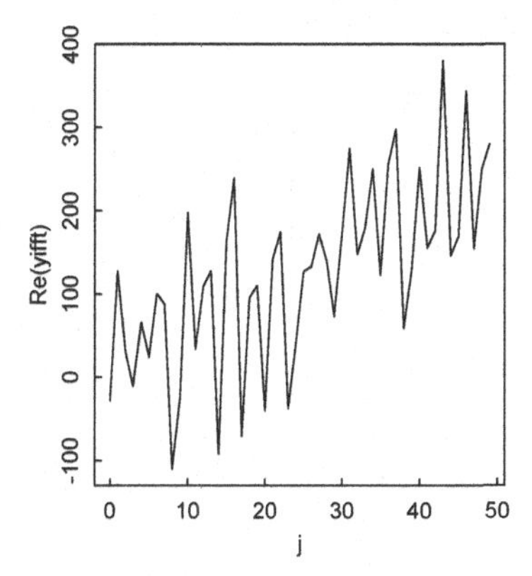

Figure 11.13: *Inverse fft of the signal in Figure 11.17.*

```
> zry = zapsmall(ry)
> ziy = zapsmall(iy)
> # Plot the real part(s)
> plot(j/(tau*N),zry,type="h",ylim=c(min(c(zry,ziy)),
+ max(c(zry,ziy))),xlab = "freq",
+ ylab ="Re(y),Im(y)", main="Fourier Components")
> # Add the imaginary part(s)
> points(j/(tau*N),ziy,type="h",lty=2)
> legend("top",legend=c("Re","Im"),lty=1:2, bty="n")
```

The frequency axis is in units of $1/(j\tau)$ with the lowest frequency being $1/(N\tau)$ and the highest meaningful frequency being $1/(2\tau)$, or 0.5 on this graph. The Fourier Components plot recovers the input frequency of 0.2; the apparent second peak at 0.8 is the result of "aliasing" as explained by the Nyquist–Shannon sampling theorem. Since the phase is $\pi/6$, both real and imaginary components of the Fourier transform are found. If the phase were 0, only the imaginary component would appear; if the phase were $\pi/2$, only the real component would appear. In both of these "pure" cases, the amplitude is $25 = N/2$; a very different normalization from that typically defined in mathematics textbooks.

11.6.2 Inverse Fourier transform

The inverse Fourier transform can be obtained with the option `inverse = TRUE` of the `fft()` function:

```
> yfft = fft(y)
> yifft = fft(yfft,inverse=TRUE)
> plot(j,Re(yifft), type="l")
```

The shape of the curve (Figure 11.13) is the same as the original function, but the normalization is different. According to the `fft` help page, "If inverse is TRUE, the

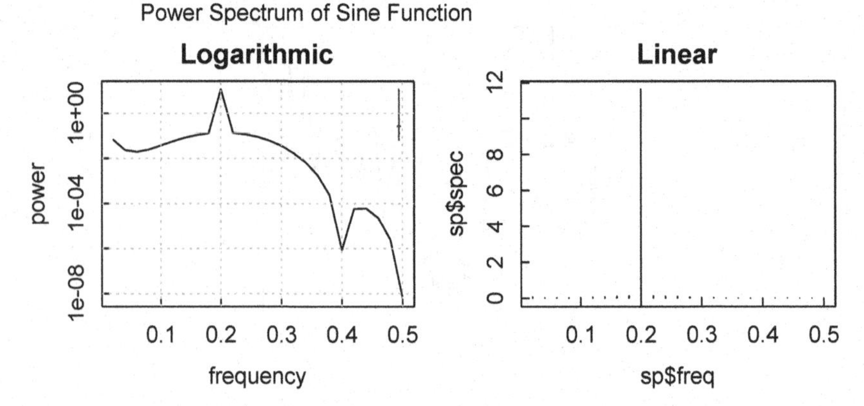

Figure 11.14: *Power spectrum of sine function.*

(unnormalized) inverse Fourier transform is returned, i.e., if y =- fft(z), then z is fft(y, inverse = TRUE) / length(y)."

11.6.3 Power spectrum: `spectrum()` *function*

Often the main quantity desired is the frequency, in which case the `spectrum()` function is appropriate, since it gives the sum of the squares of the real and imaginary components as a function of frequency, i.e., the power spectrum, in which the amplitude is plotted on a logarithmic scale. In general, only the largest values are of interest. `spectrum(y)` returns a list, from which the frequency and power components can be obtained with `$freq` and `$spec`, which enables a linear plot of power vs. frequency (Figure 11.14).

```
> # Set up for plotting two graphs with combined caption
> par(oma=c(0,0,2,0))
> par(mar=c(3,3,2,1))
> par(mfrow=c(1,2))
> # Calculate the power spectrum
> sp = spectrum(y, xlab="frequency", ylab="power",main="Logarithmic")
> grid() # To more easily read off the coordinates of the peak(s)
> # Place the combined caption
> mtext("Power Spectrum of Sine Function", side=3,line=2, adj=-2)
> # Plot the linearized power spectrum
> plot(sp$freq,sp$spec,type="h", main="Linear")
```

The `spectrum` help page states "The spectrum here is defined with scaling 1/frequency(x), following S-PLUS. This makes the spectral density a density over the range (-frequency(x)/ 2, +frequency(x)/2), whereas a more common scaling is 2pi and range (-0.5, 0.5] ... or 1 and range (-pi, pi]."

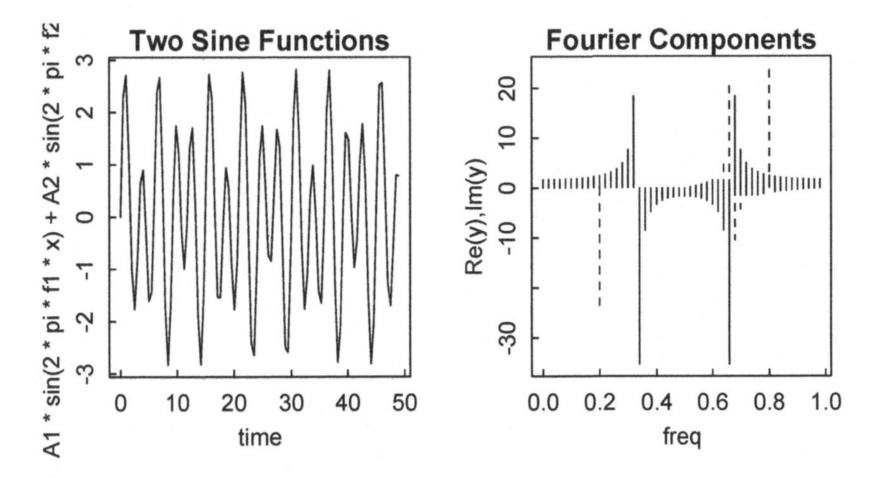

Figure 11.15: *fft of the sum of two sine functions.*

If we apply the same analysis to the sum of two sine functions, with different frequencies and amplitudes, we recover the original frequencies with approximately proportionate amplitudes with `spectrum()`. The `fft()` results, however, are not easy to interpret by inspection (Figure 11.15).

```
> par(mfrow=c(1,2))
>
> N = 50; tau = 1
> f1 = 1/5; A1 = 1; f2 =1/3; A2 = 2
> curve(A1*sin(2*pi*f1*x) + A2*sin(2*pi*f2*x),0,N-1,
+ xlab="time", main="Two Sine Functions")
>
> j=0:(N-1)
> y = A1*sin(2*pi*f1*j*tau) + A2*sin(2*pi*f2*j*tau)
>
> ry = Re(fft(y)); iy = Im(fft(y))
> zry = zapsmall(ry)
> ziy = zapsmall(iy)
>
> plot(j/(tau*N),zry,type="h",ylim=c(min(c(zry,ziy)),
+ max(c(zry,ziy))),xlab = "freq",
+   ylab ="Re(y),Im(y)", main="Fourier Components")
> points(j/(tau*N),ziy,type="h",lty=2)
```

The power spectrum (Figure 11.16) is computed and plotted from

```
> par(mfrow = c(1,1))
> sp = spectrum(y, xlab="frequency", ylab="power",
+  main="Power Spectrum 2 Sines")
> grid()
```

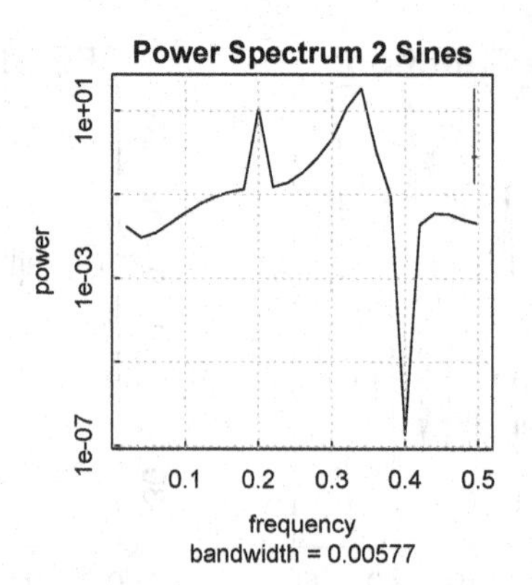

Figure 11.16: *Power spectrum of the sum of two sine functions.*

A more realistic case would be a signal consisting of two sine functions with a sloping baseline and a significant amount of random noise (Figure 11.17).

```
> par(mfrow=c(1,2))
> set.seed(123)
> N = 50; tau = 1
> f1 = 1/5; A1 = 1; f2 =1/3; A2 = 2
> j=0:(N-1)
> y = A1*sin(2*pi*f1*j*tau) + A2*sin(2*pi*f2*j*tau)
```

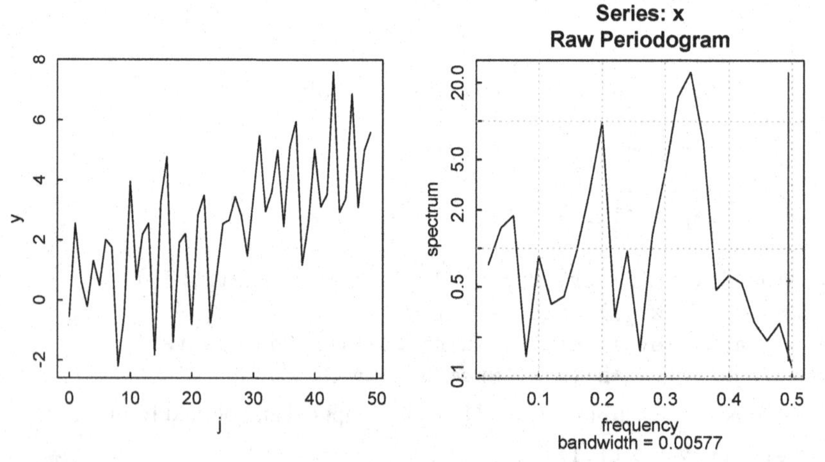

Figure 11.17: *Power spectrum (right) of the sum of two sine functions with random noise and a sloping baseline (left).*

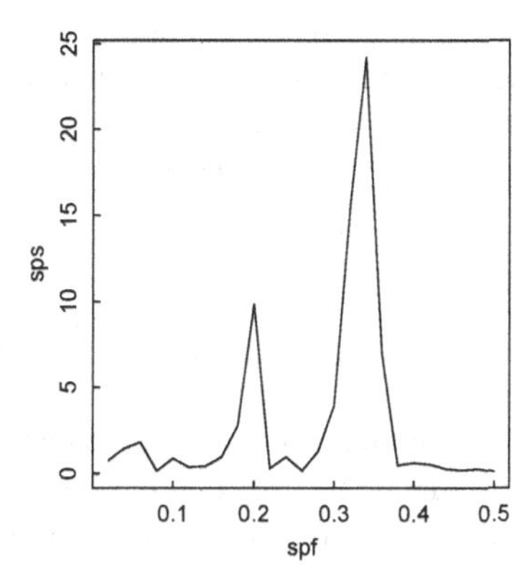

Figure 11.18: *Plot of the peaks derived from the power spectrum.*

```
> ybase = j/10 # Add a linear sloping baseline
> yrand = rnorm(N) # and some random noise
> y = y + ybase + yrand # Combine
> plot(j,y,type="l")
> sp = spectrum(y); grid()
```

Handily, `spectrum()` removes linear trends. Even with a large amount of noise, the two peaks at frequencies of 1/5 and 1/3 stand out. If the slope and intercept of the linear baseline were desired, they could be obtained from the linear fit `lm(y~j)`.

11.6.4 `findpeaks()` function

We can obtain a more precise description of the peaks in the power spectrum by using the `findpeaks()` function of the `pracma` package on the plot of the `$spec` vs `$freq` components of the `sp` list. (See Figure 11.18.) According to the `findpeaks` help page, the function "returns a matrix where each row represents one peak found. The first column gives the height, the second the position/index where the maximum is reached, the third and fourth the indices of where the peak begins and ends — in the sense of where the pattern starts and ends."

```
> spf = sp$freq
> sps = sp$spec
> plot(spf,sps,type="l")
> require(pracma)
> findpeaks(sps,minpeakheight=5)
          [,1] [,2] [,3] [,4]
[1,]  9.850003   10    6   11
[2,] 24.226248   17   13   19
> spf[c(10,17)]
[1] 0.20 0.34
```

Thus the 10th and 17th frequency values in the spectrum are 0.20 and 0.34, very close to the starting values of 1/5 and 1/3 for the pure sum of sine waves, although the heights are not in the proper ratios. If the number of sampled points had been an integral multiple of both starting frequencies (e.g., 60 rather than 50) the analysis would have yielded 0.33 for the second frequency.

11.6.5 `signal` package

According to its documentation, the `signal` package is "a set of signal processing R functions originally written for MATLAB®/Octave. Includes filter generation utilities, filtering functions, resampling routines, and visualization of filter models. It also includes interpolation functions." We confine our discussion to showing how several of the filter models can be used to approximate the underlying signal in a noisy signal.

```
> require(signal)
Loading required package: signal
Loading required package: MASS

Attaching package: signal

The following object(s) are masked from package:pracma:

    conv, ifft, interp1, pchip, polyval, roots

The following object(s) are masked from package:stats:

    filter, poly
```

11.6.5.1 Butterworth filter

The Butterworth filter is a filter designed to have as flat a frequency response as possible in the pass band. Its characteristics are plotted using the `freqz` function. By default it is implemented in signal as a low-pass filter, but it may also be high-pass, stop-band, or pass-band. Figure 11.19 shows an example.

```
> bf = butter(4, 0.1) # parameters filter order, critical frequency
> freqz(bf)
```

We use a Butterworth filter to extract a sinusoidal signal from added normally distributed random noise. Note that the pure one-pass filter introduces a phase shift, but the signal function `filtfilt` does a reverse pass and removes the phase shift, albeit at the expense of squaring the magnitude response (Figure 11.20).

```
> bf = butter(3, 0.1)  # 10 Hz low-pass filter
> t = seq(0, 1, len = 100)  # 1 second sample
> # 2.3 Hz sinusoid + noise
> x = sin(2*pi*t*2.3) + 0.25*rnorm(length(t))
> y = filtfilt(bf, x)
```

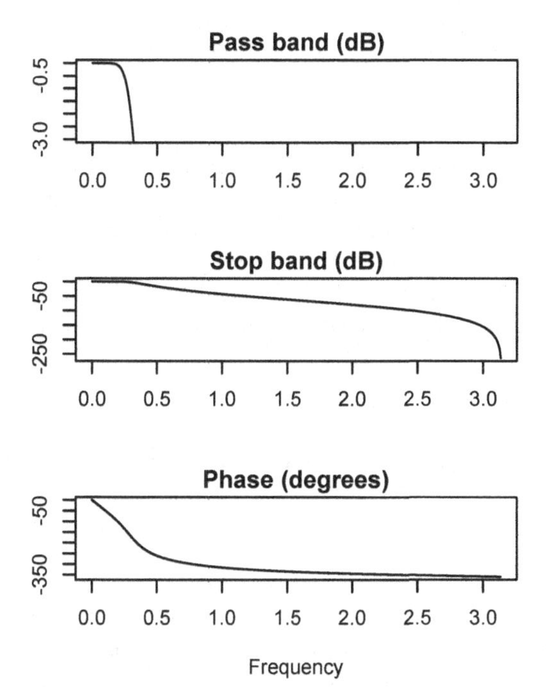

Figure 11.19: *Frequency response of the Butterworth filter* `butter(4,0.1)`.

```
> z = filter(bf, x) # apply filter
> plot(t, x,type="l", lty=3, lwd = 1.5)
> lines(t, y, lty=1, lwd=1.5)
```

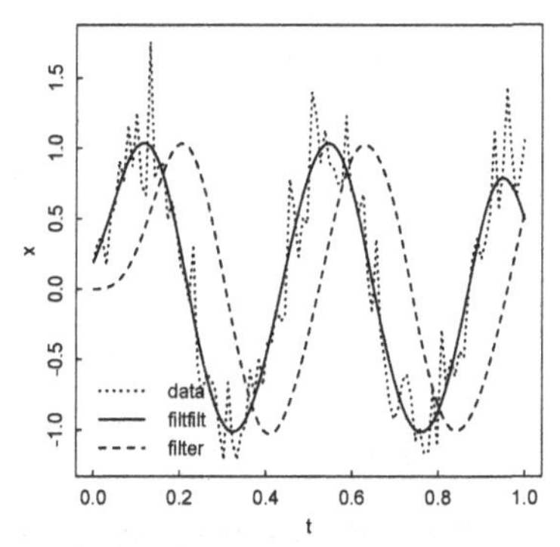

Figure 11.20: *Use of* `butter(3,0.1)` *filter to extract a sinusoidal signal from added normally distributed random noise.*

```
> lines(t, z, lty=2, lwd = 1.5)
> legend("bottomleft", legend = c("data", "filtfilt", "filter"),
+ lty=c(3,1,2), lwd=rep(1.5,3), bty = "n")
```

11.6.5.2 *Savitzky–Golay filter*

The Savitzky–Golay method performs a local polynomial fit on a set of points to determine the smoothed value for each point. It has the advantage "that it tends to preserve features of the distribution such as relative maxima, minima and width, which are usually 'flattened' by other adjacent averaging techniques (like moving averages, for example)" (Wikipedia). On the other hand, as we see from this example, it may preserve some details that were not present in the original signal (Figure 11.21).

```
> y = sgolayfilt(x)
> plot(t,x,type="l",lty=3)
> lines(t, y)
> legend("bottomleft", legend = c("data", "sgolayfilt"),
+ lty=c(3,1), bty = "n")
```

11.6.5.3 *fft filter*

The `fftfilt` function applies a multi-point running average filter to the data.

```
> z = fftfilt(rep(1, 10)/10, x) # 10-point averaging filter
> plot(t, x, type = "l", lty=3)
> lines(t, z)
> legend("bottomleft", legend = c("data", "fftfilt"),
+ lty=c(3,1), bty = "n")
```

R and its contributed packages contain many functions for analyzing time series. For more detailed and extensive views of this broad topic, see the book by Cryer and

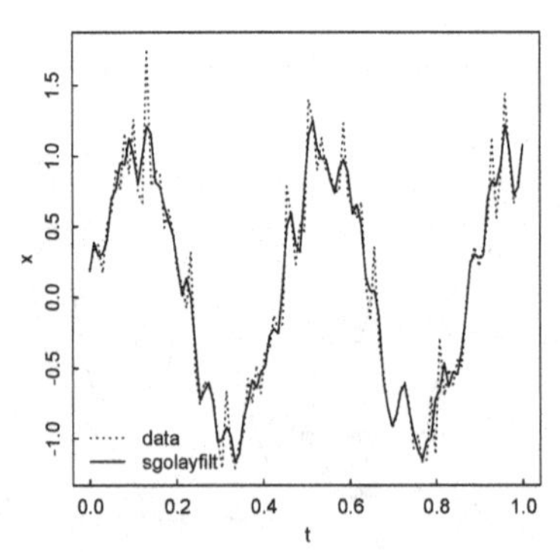

Figure 11.21: *Use of Savitzky–Golay filter to extract a sinusoidal signal from added normally distributed random noise.*

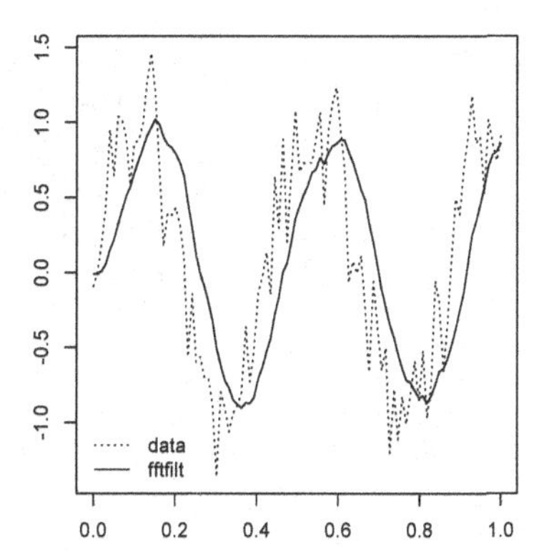

Figure 11.22: *Use of `fftfilt` to extract a sinusoidal signal from added normally distributed random noise.*

Chan (2008); Chapter 14 in Venables and Ripley (2002); and the Time Series Analysis Task View on CRAN.[1] Shorter but useful online treatments have been written by Coghlan[2] and Kabacoff,[3] among others.

11.7 Case studies

11.7.1 Fitting a rational function to data

The NIST Dataset Archives at `http://www.itl.nist.gov/div898/strd/general/dataarchive.html` contains many interesting datasets on which statistical code may be exercised. In the Nonlinear Regression subset at `http://www.itl.nist.gov/div898/strd/nls/nls_main.shtml` there are sets at three levels of difficulty: Lower, Average, and Higher. We have already used `Misra1a` and `Lanczos3` from the Lower set. Here we use `Hahn1` at `http://www.itl.nist.gov/div898/strd/nls/data/hahn1.shtml` for a dataset at an Average level of difficulty. The data are the result of a NIST study involving the thermal expansion of copper; x is the Kelvin temperature and y is the coefficient of thermal expansion. There are 236 observations, and we fit to a rational function with 7 coefficients,

$$y = \frac{b_1 + b_2x + b_3x^2 + b_4x^3}{1 + b_5x + b_6x^2 + b_7x^3}, \tag{11.4}$$

so there are 229 degrees of freedom.

We begin, as usual, by copying the data from the website, saving it to a file on the desktop, and reading it into R with `read.table()`.

```
> hahn1 = read.table(file="~/Desktop/Hahn1.txt", header=T)
```

[1] http://cran.r-project.org/web/views/TimeSeries.html

[2] http://a-little-book-of-r-for-time-series.readthedocs.org/en/latest/

[3] http://www.statmethods.net/advstats/timeseries.html

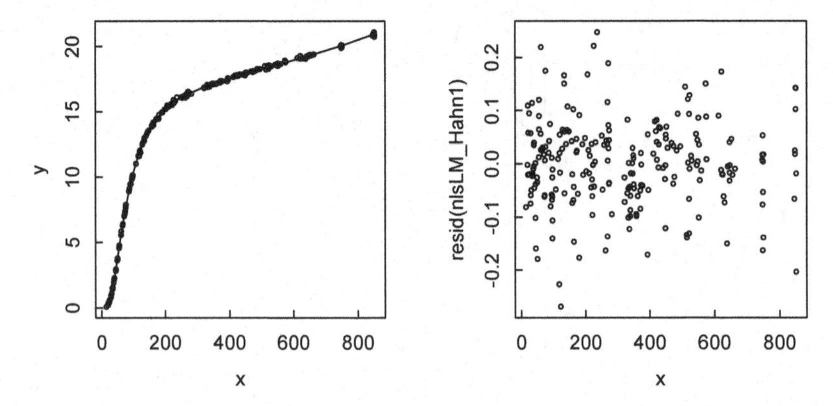

Figure 11.23: *(left) Plot of Hahn1 data and fitting function; (right) Plot of residuals.*

We extract the x and y variables, and plot the data to get a sense of its behavior. Anticipating the need to overlay the fitting function and plot the residuals, we set up a 1×2 graphics array (Figure 11.23).

```
> x = hahn1$x; y = hahn1$y
> par(mfrow=c(1,2))
> plot(x,y,cex=0.5)
```

We use the Levenberg–Marquardt approach to find the estimated best values for the coefficients in the rational function with the `nlsLM()` function in the `minpack.lm` package. As starting values we use those on the NIST website.

```
> require(minpack.lm)
> nlsLM_Hahn1 = nlsLM(y~(b1+b2*x+b3*x^2+b4*x^3)/
 (1+b5*x+b6*x^2+b7*x^3),
 start=list(b1=10, b2=-1, b3=.05, b4=-1e-5,
 b5=-5e-2, b6=.001, b7=-1e-6))
```

We get the estimated values, their standard errors, and the probabilities that they are not significant (infinitesimal in all cases) with the `summary()` function.

```
> summary(nlsLM_Hahn1)

Formula: y ~ (b1 + b2 * x + b3 * x^2 + b4 * x^3)/
(1 + b5 * x + b6 * x^2 + b7 * x^3)

Parameters:
     Estimate Std. Error t value Pr(>|t|)
b1  1.078e+00  1.707e-01   6.313 1.40e-09 ***
b2 -1.227e-01  1.200e-02 -10.224  < 2e-16 ***
b3  4.086e-03  2.251e-04  18.155  < 2e-16 ***
b4 -1.426e-06  2.758e-07  -5.172 5.06e-07 ***
b5 -5.761e-03  2.471e-04 -23.312  < 2e-16 ***
```

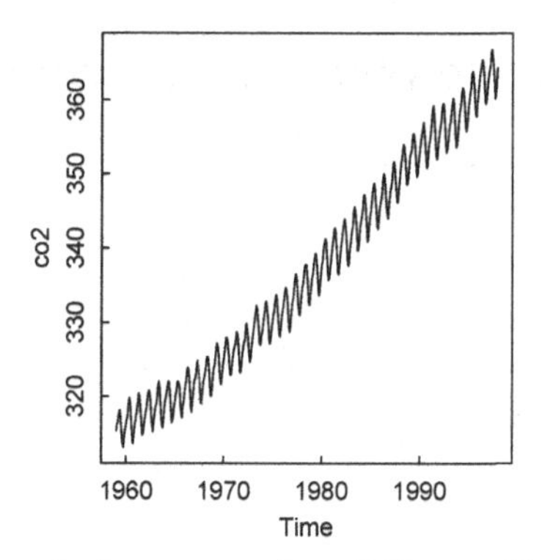

Figure 11.24: *Atmospheric concentration of CO_2 monthly from 1959 to 1997.*

```
b6  2.405e-04  1.045e-05  23.019  < 2e-16 ***
b7 -1.231e-07  1.303e-08  -9.453  < 2e-16 ***
---
Signif. codes:  0 *** 0.001 ** 0.01 * 0.05 . 0.1   1

Residual standard error: 0.0818 on 229 degrees of freedom

Number of iterations to convergence: 10
Achieved convergence tolerance: 1.49e-08
```

These values agree, to the displayed number of significant figures, with those on the NIST website. Interestingly, using starting values 10-fold lower leads to identical results.

Finally, we graphically examine the agreement between experimental and fitted values with an overlay line and a plot of residuals. Examination of the numerical data shows that the x values are not monotonically increasing, so we first sort x and y before we draw the fitted line.

```
> xsort=sort(x)
> ysort=sort(fitted(nlsLM_Hahn1))
> lines(xsort,ysort)
> plot(x,resid(nlsLM_Hahn1),cex=0.5)
```

11.7.2 Rise of atmospheric carbon dioxide

The `datasets` package included in base R contains the time series `co2`, which presents 468 monthly measurements of the atmospheric concentration of CO_2 on Mauna Loa, expressed in parts per million, from 1959 to 1997. The data can be visualized simply (Figure 11.24):

```
> plot(co2)
```

There is a clear upward trend, along with fairly regular seasonal oscillations and some random variation. The `decompose()` function separates these contributions by moving averages (Figure 11.25).

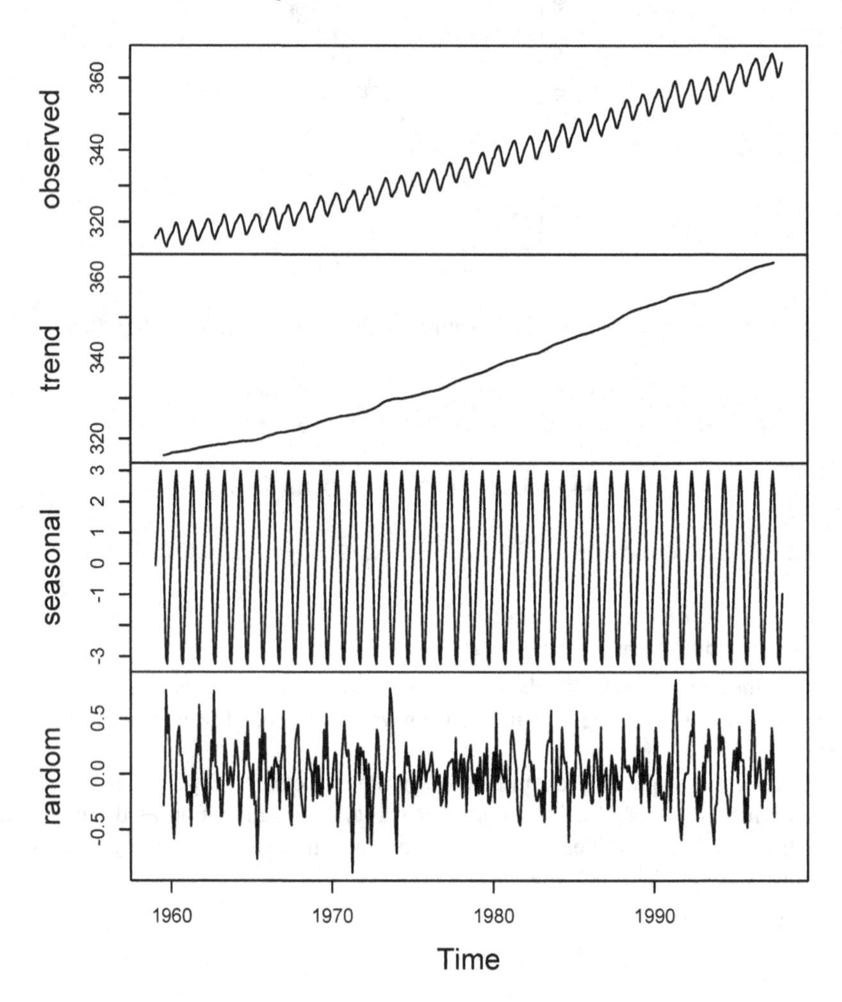

Figure 11.25: *Decomposition of* CO_2 *data into trend, seasonal, and random components.*

```
> dco2 = decompose(co2)
> plot(dco2)
```

Since the seasonal oscillations are fairly constant over time, the use of the default "additive" type is appropriate. In some other examples of time series, the amplitudes of the seasonal oscillations tend to increase or decrease. This situation is handled with the "multiplicative" option.

Bibliography

[Act90] Forman S. Acton. *Numerical Methods that Work.* Mathematical Association of America, Washington, D.C., 1990.

[Adl10] Joseph Adler. *R in a Nutshell.* O'Reilly, Sebastopol, CA, 2010.

[AS65] Milton Abramowitz and Irene A. Stegun. *Handbook of Mathematical Functions.* Dover, New York, 1965.

[Ber66] J. Berkson. Examination of randomness in alpha particle emissions. In F.N. David, editor, *Research Methods in Statistics.* Wiley, New York, 1966.

[BHS99] Bernd Blasius, Amit Huppert, and Lewi Stone. Complex dynamics and phase synchronization in spatially extended ecological systems. *Nature*, 399:354–359, 1999.

[Blo09] Victor Bloomfield. *Computer Simulation and Data Analysis in Molecular Biology and Biophysics: An Introduction Using R.* Springer, New York, 2009.

[BS00] Bernd Blasius and Lewi Stone. Chaos and phase synchronization in ecological systems. *International Journal of Bifurcation & Chaos in Applied Sciences & Engineering*, 10:2361–2380, 2000.

[BWR80] Victor A. Bloomfield, Robert W. Wilson, and Donald C. Rau. Polyelectrolyte effects in dna condensation by polyamines. *Biophys. Chem.*, 11:339–343, 1980.

[CC08] Jonathan D. Cryer and Kung-Sik Chan. *Time Series Analysis with Applications in R.* Springer, New York, second edition, 2008.

[Cha87] David Chandler. *Introduction to Modern Statistical Mechanics.* Oxford University Press, New York, 1987.

[Dal08] Peter Dalgaard. *Introductory Statistics with R.* Springer, New York, second edition, 2008.

[Gar00] Alejandro L. Garcia. *Numerical Methods for Physics.* Prentice-Hall, Upper Saddle River, New Jersey, second edition, 2000.

[HH00] Desmond J. Higham and Nicholas J. Higham. *Matlab Guide.* SIAM, Philadelphia, 2000.

[HS95] Owen T. Hanna and Orville C. Sandall. *Computational Methods in Chemical Engineering.* Prentice-Hall PTR, Upper Saddle River, New Jersey, 1995.

[JMR09] Owen Jones, Robert Maillardet, and Andrew Robinson. *Introduction to Scientific Programming and Simulation Using R*. CRC Press, Boca Raton, 2009.

[Kab11] Robert I. Kabacoff. *R in Action: Data Analysis and Graphics with R*. Manning, Shelter Island, N.Y., 2011.

[Mat11] Norman Matloff. *The Art of R Programming: A Tour of Statistical Software Design*. No Starch Press, San Francisco, 2011.

[Mit11] Hrishi V. Mittal. *R Graphs Cookbook*. Packt, Birmingham, U.K., 2011.

[MJS11] Walter R. Mebane, Jr. and Jasjeet S. Sekhon. Genetic optimization using derivatives: the rgenoud package for r. *Journal of Statistical Software. URL http://www. jstatsoft. org*, 2011.

[Mur11] Paul Murrell. *R Graphics*. CRC Press, Boca Raton, second edition, 2011.

[Pet03] Thomas Petzoldt. R as a simulation platform in ecological modelling. *R News*, 3(3):8–16, 2003.

[PTVF07] William H. Press, Saul A. Teukolsky, William T. Vetterling, and Brian P. Flannery. *Numerical Recipes: The Art of Scientific Computing*. Cambridge University Press, New York, third edition, 2007.

[Ric95] J.A. Rice. *Mathematical Statistics and Data Analysis*. Duxbury Press, Pacific Grove, CA, second edition, 1995.

[SCM12] Karline Soetaert, Jeff Cash, and Francesca Mazzia. *Solving Differential Equations in R*. Springer, New York, 2012.

[Scr12] Luca Scrucca. Ga: A package for genetic algorithms in r. *Journal of Statistical Software*, 53:1–37, 2012.

[SH10] Karline Soetaert and Peter M.J. Herman. *A Practical Guide to Ecological Modelling: Using R as a Simulation Platform*. Springer, New York, 2010.

[SN87] J.M. Smith and H.C. Van Ness. *Introduction to Chemical Engineering Thermodynamics*. McGraw-Hill, New York, 1987.

[Ste09] M. Henry Stevens. *A Primer of Ecology with R*. Springer, New York, 2009.

[Tee11] Paul Teetor. *R Cookbook*. O'Reilly, Sebastopol, CA, 2011.

[Van08] Steve VanWyk. *Computer Solutions in Physics with Applications in Astrophysics, Biophysics, Differential Equations, and Engineering*. World Scientific, Singapore, 2008.

[Ver04] John Verzani. *Using R for Introductory Statistics*. CRC Press, Boca Raton, 2004.

[VR02] W.N. Venables and B.D. Ripley. *Modern Applied Statistics with S*. Springer, New York, fourth edition, 2002.

[ZRE56] B.H. Zimm, G.M. Roe, and L.F. Epstein. Solution of a characteristic value problem from the theory of chain molecules. *J. Chem. Phys.*, 24:279–280, 1956.

Index

Printed in the United States
by Baker & Taylor Publisher Services